T0199903

Transport in Semiconductor Mesoscopic Devices
(Second Edition)

Transport in Semiconductor Mesoscopic Devices
(Second Edition)

David K Ferry

School of Electrical, Computer, and Energy Engineering, Tempe, Arizona 85287, USA

IOP Publishing, Bristol, UK

Multimedia content is available for this book from http://iopscience.iop.org/book/978-0-7503-3139-5.

ISBN   978-0-7503-3139-5 (ebook)
ISBN   978-0-7503-3137-1 (print)
ISBN   978-0-7503-3140-1 (myPrint)
ISBN   978-0-7503-3138-8 (mobi)

DOI   10.1088/978-0-7503-3139-5

Version: 20200801

IOP ebooks

British Library Cataloguing-in-Publication Data: A catalogue record for this book is available from the British Library.

Published by IOP Publishing, wholly owned by The Institute of Physics, London

IOP Publishing, Temple Circus, Temple Way, Bristol, BS1 6HG, UK

US Office: IOP Publishing, Inc., 190 North Independence Mall West, Suite 601, Philadelphia, PA 19106, USA

Contents

Preface to the second edition

It generally is regarded as being true that nanostructures may be considered as ideal systems for the study of the physics of electronic transport. Perhaps this is a self-fulfilling statement, as I have been involved in the field for my entire career. In the late 1970s, this area of research was called 'ultra-small electronics research', and the description as one of nanoscale was not applied for a few decades after that. But, it was interesting that we pursued the use of electron-beam lithography to make things small. Unfortunately, this endeavor was ended by the success of the microelectronics industry. For instance, we worked hard in the university environment to make small transistors with gate lengths on the scale of 20–50 nm. For the past few decades or so, Intel (and others, of course) has made a number something like a thousand times the population of the Earth of such devices each day, so this area of research is gone from the universities.

But this study of small structures is more complex than just the use of nano-fabrication techniques to make small transistors. Transport is the study of the motion of charge carriers in the materials of interest, mainly semiconductors in my world. This transport is characterized by a set of length and time scales. For example, velocity builds up, and decays, with a time scale governed by the scattering time that describes the interaction of the carriers with impurities and lattice vibrations. These times are on the order of a small fraction of a picosecond. Transport distances are scaled by mean free paths; e.g., the distance that a carrier travels on average between discrete scattering events. This distance can range from a few nanometers at room temperature in Si to a few microns at low temperatures in GaAs. Indeed, the distance a particle may travel before scattering has been determined to be about 180 nm in an InAs nanowire at room temperature. The time scales for transport have been probed extensively by the use of femtosecond pulse length lasers. So, the crucial length and time scales have been accessible for quite some time and this has allowed the study of the dynamic response of transport at these fundamental levels.

As a consequence of being able to access the basic length and time scales, it is possible to create structures in which the underlying physics can be probed in a meaningful manner. By lowering the temperature to easily achievable cryogenic levels, scattering can be suppressed significantly, and one can explore the physics itself. This allows us to explore quantum physics in systems which may seem to be purely classical at room temperature. The impact of these studies goes beyond a simple interest in the physics. As mentioned above, the world of the microelectronics industry has been a leader in the development of nano-transistors, with critical dimensions at the 10 nm scale, and chips at the so-called 7 nm node (in 2019) of Moore's Law. Success has been achieved here because the study of nanostructures has highlighted many important physical effects prior to their being important in the integrated circuit. Advances in the study of the underlying physics has provided important guidance for the continued reductions in device size, but the latter also provides a pull for continued study of the relevant physics. It is not a one-way street.

Rather, this field has prospered from the interplay of science and technology, and, for example, the computing power that has resulted from the latter has been crucial for continued development of the former. More importantly, however, nanoelectronics provides the driving technology for much of our high technology life today, and it holds the promise to keep driving the information growth and processing that has given us this high technology life.

As remarked in the preface of the first edition, my involvement in this area was greatly expanded when Jon Bird joined our department. But, our interest in the continued evolution of semiconductor nanoelectronics and the understanding of the relevant physics has continued even from before that time. As we move into this second edition, the understanding of what makes nanoelectronics has moved beyond just semiconductors and the corresponding materials. Now, a new component of this field has been driven by the quest to create quantum computers, which so far rely upon superconducting materials and the relevance of the Josephson tunneling junction. Consequently, in this second edition, coverage of these topics has been added to provide a more extensive coverage of the modern field on nanotechnology. Along the way, some videos from the first edition have been deleted, but those that remain for the second edition are available at https://iopscience.iop.org/book/978-0-7503-3139-5.

Some of the problems in the various chapters use tools that are available at NanoHUB.org, which is a computational science center originally founded by the National Science Foundation. All the tools are available at their website by creating an account, which is free, and then selecting 'tools' from your sign-in page. The available tools range from introductory level to quite advanced tools such as for density functional theory and nonequilibrium Green's functions. Those needed for the problems are toward the more introductory level. If problems occur, there is a reporting system online at their web page.

Finally, I should thank once more Jon Bird for continuing to collaborate and to provide information about some of the updates. I also have to thank my colleagues Steve Goodnick and Dragica Vasileska for feedback from their teaching the material. The staff at IOP Publishing have continued to provide an outstanding relationship that has existed from the commissioning of the first edition right through until the present time.

David K Ferry

Preface to the first edition

This book is the result of a project begun some years ago when Jon Bird came to Arizona State University. At that time, we had a major program on the study of quantum effects in very small semiconductor structures, an area which has become known as mesoscopic devices. Over the years, we have developed first at Arizona State University, and then subsequently when Jon moved to SUNY Buffalo, graduate level courses in this topic. The present book has grown out of several versions of lecture notes which were prepared in connection with these courses. Note that this book is intended to be a textbook for first year graduate students. Thus, many topics are not covered in the great detail that may be found for example in *Transport in Nanostructures*, from Cambridge University Press. To do so would make the book far too large for the purpose of a textbook and it would probably overwhelm the students as well.

I need to point out that, in spite of my best lamentations, Jon has continued to decline my offers to partner in the preparation of this book. Nevertheless, the book could not be possible without the input from Jon, and I owe a great debt of gratitude to him for his contributions. I also owe a debt of gratitude to Larry Cooper, long retired from the Office of Naval Research, without whom our work would not have prospered as it has. And, without the latter, I would not be in a position to write this book.

I must also thank the people who have provided various images for inclusion in the book. It is especially nice to acknowledge those who have provided the videos that are included in the ebook version. These are Jesper Nygård and the Neils Bohr Institute, JT Janssen and the National Physical Laboratory, Andrea Morelli and Andrew Dzurak and the University of New South Wales, Stuart Lindsay, Don Burgess, and Richard Akis. They all have graciously consented to do so, and I am eternally grateful for that privilege. Without these images and videos, the book would not exist.

Author biography

David K Ferry

 David K Ferry is Regents' Professor Emeritus in the School of Electrical, Computer, and Energy Engineering at Arizona State University. He was also graduate faculty in the Department of Physics and the Materials Science and Engineering program at ASU, as well as Visiting Professor at Chiba University in Japan. He came to ASU in 1983 following shorter stints at Texas Tech University, the Office of Naval Research, and Colorado State University. In the distant past, he received his doctorate from the University of Texas, Austin, and spent a postdoctoral period at the University of Vienna, Austria. He enjoyed teaching (which he refers to as 'warping young minds') and continues active research. The latter is focused on semiconductors, particularly as they apply to nanotechnology and integrated circuits, as well as quantum effects in devices. In 1999, he received the Cledo Brunetti Award from the Institute of Electrical and Electronics Engineers, and is a Fellow of this group as well as the American Physical Society and the Institute of Physics (UK). He has been a Tennessee Squire since 1971 and an Admiral in the Texas Navy since 1973. He is the author, co-author, or editor of some 40 books and about 900 refereed scientific contributions. More about him can be found on his home pages http://ferry.faculty.asu.edu/ and http://dferry.net/.

IOP Publishing

Transport in Semiconductor Mesoscopic Devices
(Second Edition)

David K Ferry

Chapter 1

The world of nanoelectronics

What we call electronics today has a confused beginning. Should it begin with the first electronic device, the metal-semiconductor junction of Braun in 1876 [1]? Or should it begin with the first suggestions of a mechanical computing machine, which dates at least to the analytical engine of Charles Babbage at the middle of the nineteenth century. If the latter, then we have to look at the theoretical work of Alan Turing in the 1930s that suggested the use of such a machine to determine if a number was computable [2]. If the former, then the follow up lies with the understanding of the vacuum diode [3, 4]. These two threads come together with the first efforts to develop computing machines at Iowa State University and the University of Pennsylvania [5, 6]. Yet, the future lay in nanoelectronics, and this became apparent with the invention of the integrated circuit in the 1950s by Jack Kilby [7] and Robert Noyce [8]. The rapid development of integrated circuits with ever smaller critical dimensions led to the evolution of what has become known as Moore's Law [9]. Here, Moore suggested that the number of transistors on an integrated circuit would continue its trend of doubling every 18 months, leading to the exponential growth that allowed first micro-electronics and then nanoelectronics to revolutionize everyday life throughout the world. The physical approach which allowed this was based upon the principles of scaling of dimensions in a manner that kept the internal electric field constant from one generation to the next [10]. Thus all dimensions and voltages of a device were reduced by the exact same factor, which led to the electrostatics of the device being unchanged. As a result, the reduced dimensionality did not, in principle, change device operation as it was downsized. This little overlooked fact meant that integrated circuits was a scalable technology, and this scalability led to the impact of nanoelectronics on the world, a transition that some have termed the information revolution due to the exponential growth in computing power [11]. For example, the iPhone 11, introduced in late 2019, uses the A13 chip based upon the ARMv8.3-A 64 bit 6 core 7 nm node technology,

doi:10.1088/978-0-7503-3139-5ch1

fabricated in Taiwan. The chip computing power is approximately 100 GFLOPS[1]. The chip itself is about 1 cm^2 and contains 8.5 billion transistors. More interesting technologically is the fact that some layers of the chip utilize extreme ultra-violet (EUV) lithography which employs 13.7 nm x-rays.

1.1 Moore's law

Since the invention of the integrated circuit, the growth in the number of transistors on the chip has been exponential. This growth has proceeded unabated now for more than half a century. Only a few years had passed after the invention of the integrated circuit when Gordon Moore recognized the important driving forces for the exponential growth. His 1965 paper became the controlling manifesto for the development of the microchip world [9]. Moore modestly suggested that 'Integrated circuits will lead to such wonders as home computers…and personal portable communications equipment'. Moore and Noyce had joined the original Shockley semiconductor company, but left with a group to found Fairchild semiconductors to pursue integrated circuits, and then eventually went on to found Intel.

The key factor in Moore's law lay in the scaling of transistor sizes. Moore noted that, if dimensions were reduced by a factor of two every so often, then the number of transistors per unit area would go up by a factor of four. This scaling is the heart of the exponential growth in transistor density in modern chips. The development of a formal scaling theory in which dimensions, dopant densities, and electric fields were all scaled according to a prescribed plan came somewhat later [10]. But, Moore recognized that there were more factors involved than merely down-sizing the transistors. He suggested that the actual size of individual chips would be increased and that gains could be made through circuit cleverness. So, the eventual factor of four increase every three years came from these three different ideas, rather than from merely reducing the size. Examples of the idea of circuit cleverness are the introduction of complementary metal–oxide–semiconductor (CMOS) technology itself, the introduction of trench capacitors in which the capacitors were buried in the silicon rather than being on the surface, and, more recently, the transition to the finFET (or tri-gate) technology.

In spite of the obvious reductions in the size of individual transistors, these are in fact results of Moore's law rather than the driving forces. The actual driving force does not derive from physical laws but from economics. In a modern integrated circuit, transistors are laid out on the chip in a planar fashion, much like the houses in a modern southwestern US city. The drive that continues the growth arises from the cost reductions associated with a given level of computing power. As the number of transistors on each chip continues to increase, the number of functional units on the chip dramatically increases which in turn increases the computing power. Because the basic cost of manufacturing a single chip has not dramatically increased over the five plus decades since the invention of the chip, the cost of computing

---

[1] Giga-floating point operations per second

power has gone down exponentially. It is this economic argument, based upon the cost of silicon real estate, that seems to drive Moore's law.

There have been glitches along the way. Intrinsic heating of the chip became a significant problem in the early 1990s. This led to an extensive industry-wide push for a new design approach based upon low power. Nevertheless, individual chip size has not increased significantly since that time, remaining at 1–2 cm^2. But, Moore's law marches onward, as new technologies are developed to solve problems. Mobility reductions were overcome with the use of strain in the transistors—n-channel devices used tensile strain while p-channel devices used compressive strain. The natural reduction of gate oxide thicknesses with scaling led to a problem of tunnel leakage through the oxide. This was overcome by the introduction of new dielectrics with high dielectric constants (currently HfO_2), so that thicker materials could be used while maintaining the gate capacitance. As devices became smaller, the physical number of dopants in the channel became quite small, which could lead to device-to-device fluctuations in dopant number [12]. This has been addressed by reducing the number of dopants considerably in the channel region itself. Hot carrier effects are ameliorated by careful design of the device. In short, as problems in the transistors are recognized, the industry finds ways to overcome these problems. The most dramatic is perhaps the move to tri-gate devices that has recently occurred. Here the problem is controlling the off/leakage current in the transistors, which has been addressed by making the transistors vertical so that a 'wrap around' gate can be used to enhance the off state resistance.

Nevertheless, it is easy to recognize that a critical dimension of 25 nm represents a chain of only 100 silicon atoms. One cannot continue forever along the path of continuing to down-size the transistor. This recognition raises possible problems with the CMOS technology itself. Another class of challenges lies in the realm of the state variable to be used for computing. Can we continue to use charge as is now done with the transistor? Or, will new paradigms arise which engender new forms of computing[2]?

The industry may be at a cross-roads for another technological change and this may involve a move to nanowires fabricated in a gate-all-around technology, where the gate wraps completely around the nanowire. This has been studied in research labs for almost two decades. Nanowires placed parallel to one another on the surface of silicon is not an attractive approach. This is because, if we are to save silicon surface area, the nanowire has to have a large diameter, and this means that it cannot compete with the finFET in saving Si area and achieving high circuit density [13, 14]. But, another approach is possible, and that is to stack the nanowires vertically [15], an approach which was followed quickly by others [16, 17]. Naturally, there have been studies to see how these devices compared with others such as SOI MOSFETs [18] and FinFETs [19, 20]. There is also a consideration whether or not to incorporate doping in the channel region, and this has been studied as well for the vertical stacked nanowires [21, 22]. The best sign that this

---

[2] But, quantum computing has not given any indication that it will solve this problem. Current qubits remain significantly larger than current Si transistors.

technology is ready for prime time applications is the entry of the major semi-conductor producers into the field [23].

What the future holds is anyone's guess, but I suspect that we will continue onward for many decades with one form or another of Moore's law. We may well discontinue reducing the size of the transistor, but perhaps we will begin to layer the transistors on the chip in the manner of lasagna, as many are suggesting already. In any of the possible scenarios, there remains a great field of research in what we call nanostructures that will underlie future advances.

1.2 Nanostructures

If we are to proceed to understand nanostructures, then it will be necessary to have a full understanding of the basic physics as it relates to those device properties. It is the objective of this book to provide just such an understanding. The first principle of nano-device physics is that it is strongly influenced by quantum effects. In this sense, it differs from the level of understanding that one requires in an undergraduate life. Here, we need to deal with a variety of effects that become quite important, such as the quantization of the electronic density of states and the implication that this has for the electronic properties of nanostructures. A direct result of this quantization effect, when we are dealing with transport that is nearly ballistic in nature, is the resulting quantization of the conductance throughout the nanostructure. These two effects go hand in hand in providing one of the most interesting observations in nanostructures—the presence of specific modes of propagation, much like a micro-wave waveguide. Indeed, it is the wave-like nature of the electrons that is being observed in these experiments. Along this same idea, the presence of quantum interference is another major observable in nanostructures. That is, the wave can propagate along a pair of trajectories, much like a two-slit experiment, and the resulting wave interference is clearly observable in experimental studies. The tunneling, which can arise from the wave properties of the electrons, is also easily observable and has led to a number of interesting devices.

To create the devices in which our quantum effects are to be studied, one can proceed by at least two different approaches. These approaches will be discussed further below. First, however, we want to note some important points. It is possible to create nanostructures using top-down microfabrication which follows the same procedures as used in the microelectronics industry. One difference, however, is the preferred use of electron-beam lithography for the ease with which it can make relatively small structures [24]. With such lithography, one can routinely approach dimensions in the 10–30 nm range, and with some extra effort, lines approaching 5–7 nm have been fabricated [25]. Thus, the standard processes of lithography and etching can be used to prepare a large number of different nanostructures. On the other hand, bottom-up processing can be used to create even smaller structures. Typical approaches utilize either molecular self-assembly [26] or, for example, the deposition of small quantum dots through strain relaxation of a very thin epitaxial layer [27]. Then there are entirely new approaches which can arise from the layer compounds, such as graphene, which redefine some of our ideas about whether to go top-down or bottom-up.

Hence, there is a vast range of fabrication tools which can be brought to bear to create a specific nanostructure needed to study an interesting point in physics. We will discuss a few of these methods in a later section. But, the field of nanostructures is truly enormous, covering a wide range of disciplines that range from fundamental physics, to materials growth and development, and to electronic and optoelectronic investigations. It is impossible to cover this entire range of topics within a single book, and even concentrating on transport can lead to a large book [28]. A rather narrower view will be taken here, and we will focus on a discussion of the basic effects that are associated with the germane quantum transport effects that can be found in the electronic properties of various nanostructures. The short term application of some of these effects is not the primary aim, especially as they relate to today's CMOS microelectronics. Rather, the field of mesoscopic physics and devices goes beyond today's transistors, and is wide enough to keep us busy throughout this book. It is the purpose here, however, to try to take a coherent journey through this field to highlight the common physics and understanding that underpins this field.

An important point about nanostructured devices is that the critical dimensions of the structure are comparable to the corresponding de Broglie wavelength of the electrons. This allows their properties to be strongly influenced by quantum mechanical effects. For example, if we have an electron in GaAs at a Fermi energy of, for example, 10 meV, then this corresponds to a momentum of approximately 1.4×10^{-26} kg m s^{-1} or a wave vector of about 1.3×10^{8} m^{-1}. This now corresponds to a de Broglie wavelength of almost 50 nm. We will see later that this corresponds to a two-dimensional density of just over 10^{11} cm^{-2}, which is easily obtained in high quality GaAs heterostructure layers. One of the nicest demonstrations of the observability of de Broglie waves was that of the quantum corral by Don Eigler *et al* at the IBM Almaden Research Center [29], as shown in figure 1.1. Using a

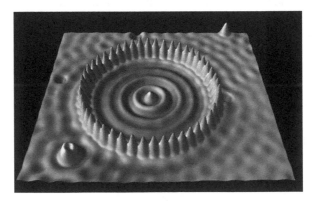

Figure 1.1. Scanning tunneling microscope image of 48 Fe atoms (the sharp peaks) forming a ring on the surface of copper [29]. Within the ring, standing waves from the confined copper surface electrons are clearly visible. The image is used with permission from IBM research.

scanning tunneling microscope (STM), they arranged iron atoms on the surface of Cu in the shape of a ring approximately 14.6 nm in diameter. Within the ring, they could then image the square magnitude of the wave function of electrons on the Cu surface, even though the relatively higher energy of the electrons at the Cu Fermi energy have a much shorter wavelength than those in the low density GaAs layer. Nevertheless, it is absolutely clear that the wave nature of the electrons is exceedingly important in these nanostructures.

As the preceding example indicates, when we confine electrons on the scale of their wavelengths, or even on larger scales when the motion is coherent, these electrons are subject to quantization by their confinement. This quantization gives rise to a dramatic difference in their density of states from that expected in classical bulk material. One reason for this is that the energy spectrum of the electrons becomes very different from the quasi-continuous one expected in bulk materials. The presence of this quantization gives opportunities to probe new and different physics and applications, some of which might be useful for future device applications. Moreover, the interaction of the electron with defects and impurities becomes a much more singular process, in that individual scattering events become important processes in the transport of the electrons. This disorder, arising from the impurities or the defects, can introduce new observables in the transport conductance, which may not be small changes. To understand this, we want to consider some basic ideas of length and time scales.

In large, bulk conductors, the resistance that exists between two contacts is related to the bulk conductivity and to the dimensions of the conductor, as expressed by

$$R = \frac{L}{\sigma A},$$ (1.1)

where σ is the conductivity and L and A are the length and cross-sectional area of the conductor, respectively. If the conductor is a two-dimensional conductor, such as a thin sheet of metal, then the conductivity is the conductance per square, and the cross-sectional area is just the width W. This changes the basic formula (1.1) only slightly, but the argument can be extended to any number of dimensions. Thus, for a d-dimensional conductor, the cross-sectional area has the dimension $A = L^{d-1}$, where L must be interpreted as a 'characteristic length'. Then, we may rewrite equation (1.1) as

$$R = \frac{L^{2-d}}{\sigma_d}.$$ (1.2)

Here, σ_d is the d-dimensional conductivity. Whereas one normally thinks of the conductivity, in simple terms, as $\sigma = ne\mu$, the d-dimensional term depends upon the d-dimensional density that is used in this latter definition. Thus, in three dimensions, σ_3 is defined from the density per unit volume, while in two dimensions σ_2 is defined as the conductivity per unit square and the density is the sheet density of the carriers. The conductivity (in any dimension) is not expected to vary much with the characteristic dimension, so we may lake the logarithm of the last equation. Then, taking the derivative with respect to $\ln(L)$ leads to

$$\frac{\partial \ln(R)}{\partial \ln(L)} = 2 - d \tag{1.3}$$

This result is expected for macroscopic conducting systems, where resistance is related to the conductivity through equation (1.2). We may think of this limit as the bulk limit, in which any characteristic length is large compared to any characteristic transport length.

In mesoscopic conductors, it has been suggested that the above would no longer be valid, primarily under the assumption that disorder effects would be proportionally larger in small structures. Let us first consider how this might appear. We have assumed that the conductivity is independent of the length, or that σ_d is a constant. However, if there is surface scattering, which can dominate the mean free path, then one could expect that the mean free path is $l \sim L$. Since $l = v_F \tau$, where v_F is the Fermi velocity in a degenerate semiconductor and τ is the mean free time between collisions, this leads to

$$\sigma_d = \frac{n_d e^2 \tau}{m^*} = \frac{n_d e^2 L}{m^* v_F}, \tag{1.4}$$

Hence, the dependence of the mean free time on the dimensions of the conductor changes the basic behavior of the macroscopic result (1.3). This is the simplest of the modifications. For more intense disorder or more intense scattering, the carriers may well be localized because the size of the conductor creates localized states whose energy difference is greater than the thermal excitation, and the conductance will be quite low. In fact, we may actually have the resistance increasing exponentially with length according to [30]

$$R = e^{\alpha L} - 1, \tag{1.5}$$

where α is a small quantity. We think of the form of equation (1.5) as arising from the localized carriers tunneling from one site to another, and the last term (-1) is required to recover zero resistance with zero length. Then, the above scaling relationship (1.3) is modified to

$$\frac{\partial \ln(R)}{\partial \ln(L)} \approx \alpha L. \tag{1.6}$$

In this situation, unless the conductance is sufficiently high, the transport is localized and the carriers move by hopping. The necessary value has been termed the minimum metallic conductivity [30], but its value is not given by the present arguments. Here we just want to point out the difference in the scaling relationships between systems that are highly conducting (and bulk-like) and those that are largely localized due to the high disorder.

In a strongly disordered system, such as that discussed above, the wave functions decay exponentially away from the specific site at which the carrier is present. This means that there is no long-range wavelike behavior in the carrier's character. On the other hand, by bulk-like extended states we mean that the carrier is wavelike in nature and has a well-defined wave vector k and momentum $\hbar k$. Most semi-conducting mesoscopic systems have sufficiently weak scattering that the carriers do not have localized behavior. Thus, when we talk about diffusive transport, we generally mean almost-wavelike free states with relatively low scattering rates. Here, we tend to mean that the concept of an electron mobility is quite valid and that the scattering occurs often enough to make this the case.

We will examine the idea of localization later (in chapter 5), but here we assume that the entire conduction band is not localized. Rather, it retains a sufficiently large region in the center of the energy band that has extended states and a nonzero conductivity as the temperature is reduced to zero. For this material, the density of electronic states per unit energy, per unit volume, is given simply by the familiar (we will return to derive the density of states in section 2.2) dn/dE. Since the conductor has a finite volume, the electronic states are discrete levels determined by the size of this volume. These individual energy levels are sensitive to the boundary conditions applied to the ends of the sample (and to the 'sides') and can be shifted by small amounts on the order of $\hbar/\tau$, where τ is determined here by the time required for an electron to diffuse to the end of the sample. In essence, one is defining here a broadening of the levels that is due to the finite lifetime of the electrons in the sample, a lifetime determined not by scattering but by the carriers' exit from the sample. This, in turn, defines a maximum coherence length in terms of the sample length. This coherence length is defined here as the distance over which the electrons lose their phase memory, which we will take to be the sample length. The time required to diffuse to the end of the conductor (or from one end to the other) is L^2/D. where D is the diffusion constant for the electron (or hole, as the case may be) [31]. The conductivity of the material is related to the diffusion constant (we assume for the moment that the temperature $T = 0$) as

$$\sigma_d(E) = \frac{n_d e^2 \tau}{m^*} = e^2 D \frac{dn}{dE}, \tag{1.7}$$

where we have used the fact that $n = (2/d)(dn/dE)E$, d is the dimensionality, and $D = v_F^2 \tau / d$. If L is now introduced as the effective length, and τ is the time for diffusion, both from D, one finds that

$$\frac{\hbar}{\tau} = \frac{\hbar \sigma}{e^2 L^2} \frac{dE}{dn}. \tag{1.8}$$

The quantity on the left-hand side of equation (1.8) can be defined as the average broadening of the energy levels ΔE_a, and the dimensionless ratio of this broadening to the average spacing of the energy levels may be defined as

$$\frac{\Delta E_a}{dE/dn} = \frac{\hbar \sigma}{e^2 L^2}. \tag{1.9}$$

Finally, we change to the total number of carriers $N = nL^d$, so that

$$\frac{\Delta E_a}{dE/dn} = \frac{\hbar\sigma}{e^2}L^{d-2}.$$ (1.10)

This last equation is often seen with an additional factor of 2 to account for the double degeneracy of each level arising from the spin of the electron. Nevertheless, this last result agrees with the expected scaling of the resistance in equations (1.2) and (1.3).

It is now possible to define a dimensionless conductance, called the Thouless number by Anderson *et al* [32], in terms of the conductance as

$$g(L) = \frac{2\hbar}{e^2}G(L),$$ (1.11)

where $G(L) = \sigma L^{d-2}$ is the actual conductance. These latter authors have given a scaling theory based upon renormalization group theory, which gives us the dependence on the scale length L and the dimensionality of the system. The details of such a theory are beyond this book. However, we can obtain the limiting form of their results from the above arguments. The important factor is a critical exponent for the reduced conductance $g(L)$ which may be defined by

$$\beta_d \equiv \lim_{g\to\infty} \frac{d[lng(L)]}{dln(L)} \to d - 2,$$ (1.12)

which is just equation (1.3) rewritten in terms of the conductance rather than the resistance. By the same token, one can rework equation (1.6) for the low conducting state to give

$$\beta_d \equiv \lim_{g\to0} \frac{d[lng(L)]}{dln(L)} \to -\alpha L.$$ (1.13)

What the full scaling theory provides is a connection between these two limits when the conductance is neither large nor small.

For three dimensions, the critical exponent changes from negative to positive as one moves from low conductivity to high conductivity, so that the concept of a mobility edge in disordered (and amorphous) conductors is really interpreted as the point where $\beta_3 = 0$. This can be expected to occur about where the reduced conductance is unity, or for a value of the total conductance of $e^2/\pi\hbar$. In less than three dimensions, there is no critical value of the exponent, as it is by and large always negative, approaching 0 asymptotically. That is, it has been suggested that all states are localized in one and two dimensions [32], although experiments tell us otherwise. What the results (1.12) and (1.13) is that the curve can cross over the asymptote (1.12) and approach 0 from the positive side for two dimensions. In studies of high mobility silicon MOS field-effect transistors (MOSFETs) at low temperatures, it has been demonstrated that there is a phase transition between a localized regime and a diffusive regime [33–35]. Subsequently, this phase transition has been observed for a variety of semiconductor heterostructures and for both electrons and holes. It is clear that the critical density for the onset of diffusive transport is affected by the impurities in the semiconductor, as measurements on very high purity, undoped GaAs/AlGaAs structures suggest that the critical electron density is below 2.6×10^9 cm^{-2}. Thus, it seems to be clear that any one parameter scaling theory fails to uncover the crucial physics of any metal–insulator transition behavior in high quality mesoscopic devices, except in the $d = 3$ case. Yet, the ideas of the scaling theory can be a useful guide to what may generally be expected in any dimensionality, although the exact details may be surprising as in the structures mentioned.

1.3 Some electronic length and time scales

We have already encountered a number of appropriate length scales with the device characteristic length L and the mean free path $v_F \tau$, where τ is the mean free time between collisions, and v_F is the Fermi velocity. These are connected with the ideas of mobility $\mu = e\tau/m^*$, where m^* is the effective mass of the electron in the semiconductor, and the diffusivity D ($= \mu k_B T/e$), which is given in terms of the Fermi velocity as $v_F^2 \tau/d$, and d is the dimensionality as above. We have also discussed briefly the idea of a coherence time, or phase-breaking time τ_φ which describes the time over which the wave function retains its coherence. While this is a vague meaning and description, a better understanding can only arise as we describe the effect of this phase coherence in the scenario of a number of experiments, in which the quantum behavior remains well observable for a time scale on the order of this quantity. This phase-breaking time allows us to connect to a phase-breaking length through the diffusivity via $l_\varphi = D\tau_\varphi$. We can think of this phase-breaking length as the average distance which the electrons diffuse before their phase is disrupted through various scattering events. Naturally, we desire that this length be larger than, or comparable to, the size of the mesoscopic device under investigation, which usually implies that the measurements are to be performed at cryogenic temperatures. Some of the earliest measurements of the phase-breaking length were performed in thin metallic wires [36, 37], which tend to have more disorder than semiconductors. In most cases, the phase-breaking time and length were inferred by fitting to the dependence of the weak localization in these wires on the applied magnetic field. We will turn to a discussion of this effect in chapter 3.

Other important lengths are the Fermi wavelength, which was introduced earlier, and the thermal length. This latter is a length that is a little more difficult to grasp, as it connects the diffusivity with a time defined by the thermal broadening of a typical energy level. This broadening is used to define a time scale via $\hbar/k_B T$, and this connects the diffusivity to the thermal length as $l_T = \sqrt{\hbar D/k_B T}$. Again, we will encounter this length in some detail in the discussion of weak localization and conductance fluctuations in chapter 3.

Another important time for nonequilibrium systems is the energy-relaxation time τ_E, which describes the time scale over which the energy per carrier in the system returns to its equilibrium value. Usually, this time describes the particular electron–phonon interactions by which energy is transferred from the electrons to the lattice, and so is dominated by inelastic interactions. This raises another issue over the various time scales. Those discussed in the preceding paragraphs are generally defined only in equilibrium conditions. For example, the relationship given above between the diffusion constant and the mobility is only valid in equilibrium (and, as stated, only for non-degenerate conditions). When the system goes out of equilibrium, there is no such relationship, and, quite generally, there is no fluctuation-dissipation theorem from which this relationship is derived [38]. In this case, however, the thermal length and the phase-breaking length should more likely be defined in terms of the thermal velocity $v_T = \sqrt{2k_B T/dm^*}$, and the mean free time or the phase-breaking time, respectively.

1.4 Heterostructures for mesoscopic devices

In this section, I would like to describe some structures and materials which are commonly used for mesoscopic devices. First, we treat the two most common device types—the MOSFET, and the high-mobility, heterostructure device. The most common type of the former is the Si MOSFET while the most common form of the latter is the AlGaAs/GaAs heterostructure. Both approaches have been extensively used to study mesoscopic physics, and our goal here is to describe the two device structures and discuss a few of their key features, as well as some ideas on how these structures are fabricated. Then, we turn to superconductors, which have become of great interest in recent years for their use in possible structures for quantum computing devices.

1.4.1 The MOS structure

The field effect transistor (FET) is actually the oldest known form of transistor, as it was originally patented by Lilienfeld in 1926 [39]. But, it was not until 1959 that a working device was demonstrated [40], as control of the surface was a serious problem which held up development until after the bipolar transistor. Within a few years, the MOSFET was the preferred device for the integrated circuit due both to its planar technology and its generally lower power dissipation. The rest, so they say, is history. However, it is not generally appreciated that it is also a good device in which to study mesoscopic physics. Yet, the quantum Hall effect was discovered in a Si MOSFET [41]. In addition, some of the most extensive early work on conductance fluctuations [42] and on the disorder induced metal–insulator transition [43] were both performed in Si MOSFETs. In addition, InSb, GaAs, and Ge have been studied using mylar as the gate insulator [44], and a wide range of semiconductors with anodic oxides [45] and deposited oxides [46] have been investigated. So, it seems clear that the MOSFET is one of the major devices that has been studied for mesoscopic phenomena.

In figure 1.2, we display the conduction band energy profile for a typical planar MOSFET. A positive voltage is applied to the drain relative to the source, which is typically grounded. Between the two is the *p*-type substrate (the source and drain are *n*-type). The oxide is usually silicon dioxide, although these days it is mostly a high dielectric constant oxide such as hafnium oxide (or hafnium silicate). Between the

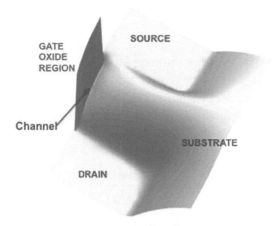

Figure 1.2. The conduction band energy profile for a typical planar MOSFET.

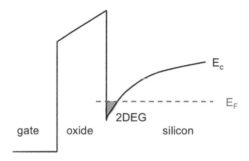

Figure 1.3. Sketch of the band bending in the silicon for a MOS structure under gate bias.

oxide and the *p*-type regions lies the inversion channel, indicated by the arrow in the figure. This is an active *n*-channel which connects the source and drain and is the key part of the transistor. Normally, one describes the transistor action via classical statistics, but this is not correct, as we will show. Indeed, the channel is a quantum object, but this is not very important at room temperature as the quantization is normal to the current flow, so is a second-order effect. At low temperatures, it is the quantized channel that is of great interest to researchers.

The inversion channel electrons reside between the potential barrier introduced by the oxide–semiconductor interface and the confining potential represented by the conduction band in the silicon. A cut of this confinement is shown in figure 1.3. In the classical model, the density in the conduction band is related to the separation of the Fermi energy from the conduction band edge as

$$n = N_c \exp\left(-\frac{E_c - E_F}{k_B T}\right), \tag{1.14}$$

where N_c is the effective density of states, a number on the order of 10^{19} cm^{-3} [46]. As shown in the figure, the conduction band edge is a function of position, so this makes the density a function of position, and a maximum at the oxide–semiconductor interface. From equation (1.14), it is clear that we can write the decay of the density into the semiconductor in the form

$$n(x) = n(0)\exp\left(-\frac{x}{L}\right),$$ (1.15)

where L is a characteristic decay length [47]. In fact, the exponential decay is given by defining the surface potential $\varphi_s(x)$ as the variation of the conduction band away from its bulk equilibrium value. Then, the decay is determined by the thermal voltage in equation (1.14), and

$$n(x) = n(0)\exp\left(-\frac{\varphi_s(x)}{k_B T}\right).$$ (1.16)

Hence, the density has an effective thickness that corresponds to the surface potential falling by $k_B T$. But, the surface electric field is given as

$$E_s = -\frac{\partial \varphi_s(x)}{\partial x}\Big|_{x\to 0} = \frac{en_s}{\varepsilon_s} \sim \frac{en(0)L}{\varepsilon_s}.$$ (1.17)

We can put all of this together to obtain a good estimate of thickness of the inversion layer as

$$d_{eff} = \frac{\varphi_s(0)}{eE_s} \sim \frac{k_B T}{e}\frac{\varepsilon_s}{en_s}.$$ (1.18)

If we take an inversion density of 5×10^{11} cm^{-2} at room temperature, then we find that the effective thickness of the inversion layer is about 3.3 nm.

But, quantum mechanically, we need to ask what the corresponding de Broglie wavelength is for an electron at the Fermi energy in the inversion layer. Suppose we assume that the average energy of the carrier is just the thermal energy. Then, the de Broglie wavelength is given as

$$\lambda_d = \frac{h}{p} = \frac{h}{\sqrt{2m*k_B T}}.$$ (1.19)

Using the transverse mass for transport along the channel, we find that this wavelength is 18 nm. Now, there is just no way to stuff an 18 nm Marshmallow into a 3.3 nm hole. The classical idea of band bending is not valid in this quantized inversion layer, where the potential must be solved in a self-consistent manner. Before addressing this, let us talk about the phrase 'transverse mass'. Silicon has a complicated band structure. The minimum of the conduction band lies along the line from Γ to X in the Brillouin zone, and is located about 85% of the way to X. Because of the symmetry of the Brillouin zone, there are six equivalent minima, as shown in figure 1.4. Each of the six ellipsoids has a longitudinal axis and two transverse axes, and corresponding values for the mass. In Si, it is generally felt that the effective mass values are $m_L = 0.91\ m_0$, $m_T = 0.19\ m_0$.

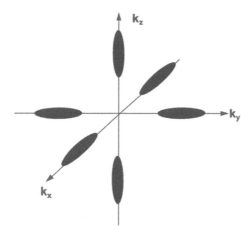

Figure 1.4. A constant energy surface near the minima of the conduction band in silicon consists of six equivalent ellipsoids oriented along the lines from Γ to X.

Silicon MOSFETs are usually fabricated with the surface normal along a (0,0,1) direction. Then, the quantization has a beneficial result of splitting the six ellipsoids into two sets. One pair of ellipsoids has the longitudinal mass normal to the interface, and this gives a lower-energy set of quantum levels. The other four ellipsoids show the transverse mass in the direction normal to the interface and, as this is the smaller mass, will have higher lying quantum levels for motion normal to the surface. The set of levels corresponding to the two-fold valleys is generally denoted as E_0, E_1, E_2, ... while the set of levels for the four-fold set of valleys is denoted with a prime on each level. The advantage is that the two-fold set of valleys now shows the smaller transverse mass in the transport direction, which gives a higher mobility. The introduction of strain about a decade ago in the industry was done for the same reason—to separate these valleys and gain a higher mobility.

In figure 1.5, the energy levels for the lowest two states in the two-fold set of valleys and the lowest energy level for the four-fold set of valleys are plotted for a range of inversion densities for a silicon substrate doped to 10^{17} cm^{-3}. The doping has an effect on the potential, as band bending depletes the substrate so that there is a contribution to the surface field from this charge. In fact, the surface field has been estimated to be [48]

$$E_s = \frac{e}{\varepsilon_s}\left(N_A x_p + \frac{n_s}{2}\right), \qquad (1.20)$$

where N_A is the substrate doping density and x_p is the depletion depth of this region. The factor of ½ arises from the fact that the electric field appears on both sides of the inversion charge, while it only appears on the oxide side of the bulk depletion charge. So the results in figure 1.5 are doping-/dependent. These energy levels were computed using the self-consistent Poisson–Schrödinger solver for the silicon system developed by Vasileska, and they are referenced to the conduction band minimum at the oxide–semiconductor interface. The simulation package is called SCHRED 2.0,

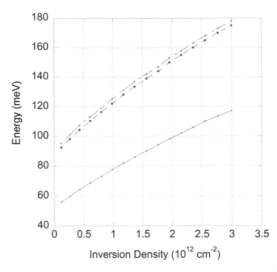

Figure 1.5. The lowest three energy levels, or subbands, for the silicon-silicon dioxide interface with an acceptor doping of 1017 cm^{-3}. The energy level E0 is shown as the solid red curve, while E1 is depicted by the upper dashed red curve. The lowest energy level for the four-fold valleys is E_0' and is indicated by the blue line and symbols.

and is available at NanoHUB.org for anyone to use [49]. Surprisingly, the lowest state has a 'thickness' of about 2.8 nm, which is quite close to the classical value found above. However, only about 49% of the electrons are in this subband at room temperature. Some 8.3% are in the second subband of the two-fold valleys, while the remainder are in the upper, four-fold set of valleys. Of course, at low temperatures, we expect all of the carriers to be in the lowest energy state, corresponding to a single highly degenerate subband.

With the momentum normal to the interface quantized, the transport is con-strained to lie in the plane of the interface. Hence, these electrons form what is known as a quasi-two-dimensional electron gas (2DEG) [48]. The mobility of these electrons is limited by scattering from a variety of sources, but primarily by scattering from ionized impurities, such as the acceptors in the bulk Si. This impurity scattering is largely an elastic process which does not dissipate energy, especially at low temperatures. However, the scattering is screened by the electrons in the inversion layer to some extent. Additional scattering in the silicon system comes from the roughness at the interface. While this interface is quite good, with smoothness nearly on the atomic scale, it is random enough to provide significant scattering of the electrons [50, 51]. Additional scattering processes arise from the phonons, but the optical phonons are not important at low temperatures, and the acoustic phonons provide only a weak scattering. In figure 1.6(a), we plot the mobility as a function of the effective electric field for various temperatures [52]. This effective field is given by the inversion density, as in equation (1.20). Surface roughness scattering varies as the square of this field, as in normal perturbation theory, and the behavior at low temperature agrees well with this interpretation.

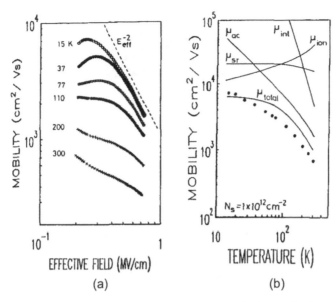

Figure 1.6. (a) Variation of the mobility with the effective field, given by (1.20). The decay with the square of the field at low temperatures signals surface roughness scattering. (b) Variation of the mobility with temperature. Various scattering processes are indicated in the plot. (Reprinted with permission from [52]. Copyright 1992 IOP Publishing.)

The contributions from several scattering mechanisms are shown in panel (b) for a fixed density of 10^{12} cm^{-2}. Again, it is fairly clear that ionized impurity scattering dominates the mobility at low temperature, but that surface roughness scattering plays a role in limiting the mobility.

1.4.2 Fabricating the MOSFET

While this book is not about semiconductor devices per se, one does encounter the problems of nano-fabrication when trying to create mesoscopic structures with which to study the physics. In order to fully comprehend the broad range of problems one encounters in this task, it is useful to actually go through the levels of detail that are encountered in creating a MOSFET, although this is only one type of device. Most of the necessary processing is quite similar in any type of semiconductor device, and this should be kept in mind while reviewing the topic. To do this, we follow a detailed process flow for a MOSFET with an effective 25 nm gate length. Although this process is about a quarter of a century old, much of it is still used in detail today, so it remains surprisingly relevant [53]. Now it is also important to understand that the end result may not be the proper one. That is, the device that has been built may not be the one that was either desired or designed. One would hope that it is quite close, especially if we want to say that it we understand the various processes from beginning to end.

The first step is isolation of the individual device, which begins with oxidizing the bare silicon wafer, particularly in the region where the device is to be located, as

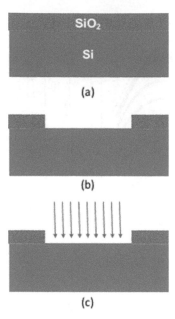

Figure 1.7. Isolation and forming the channel. (a) The oxidized silicon waver. (b) A hole is opened by lithographic techniques (see text). (c) Dopants are implanted into the waver at the desired location.

shown in figure 1.7. Shown in panel (a) is the oxidized wafer. After this, the sample is coated with a positive photoresist. With a positive photoresist, the molecules are usually long-chain molecules for which the light breaks up these chains. Then the developer dissolves these short chains leaving a hole where the light was projected. Then the oxide can be etched away leaving a hole, as in panel (b). The remaining oxide isolates this device from others. To create the p-type layer that will be the channel region, boron is implanted through the hole at an energy of 30 keV and a dose of 3.6×10^{13} cm^{-2}. The energy determines the depth of the implant while the dose will determine the final doping density in the p-type region. Now, we have to correct a problem. When the Si is implanted, the energetic ions severely damage the crystal structure of the Si. To repair this damage, the structure must be annealed. But, the annealing proceeds via an oxide regrowth process. In Si, the fast growth direction is the (001) direction, so the wafer we use has to be one in which the surface normal is the (001) direction. If it is any other direction, the fast growth direction will not be normal to the interface, and we will not be able to anneal out all of the damage. This follows as the growth proceeds in the (001) direction, which if it is not the surface normal causes many growth fronts to interfere, leading to dislocations and grain boundaries. In the case under discussion, the annealing is carried out at 1000 °C for 2 h. Of course, the implanted atoms will diffuse during the anneal procedure, and this has to be accounted for during the 'design' of the device. As a result of this thermal process, the oxide is regrown over the implanted region and the surface planarized for the next step.

The next step is to form the central gate oxide. Once more a positive photoresist is deposited on the oxidized wafer, and exposed and developed. This time, however, the new hole is smaller than the previous one and must be located in the center of the p-type layer formed with the previous hole. This new hole is shown in figure 1.8(a), and is expanded in panel (b) (red dashed lines indicate the expansion). Once the hole is opened in the resist, the oxide is etched away so that the surface of the p-region is exposed. Now, the gate oxide is grown at 800 °C for 8 min, producing a 3 nm thickness. Over the top of this is deposited/grown polysilicon which will serve as the actual gate material (dark blue in figure 1.8(b)). This polysilicon is heavily phosphorous doped to make it n-type and to have a low resistivity. At this point, it should be pointed out that the shift to an alternate gate material, of high dielectric constant, was made a few years ago. Typically, this new 'oxide' is hafnium oxide. The technology was developed at Intel (and elsewhere) [54]. The process was introduced in the 45 nm 'node', where the effective gate length is about 25 nm, in 2007–8. In the process flow, the deposition of the polysilicon gate is considered to be a dummy, in that it will be removed later and replaced with the new oxide. After the source and drains and stressor components have all been introduced (discussed below), the polysilicon gates and the gate oxides are removed. Then, the HfO$_2$ is deposited by atomic layer epitaxy, which basically is a set of chemical reactions that leave the desired material layer in place [54, 55]. The higher dielectric constant means that the same gate capacitance can be achieved with a thicker oxide, which is necessary to prevent tunneling through the thin oxide. The downside is that the gate itself is no longer polysilicon, but is another material which is chosen to have the proper work function for the device design. Typically, now different gate materials are used for the n-channel and the p-channel.

Now that the gate material is deposited, it is time to pattern it to the actual short gate length desired. This time, a negative photoresist is deposited on the gate material. With a negative photoresist, we desire the exposing light to cross-link the material which is dominantly short molecules. Hence, when developed, the exposed pattern is all that remains. This exposure is typically carried out with an excimer laser, but even this has a wavelength that is too large to make the small gate desired.

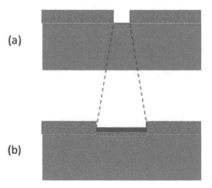

(a)

(b)

Figure 1.8. (a) The gate region opening is produced in the oxide. (b) The thermal oxide gate is then grown, followed by deposition of the heavily doped polysilicon gate.

So, the photoresist that remains, shown in figure 1.9(a) is thinned in an oxygen plasma (figure 1.9(b)). The resist is thinned until it is 40 nm in length in the source-drain direction. Hence this is called a 40 nm *drawn* (lithographically defined) gate. This pattern is then used to protect the desired gate while the rest of the polysilicon is etched away with a reactive ion etching process leaving the desired gate in figure 1.9(c).

It should be remarked that the gate is oriented so that the carriers (in this case, electrons) will move in the (110) direction in the surface layer. From figure 1.4, and the discussion about it, it may be recalled that the minimum of the conduction band lies along the Γ to X line, which is the (100) direction, and lies about 85% of the way to X. Because of the crystal symmetry, there will be 6 such minima, all of which are equivalent. A constant of energy surface near this energy will be an ellipsoid of revolution (a cigar-like shape). The long axis of the ellipsoid is parallel to one of the (100) directions. Hence, in our device, 4 of these valleys lie in the surface plane, and 2 of them are oriented with the long axis normal to the plane. Now, the (110) direction makes an equal angle with all 4 of the ellipsoids that lie in the plane, so that the transport in the device does not see differences in the 4 ellipsoids. Hence, the transistors are almost always oriented in the (110) direction.

To proceed, we have to face another problem. The natural approach is to now to implant the source and drain dopants, in order to convert the *p*-type layer to a heavily doped *n*-type layer (creating the *n-p-n* doping profile down the channel). If we do this directly at this point, using the gate as a mask to keep the implanted ions from the region under the gate, these dopants will diffuse into this masked region during the annealing process after the implant. This will do away with the *p*-layer entirely and ruin the transistor. We could place a spacer layer on either side of the

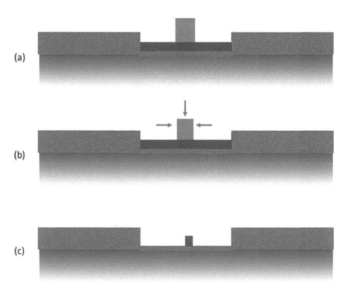

Figure 1.9. (a) The negative photoresist after exposure and development. (b) This resist is then thinned in, for example, an oxygen plasma. (c) The final gate, after the resisit is stripped away.

gate to account for the diffusion but this makes the size of this spacer hyper-critical to the entire process. However, this is the approach we shall follow, but the spacer will be made larger than the anticipated diffusion along the surface, and it will be made of a special material. In this case, the spacer is a phospho-silicate glass (PSG). It is deposited, and again patterned by a lithographic process to result in a structure such as that of figure 1.10(a), where the PSG is shown in green. This will serve two purposes. First, it protects the channel area from diffusion of the implanted atoms. Secondly, it will provide a source of phosphorous atoms that will diffuse out of the PSG during the anneal and these will form a shallow doped layer that extends from the source/drain regions to the channel. These regions are known as source(drain)-extensions. To create the proper source and drain regions, As atoms are implanted at 30 keV with a dose of 5×10^{15} cm^{-2}. The PSG has a phosphorous concentration of about 3×10^{21} cm^{-3}, and provides a relatively constant source of this dopant during the anneal. The anneal itself is done by rapid thermal annealing (RTA) at 1000 °C for 5–10 s. The implantation and anneal of the As gives source and drain regions that are doped at approximately 5×10^{19} cm^{-3}, and extend about 70 nm below the surface. The extensions created by the phosphorous out-diffusion from the PSG, are doped much higher and extend some few nm from the surface, as shown in figure 1.10(b), which displays what is essentially the completed transistor, lacking only the metallization layers. One should note that even the shallow phosphorous diffusion lets the atoms penetrate into the region under the gate. This gives the distance between the two (red) extensions as about 25 nm, which is called the *effective* gate length.

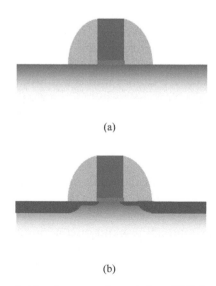

(a)

(b)

Figure 1.10. (a) The addition of sidewall spacers, formed from PSG (in green), helps in the transistor fabrication and also helps keep the implanted source/drain atoms from diffusing into the channel regions. The polysilicon gate is the top blue region, while the narrow grey region is the gate oxide. (b) The finished transistor. The red regions represent the heavily doped n-type areas that form the source and drain regions of the transistor.

Now, there is an additional detail that has appeared in recent years, and that is the use of strain to change the effective mass, as was pointed out in section 1.4.1. To accomplish the tensile strain in the n-channel device, a material such as Si_3N_4 is laid over the gate stack (the gate itself and the sidewall spacers in figure 1.10) at high temperature. As it cools, it expands and imposes the tensile strain on the channel material. For the p-channel device, we need compressive strain. To achieve this, we do not implant the source and drain as described above. Instead, we use another lithographic step to open holes in the p-Si layer where the source and drain are desired to sit. Then, acceptor doped SiGe is regrown in these holes. With the large lattice constant of this latter material, the expansion imposes compressive forces upon the channel material between the source and drain.

1.4.3 The GaAs/AlGaAs heterostructure

The most popular material system for mesocopic devices is the GaAs/AlGaAs heterostructure system. In GaAs, the zinc-blende lattice is a face-centered cubic (fcc) lattice in which each lattice site is occupied by a diatomic molecule of one Ga atom and one As atom. If the Ga atom sits on the lattice site, the As atom is displaced one quarter of the distance across the cube in the body diagonal direction; e.g., (a/4) (111), where a is the edge of the fcc cube. These two atomic sites are known as the A site and B site. GaAlAs is an example of a ternary alloy system in which some fraction of the A site Ga atoms are replaced with Al atoms. Thus, if 30% of the Ga atoms are randomly replaced with Al atoms, the alloy is referred to as $Ga_{0.7}Al_{0.3}As$, or $Ga_xAl_{1-x}As$ with $x = 0.7$. It is important to point out that this is a totally random alloy in principle, so that there is no clustering or precipitation of various compounds within the crystal. One reason for the ternary is that the band gap increases as the percentage of Al increases, thus one can create a series of quantum barriers and wells with multiple layers of the two compounds. However, AlAs is an indirect material, and the alloy becomes indirect at around $x = 0.55$. One usually stays with the more Ga-rich alloys, especially for optical applications, so that the band gap is direct at the Γ point. Another important issue is that the lattice constants of AlAs and GaAs are nearly the same. Hence, the alloy can be grown on GaAs with almost zero strain in the crystal, and this results in almost atomically sharp interfaces [56].

The band gap of AlGaAs is larger than that of GaAs, and this difference must be taken up by band bending at the interface. Part of the band discontinuity is taken up in the conduction band and part in the valence band, as shown in figure 1.11. Currently, it is felt that the conduction band discontinuity is about 63% of the total energy band discontinuity [57]. Prior to bringing the two materials together (conceptually), the Fermi level will be set in each material by the corresponding doping. Usually, the GaAs is undoped so that the Fermi level is near mid-gap and thought to be set by a deep trap level in this material. Once the interface is formed, however, there must be a single Fermi level that is constant throughout the material, as no current is flowing. This leads to the band bending as shown in the figure, and the band discontinuities provide certain offsets that are shown in the figure. In the

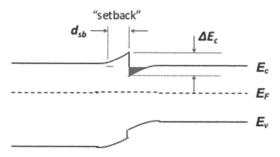

Figure 1.11. The band lineup at the hetero-interface between GaAs (on right) and AlGaAs (on left). The setback of the dopants is also indicated.

early days, it was common to uniformly dope the GaAlAs, but this is no longer done. Rather, a single layer of dopant atoms is placed in the GaAlAs a distance d_{sb} from the interface [58]. Regardless of the doping method, electrons near the interface will move to the lower-energy states in the quantum well on the GaAs side of the interface. For the δ-doping case (a single layer of dopants), all of the electrons will move to the GaAs. In the uniform doping situation, only a small fraction will move. This technique of getting the electrons into the GaAs is known as modulation doping [59]. In this approach, the actual ionized dopants are set some distance from the electrons, so that the Coulomb scattering potential is weakened. Moreover, the electrons themselves work to screen this scattering potential. These effects lead to very high mobilities for the electrons in the GaAs. In fact, mobilities above 10^7 cm^2 Vs^{-1} can be obtained for the electrons in the GaAs at low temperatures [60] and this is thought to be limited by scattering from the dopants themselves [61]. The usual dopant for the GaAlAs is silicon, which acts as a donor. However, it can also form a complex which leads to a trap level, known as the DX center [62]. This trap can be avoided if the composition of the GaAlAs is kept below about $x = 0.25$. So this sets additional limits on the heterostructure.

The GaAs/GaAlAs heterostructure is typically grown on a GaAs semi-insulating substrate. First, a superlattice formed of thin layers of GaAlAs and GaAs is grown. This has the double effect of smoothing the surface and trapping dislocations within the superlattice. Then, a thick undoped GaAs layer is grown, followed by the GaAlAs layer. For the latter layer, the growth is interrupted to deposit the dopant layer a desired distance from the interface, and then the additional ternary is grown to the desired thickness. For the highest mobility layers, d_{sb} can be more than 20 nm thick. The thicker this set-back layer is, the higher the mobility will be and the lower the inversion layer density will be. Finally, a heavily doped GaAs 'cap' layer is grown on the surface, which serves two purposes. First, it prevents unwanted oxidation of the GaAlAs surface. Second, it provides a layer to which it is easier to make ohmic contacts. The preferred growth method for high quality material is with molecular beam epitaxy, a growth technique achieved in an ultra-high vacuum. The atomic constituents are provided by heated sources, known as Knudsen cells, in which the atomic species are individually vaporized (within the Knudsen cell) and shutters are used to turn on and off the flow of atoms from the cell. Careful control

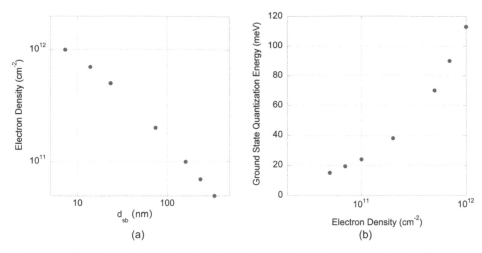

Figure 1.12. (a) The electron density as a function of the setback distance of the δ-doping layer. (b) The energy of the lowest quantum level as a function of the electron density.

of the deposition rates and the substrate temperature can lead to atomic layer epitaxy and the ultimate control of the growth. However, this process leads to low growth rates, below a micron per hour, so it is not conducive for thick structures. This growth process, and slow grow rate, allows for the design of semiconductor multi-layers with very precise control of the overall structure for quite specific applications. Quantum well structures are created by sandwiching a GaAs layer between two GaAlAs layers, and the bound states in the well can be precisely tuned by careful control of the thickness of the GaAs layer and the effective barriers formed by the band offsets arising from a precisely controlled alloy composition. This process has led to a wide variety of emitters and detectors of radiation over a wide spectral range.

In figure 1.12, we illustrate the density and quantization energy in the inversion layer for a GaAlAs/GaAs heterostructure. This is an estimate using the 'bound state calculation lab' at NanoHub.org [63]. The tool was used to find the quantization energy for a given density, from which the Fermi energy could be found. We used a δ-doped layer of 10^{12} cm^{-12} which was set back a distance d_{sb} and assumed a conduction band offset of 0.25 eV. It can be seen that the larger one makes the setback distance, the lower the electron density becomes. At the same time, this will increase the mobility by moving the ionized dopant atoms further from the electron layer.

1.4.4 Other important materials

If we consider just the binary III–V materials, there are already a large number of possible heterostructures that can be grown. In figure 1.13, we plot the energy gaps as a function of the lattice constant for the group IV and III–V compounds. Also shown are some rough connector lines for a few ternary alloys. One popular substrate is InP, not the least because it is lattice matched to In$_{0.53}$Ga$_{0.47}$As which

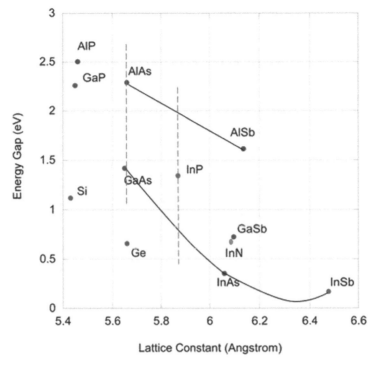

Figure 1.13. A wide range of possible materials can be chosen for specific properties of energy gap and lattice constant to fit desired characteristics and a convenient substrate.

has an energy gap well suited to match the minimum dispersion in quartz fibers, a match very important for long distance fiber communications. The quaternaries InGaAsP or InGaAsSb provide the ability to both lattice match InP as a substrate and to vary the band gap over a very wide range. The ability to thus choose a material to provide desired characteristics is quite important in these materials. InAs has become important for THz high-electron-mobility transistors (HEMTs) as well as with AlSb for spin applications.

The group III nitrides provide another set of materials with a wide range of attributes. For example, GaN has a band gap of 3.28 eV and an *a*-plane lattice constant of 0.316 nm. In AlN, these values are 6.03 eV and 0.311 nm, while in InN, these values are 0.7 eV and 0.354 nm. Thus, alloys of these materials can span the entire visible range, and GaN-based systems have found a home in blue and blue-green lasers. They are also being pursued for high power HEMTs. While one normally tries to lattice match the various layers in a heterostructure, a controlled mismatch can be used quite effectively. The group III nitrides tend to be ferro-electric, which means that they have a built-in polarization in the lattice. The discontinuity in this polarization at an interface can be used effectively to induce the inversion layer charge to form without the need for dopants [64].

We cannot ignore the newer two-dimensional materials. The world is full of real two-dimensional semiconductors, which are called layered compounds. The best known on is graphite, in which layers of atomically thin carbon are weakly bound together to produce the bulk. The single layer of carbon atoms is called graphene, and it has been isolated only recently [65]. In this single layer graphene, the carbon atoms are arranged in a hexagonal lattice, which has two atoms per unit cell. The single layer of graphene is exceedingly strong, and has been suggested for a great many applications. There are other layered compounds, which have similar structure, and a well-known group is the transition metal di-chalcogenides (TMDC), such as MoS_2 and WSe_2 [66]. In the TMDC, the layer is actually an atomic tri-layer, with the metal atoms forming the central layer, and the chalcogenide atoms forming the top and bottom layers. Each metal atom has a triangle of three chalcogenide atoms above it and below it. Yet the basic structure still consists of hexagonal coordination with two so-called atoms per unit cell—these consist of a metal atom and an up and down pair of chalcogenide atoms, which sit above one another to form a pseudo-atom. As a result, both graphene and the main TMDCs have a quite similar band structure. In the graphene case, the conduction and valence bands are composed of the p_z orbitals of carbon [67]. In the TMDC case, the conduction and valence bands are determined by the metal d orbitals [66], as these bands lie in the gap between the bonding and anti-bonding sp hybrids. The TMDC materials have some interesting properties, particularly with their spins, as they basically lack inversion symmetry in the plane of the layer. As a result, the spin–orbit interaction can lead to unique effects such as the spin Hall effect. We will deal with these materials in subsequent chapters—graphene in chapter 4 and the TMDCs in chapter 5.

1.5 Superconductors

Superconductors have been an interest set of materials for a great many years. Kammerlingh Onnes discovered superconductivity in 1911, while examining the properties of metals at low temperatures [68]. He had succeeded in liquefying He only three years earlier and was anxious to use this new cryogen. As he cooled mercury below 4.2 K, he observed a strange phenomenon—the resistance dropped almost to zero abruptly at this temperature. By the time it had been cooled to 3 K, the resistance was less than 10^{-6} Ω. This phenomenon has been observed in more than one quarter of the elements of the periodic table. And, it has been observed in a number of compounds at higher temperatures (today referred to as high-temperature superconductivity). The temperature at which the transition begins is known as the *transition temper*ature T_c, and the transition itself is recognized as a thermodynamic phase transition. Nothing much was done with this phenomenon, primarily due to the scarcity of liquid helium, until after the Second World War. Subsequently, the needed cryogen began to be available in quantity and cheaply, which led to a relative boom in studies. Then, these materials became of use in producing magnets, both for experimental systems for science and for their application in motors and generators. Perhaps the largest applications are for measuring derivatives of the magnetic field using superconducting quantum interference devices (SQUIDS) by the military, and

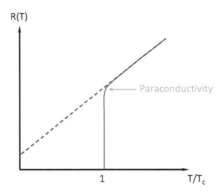

Figure 1.14. Superconductivity appears below the transition temperature T_c. There is a fluctuation induced lowering of the metallic resistance just above this temperature, which is called paraconductivity.

now for the burgeoning field of quantum computing. We will deal with the latter in detail both below and in chapter 8.

The most easily observed characteristic of superconductors is the *transition temperature T_c*, below which superconductivity occurs in the material. One can readily observe the transition temperature when one plots the resistance versus the temperature, as in figure 1.14. The temperature scale in the figure is normalized to the transition temperature, T_c. There is a slight rounding of the curve just above T_c, and the enhanced conduction in this region is often called *paraconductivity*, in analogy with paramagnetism, or the enhancement of magnetism. This enhanced conductivity is attributed to thermal fluctuations at the transition temperature, and we can think of them as very small regions, which are beginning to exhibit superconducting behavior. Superconductivity of the entire sample does not occur, since the small regions are superconducting for only short periods of time and are generally unconnected. In a sense, these are like the domains exhibited in magnetic field. Only a few of the domains are superconducting, and then only for short periods of time. The drop in resistance at $T = T_c$ is exceedingly sharp, and occurs over a small fraction of a degree, so that the onset of superconducting behavior is easily observed. The transition temperature is a fundamental property of superconductors.

As we mentioned above, only about one-fourth of the elements of the periodic table exhibit superconductivity. Such elements as silver and copper are not super-conductors at the lowest temperatures measured. In fact, it seems that the higher the conductivity of a material in the normal state, the poorer this material is as a superconductor. This implies that if a material is a good conductor in the normal state $(T > T_c)$, it probably has a very low transition temperature or does not exhibit superconductivity at all. We can understand this effect by thinking about the dominant interaction mechanism leading to superconductivity. This interaction, at least in the pure compounds, is generally conceded to be the electron-lattice interaction, that is, the interaction between two electrons of opposite spin mediated by interactions with the lattice. The stronger this interaction is, the more likely a material is to be a superconductor. However, this is related to the dominant scattering mechanism in normal metals. If the electrons are strongly scattered and the conductivity is low, this interaction is strong, and the conductivity is low. Hence the material is a poor conductor.

An exceedingly useful property in superconductors is the presence of *persistence currents*. Consider, for example, a superconducting ring. If a current is induced in the material in the normal state, and the material is then cooled below T_c, the current in the ring will persist for a very long time. Since the material has no resistance in the superconducting state, there is no decay mechanism for the current. Such a mechanism is exceedingly useful in superconducting magnets. If a fixed value of magnetic field is required, the superconducting solenoid is energized to the proper current value by an external power source. Then the current leads are shorted together at the magnet by a second superconductor. This forms a superconducting loop, and the external supply can be turned off, while the circulating current in the loop remains unchanged.

1.5.1 The Meissner effect

A magnetic field can be used to destroy superconductivity. If a magnetic field is applied to a superconductor, then, for $H > H_c$, a critical value of the magnetic field which is different for different superconductors, the superconductivity is destroyed and the material reverts to the normal state. It is generally found that H_c is a function of temperature as

$$H_c = H_{c0}\left[1 - \left(\frac{T}{T_c}\right)^2\right],\tag{1.21}$$

where H_{c0} is the *critical field at absolute zero*. The critical field at $T = 0$ K is a function of the critical temperature. The additional magnetic energy of the electrons serves to break the superconducting bond interaction, just as thermal energy does. The effect of both types of additional energy terms leads to a result like equation (1.21). The critical magnetic field will also limit the amount of current which a superconductor may carry. Since the current itself gives rise to a magnetic field, the current carried by a superconductor must be less than that which would produce H_c. For a higher current than this, the magnetic field produced by the current destroys superconductivity and the material reverts to the normal state. Thus, superconducting magnets utilize materials with a high critical temperature, such as Nb_3Sn, where $T_c = 18.05$ K.

The onset of superconductivity leads to a phenomenon called the *Meissner effect*. If a superconductor is cooled below T_c, the magnetic flux lines are expelled from the material. Thus, within a superconductor, the magnetic flux density is zero, $B = 0$. Hence, superconductors exhibit perfect diamagnetism, $\mu_r = 0$. The magnetization which results from the applied field must completely oppose any applied field. This occurs up to the critical magnetic field in type I superconductors. But, in many compound superconductors, there is a lower critical field, H_{c_1}, where the Meissner effect begins to fail. Thus, there is incomplete expulsion of the flux, and this gets weaker and weaker as the external field intensity increases, up to an upper critical field, H_{c_2}, where the superconductivity vanishes.

A somewhat different behavior is observed if the superconductor is fabricated as a thin film. If the films are very thin, the material does not exhibit perfect diamagnetism. As we shall see in the next section, the film must be of a certain thickness before it exhibits the properties of bulk superconductors, such as perfect diamagnetism. In addition, in a thin film, the magnetic field has very little effect and the critical field is much higher than for the bulk material.

1.5.2 The BCS theory

A number of the preceding experimental observations are consistent with the existence of an *energy gap* associated with superconductivity. Theoretically, the electrons which contribute to superconductivity do so as paired electrons. This set of two electrons is known as a *Cooper pair*, and is formed from two electrons which have opposite spin angular momentum [69]. Thus, if one of the electrons is spin up, the other is spin down. Normally, electrons repel one another because of Coulomb forces. But at low temperatures, an additional interaction between the electrons, in which they interact with the lattice, leads to a weak attractive force between the electrons. This leads to a pairing of the electrons. This pairing of the electrons results in a lower energy state than would result from the electrons remaining unpaired. This, in turn, results in an energy gap of the electrons [70]. The observed energy gap, though, would correspond to $E_G = 2\Delta$, since each of the paired electrons must receive an additional energy of Δ in order to break the pairing bond, as shown in figure 1.15.

At absolute zero, all the conduction electrons are paired. At higher temperature, some of the pairings are broken by thermal agitation, so that some normal electrons are excited across the energy gap in much the same way that electrons are excited across the band gap in semiconductors. For temperatures slightly above T_c, all the pairs are broken and the material exhibits normal resistivity. In the superconducting state, a mixture of normal electrons and superconducting paired electrons is present in the material. The normal electrons may be observed in tunneling experiments, such as we discuss in a later section of this chapter. The energy gap also varies with

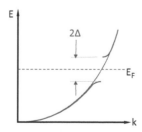

Figure 1.15. An energy gap 2Δ opens at the Fermi energy and this gap separates the superconducting electrons from the normal electrons. At absolute zero of temperature, all the electrons form Cooper pairs. This has the unusual requirement that the electron number is an even quantity.

temperature, decreasing with increasing T until, at T_c, the energy gap is zero. The general behavior of the gap near T_c is given by

$$E_G = E_{G0}\sqrt{1 - \frac{T}{T_c}}, \tag{1.22}$$

where E_{G0} is the value of the gap at zero temperature. As a first approximation, the energy gap at zero temperature is related to the energy gap at the transition temperature as $E_{G0} = 3.5k_B T_c$. This result is not exact, but is approximately correct for a wide range of superconductors.

The basic developments of the BCS theory show that an attractive force between the electrons can lead to a lower energy state at low temperatures, and, hence, can lead to an energy gap, as discussed above. For the attractive force, their theory suggests a second-order interaction between electrons and the lattice. One electron interacts with the lattice; this interaction deforms the lattice slightly in the neighborhood of the electron. This deformation is much like the modification of the density of the electrons near an ionized impurity that leads to electron screening effects over a length of the order of the Debye length. In the BCS theory, a similar effect occurs between the electron and the deformed lattice. A second electron encounters the deformed lattice and interacts with it. Thus, in effect, the second electron interacts with the first, but does so through the lattice interaction. Effectively, one could say that the deformed lattice shields the electron's repulsive force (which is weaker for electrons with opposite spin) so that the second electron is attracted to the deformed area, and hence to the first electron. An important aspect is that the two electrons must have opposite spin, otherwise the Coulomb interaction between them is just too strong for them to form a pair. The attractive force between these two electrons, with opposite spin, *lowers* the energy of the pair. Thus, it costs energy to break the pair and a gap opens between the paired electrons and the un-paired electrons, as shown in figure 1.15. The details of the BCS theory are beyond the scope of this book.

Normally, when a gap opens, the electrons cannot carry any current as they cannot gain any energy due to the fact that the states below the gap are completely full. This was important in the band theory of semiconductors. In fact, super-conductors carry current *without dissipation*, so there is no need for the paired electrons to gain energy. In fact, the pair will hold together until a large amount of energy, given by the gap, is injected into the pair by some mechanism. Hence, there must be a critical minimum velocity, which corresponds to the presence of an energy gap. If the velocity of the electrons is less than the value of the critical velocity, no dissipation, or energy loss, occurs, and the material has zero resistance. However, for sufficient energy, the pair is broken and normal electrons occur.

The existence of the energy gap can be readily demonstrated by measurements at microwave or infrared wavelengths [71]. At absolute zero, photons with energy of $\hbar\omega < E_G = 2\Delta$ are not absorbed, while those with by $\hbar\omega > E_G = 2\Delta$ cause transitions across the energy gap and result in absorption of photons. The photon energy is absorbed by the electron pair, and this additional energy is sufficient to break the bond of the pair. For temperatures above zero, some absorption occurs due to the presence of a few normal electrons. Since these normal electrons are already excited across the gap, they are free to absorb any photon that comes along.

1.6 Bits and qubits

When the ENIAC was first developed at the University of Pennsylvania at the end of the Second World War, it computed with base-10 circuits [72]. The simplest operation, the addition of two numbers took 0.2 ms. But, calculation in base-10 is somewhat wasteful of computer resources. It was John von Neumann who suggested at the time that the binary system should be used (and he also suggested the stored program approach, which along with binary arithmetic, formed what is called today the von Neumann architecture [73], the EDVAC was the successor of the ENIAC). Essentially, all computers from this time forward have used the binary system, as it is eminently suitable to two-state bits. In this space, we define the two unit values for a bit as the states 0 and 1. Then, using quantum notation, we define the state of a bit in this space as

$$|\psi\rangle = a|0\rangle + b|1\rangle, \tag{1.23}$$

where

$$a, b \subset (0, 1), \quad a + b = 1. \tag{1.24}$$

Hence, a and b can be either a 0 or a 1, but the two must be different.

In the computer itself, a particular state is defined by the values of the bits used to describe the state. Then, one can define a state transition diagram, in which each possible state is a node, and transitions between these nodes indicate the action taken when a control signal is applied [74]. Various outputs arise from the set of transitions and the state of the machine after the transition. Hence, an algorithm is a set of instructions that guide the system through a set of transitions, changing the various bits under the guidance of the control signals, to a desired result. Good algorithms lead to the machine halting at the end of the algorithm and yield a desired output (and with bad algorithms, the machine sometimes never stops). In fact, the stop state was required by Turing, as he showed that a number was computable only if the machine stopped [2]. This same operational model will exist also in the quantum computer. That is, the qubits will define a state, state transitions will depend upon control signals.

Let us diverge a bit, as ever so often, someone suggests that the quantum computer can be reversible and therefore not dissipate energy. But, this is a perpetual motion machine, even if it is superconducting. Above, we designated the bit (or qubit) by $|\psi\rangle$. Suppose we have n bits in the machine. Then, there will be 2^n combinations of the qubits and this many actual states of the machine. Now we can express the value of the state by a many-body state with n 0's or 1's total. With this designation, the state transition matrices will be $n \times n$ matrices, each selected by a particular input. However, we can also express the state by a many-body matrix with $2^n - 1$ zeroes and a single 1 [75]. This single 1 designates in which state the machine exists. Now, if the machine is to perform reversible logic, each state must have a single successor state and a single predecessor state; e.g., the mapping must be 1:1 in each direction—no fan out or fan in. In this case, the state transition matrix is such that each row has a single 1 and each column has a single 1, and the matrix is full rank. It turns out this type of matrix is a characteristic matrix of the cyclic permutation group of order 2^n. Thus, each state lies on a ring upon which it cycles forever. This machine has no stop state and cannot do computation! And, it is a perpetual motion machine. Hence, a machine such as this does not meet the general requirements for computation [76].

When we move from the bit to the qubit, the change is rather subtle. We still have the two values 0 and 1, which can considered to be coordinate axes representing these values. But, now the wave function is analog and can take any value on the unit circle. Hence, equation (1.23) is still valid, but the coefficients now must satisfy

$$|a|^2 + |b|^2 = 1. \tag{1.25}$$

Thus, the two coefficients are complex numbers, and equation (1.25) tells us that the net wave function must have a magnitude of 1, as required by any quantum wave function. Now, the problem that arises is that this formulation is overly *restrictive* to the actual wave function for the quantum qubit. The two states 0 and 1 can be thought of as representing real energy levels of a two level atom (we will return later to why we use the atom model, and connect the argument with spin in chapter 7). For convenience we assume these two states are aligned along the z-axis in spherical coordinates. From the latter concept, we then tend to write the Hamiltonian representing this two level system with the Hamiltonian

$$H = \frac{\Delta}{2}\begin{bmatrix} 1 & 0 \\ 0 & -1 \end{bmatrix}, \tag{1.26}$$

and the wave function (1.23) is expressed as

$$\psi = \begin{bmatrix} a \\ b \end{bmatrix} = \begin{bmatrix} |a|e^{i\varphi_1} \\ |b|e^{i\varphi_2} \end{bmatrix}. \tag{1.27}$$

Hence, we have mapped the (0,1) states into the states $(-1,1)$, in terms of $\Delta/2$, with Δ the energy difference between the two states. Hence, the upper level is taken to be the 1 state and the lower level the -1 state.

It is convenient to write the wave function in a slightly different form which characterizes the system, but make the two phases somewhat less arbitrary. Hence, we write the two-level system in terms of the density matrix

$$\rho = \begin{bmatrix} |a|^2 & ab* \\ a*b & |b|^2 \end{bmatrix}, \tag{1.28}$$

from which we infer that equation (1.25) requires the trace of the density matrix to satisfy

$$Tr\{\rho\} = 1. \tag{1.29}$$

In this approach, we introduce the polarization $\boldsymbol{P}$, and write equation (1.28) as

$$\rho = \frac{1}{2}(I + \boldsymbol{P} \cdot \boldsymbol{\sigma}), \tag{1.30}$$

where I is the 2×2 unit matrix and the vector $\boldsymbol{\sigma}$ has the components σ_x, σ_y, σ_z. These so-called spinors are each a 2×2 matrix, which will be discussed in chapter 7. The important point here is the polarization, which is related to the coefficients in equation (1.25) as

$$
\begin{aligned}
P_x &= 2\mathrm{Re}(a^*b) \\
P_y &= 2\mathrm{Im}(a^*b) \,. \\
P_z &= |a|^2 - |b|^2
\end{aligned}
\tag{1.31}
$$

The language in the above description of the density matrix gives the polarization in the real three dimensional space. But, we can think of the polarization itself as a vector in an abstract three dimensional Euclidean space. In this space, the state 0 is a unit vector pointing in the $+z$ direction, while the state 1, is a unit vector pointing in the $-z$ direction. The polarization itself resides on the unit sphere in this space; this unit sphere is called the *Bloch sphere*. Such a Bloch sphere is shown in figure 1.16, and the two angles are defined as the polar angle and the azimuthal angle, where the former is the polar angle measured away from the $+z$-axis, and the latter is the azimuthal angle and lies in the (x,y) plane and is measured from the x-axis, as is normal for spherical coordinates. Thus, many qubit operations are easily expressible as rotations around the three axes of the Bloch sphere.

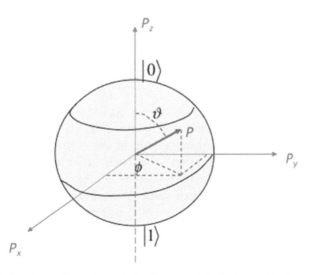

Figure 1.16. The Bloch sphere allows the polarization to rotate in three directions. Then, many qubit operations are expressible as rotations around one of the three axes.

To examine the nature of the polarization itself, let us make some angular definitions of the various components of the wave function (1.23), as

$$a = e^{i\gamma}\cos(\vartheta/2)$$
$$b = e^{i\gamma-\phi}\sin(\vartheta/2)$$
(1.32)

so that

$$P_x = \cos(\phi)\sin(\vartheta)$$
$$P_x = \sin(\phi)\sin(\vartheta)\ .$$
$$P_x = \cos(\vartheta)$$
(1.33)

In the above discussion, we talked about the qubit as a two-level system, specifically a two-level atom. This could be two specific levels of a real atom, such as in the ion trap system, or in an artificial atom such as a quantum dot. The important point is that the phrase 'two level atom' is often a 'code word' for the qubit system. This is because the energy levels in an atom are nonlinearly spaced; that is, they vary roughly as $1/n^2$. Hence, the two desired levels are separated by an energy that is not the same as the separation of any other two levels. If the levels were linearly spaced, as in a normal harmonic oscillator, one has a problem. For example, if the lower two levels are those of interest, they are separated by exactly the same energy as any other two adjacent levels. This means that excitation of the lower two levels can also be absorbed by any other two levels, and this constitutes a sizable source of decoherence in the qubit. Hence, one desires the energy levels to be nonlinearly spaced, so that the two levels of the qubit do not couple to other unused energy levels. Since, the atom has this property, the phrase 'two level atom' is often applied to any qubit system to indicate it has the desired properties.

One of the most important aspects of the quantum processor is *entanglement*. Indeed, Erwin Schrödinger called entanglement the most important aspect of quantum mechanics [77]. This paper was his response to the Einstein–Podolsky–Rosen paradox published earlier in the year [78]. His point was that once the two particles (envisioned by the latter authors) interacted, they were no longer independent from one another. Rather, they must now be described by a single entangled wave function incorporating both particles. And, it was this entanglement that differentiated the quantum system from the classical one. Needless to say, it is entanglement that has become the crucial ingredient in quantum computing that gives rise to the possibility of major speedup in the computation. In the quantum computer, we want qubits to interact with each other under controlled circumstances. When two qubits interact, they become entangled, just as the two particles in EPR. They remain entangled until some mechanism gives rise to decoherence, which breaks up the entanglement. As a result, one needs long decoherence times in the qubits.

In general, the quantum computer works on controlled gates, as these gates provide the interaction between the qubits. As an example, we take the so-called CNOT gate, or controlled NOT. In the classical system, any transistor (or CMOS gate) is naturally a NOT gate. This is because the transistor inverts its input. If we take a high voltage as a logical 1 and a low voltage as a logical 0, then a 1 input to the gate causes the transistor to turn on, which lowers the output voltage, and this gives a 0 output. Similarly, when the input voltage is low, the transistor is in the off state and the output voltage is high. Thus, the input is inverted, which means that the logical output is NOT(input). The CNOT differs from this slightly, in that the action of the transistor gate is controlled in a manner that makes the classical XOR gate. Now, the XOR gate has two inputs, which are called a and b. If either a or b is high (a 1), then the output is also high. But, if both inputs are either high or low, the output of the gate is low. In other words, the output of the XOR gate is 1 if either a or b is high, but not when both are high. So, if we take input b as the control input, when b is high, a is inverted, and we thus have the CNOT. We can write the output as a wave function, in which the state of the control bit is the first signal and the state of x is the second variable:

$$\text{output} \rightarrow |\psi\rangle = \frac{1}{2}(|0\rangle|1\rangle \pm |1\rangle|0\rangle) = \frac{1}{2}(|01\rangle \pm |10\rangle). \tag{1.34}$$

That is, if the control bit (b) is 0, y is 1 only if x (a) is 1, and if the control bit is 1, y is 1 only if x is 0. The wave function (1.34) is entangled. That is, if we take the control bit as a two state Hilbert space, and the input bit as a two state Hilbert space, then the entangled output lies in a tensor product Hilbert space, denoted as $H^2 \otimes H^2$. But note that the wave is not a simple product, and it cannot be separate into parts that lie in only one of the individual Hilbert spaces. This is what defines the entanglement. Hence, the wave function in (1.34) is an entangled wave function. The plus/minus sign gives two possibilities, which both give the same value of 1 for the square magnitude of the wave function (the factor of ½ provides proper normalization). The introduction of the control concept is really accomplished by the manner in which the interaction between qubits is accomplished and managed. Hence, any algorithm tells us how to manipulate the control signals, which change the interactions so that entanglement is created, manipulated, or destroyed. This is no different than the classical computer where bit strings are pushed around, manipulated, and then stored or erased. Here, however, we presumably use entanglement to increase the power of individual qubits.

1.7 Some notes on fabrication

In creating mesoscopic devices, one generally combines the growth of a specific material or heterostructure with various processing steps such as those described above in section 1.4.2. As can be observed there, the two dominant processes that are universally used are lithography and etching to define the specific structure that is desired. We cannot do a thorough job of covering these processes, as that would require a book unto itself. Nevertheless, we give only introduction to the various processes that can be used to shape mesoscopic structures.

1.7.1 Lithography

Lithography is primarily the same process as photography in which an image is exposed and then developed. It does not matter whether one is carrying out optical lithography, electron-beam lithography, or something more exotic, the approach is much like the early days of photography. Anyone who has done darkroom work understands that the developed image results from a combination of the exposure (amount of light) and the developer. The wafer is our equivalent of the glass plate or the final print of modern photography. The first step is to spread some 'goop' on the surface of the substrate. This goop is a resist which is spun onto the surface. That is, the wafer is attached to the rotating plate of a machine which will then spin it at a high rate of revolution. The liquid containing the resist is dropped onto the spinning surface and the spinning action will spread the liquid to a nearly uniform thickness. Then, the wafer is baked to remove the liquid solvent leaving a hard resist coating. The progress that has been made since Matthew Brady's time is that the tent has been replaced with a cleanroom. Then, the resist is exposed, either by photons or electrons, or some other more exotic form of energy deposition. The developer then removes unwanted parts of the 'goop', leaving the desired image.

In the industry, the exposures discussed in section 1.4.2 have, in recent days, been produced using excimer lasers as the source, and hard masks in a projection tool which gives something like a 10:1 reduction in the image size. These tools step across the wafer to make a great many identical exposures on the typical 300 mm diameter wafer. Only quite recently has the XUV exposure tool, which uses 13.7 nm x-rays, been introduced with the new chip in the iPhone 11. Nevertheless, both of these approaches using large stepping machines are far too expensive for use in research labs where only 1 or a few devices are desired. Here, one typically uses electron-beam lithography, or ion-beam lithography, which differ from optical lithography in only a couple of ways. First, electron-beam lithography uses electrons rather than photons to deposit the energy in the resist [79]. Similarly, ion-beam lithography would use ions rather than electrons or photons [80]. The second difference is that one generally exposes a single pixel at a time in electron-beam, or ion-beam, lithography so that this is a serial writing method where the beam is raster scanned across the image area. In optical lithography, the light beam is a large area beam which usually exposes a chip at a time, and then is stepped across the wafer, as mentioned. In optical lithography, a photomask is used, and this is generally a metal coating on a glass plate, in which areas of the metal have been removed, corresponding to those areas of the chip that are to be exposed to the photons. In electron-beam lithography, or ion-beam lithography, the beam is turned on or off for each pixel (technically, the beam is 'blanked'), and the size of the pixel is set by the size of the beam spot. Higher resolution requires a smaller spot size.

There are basically two kinds of resists, positive and negative. In a positive resist, the electron beam breaks long polymer bonds, making this exposed material more able to be dissolved in a developer. In a negative resist, the electron beam is used to cross-link the material into a polymer that resists the dissolving action of the developer. The most common electron beam resist in the mesoscopic world is poly-methyl methracrylate (PMMA), and it is usually used as a positive resist that yields very high resolution. Negative resists are used, for example, in isolating a local region, such as a mesa, on the chip. Only a small part of the resist over the entire chip has to be exposed, and, after development, protects the mesa as the surrounding material is etched away. Positive resist is used where one wants to deposit a metal wire, so only the position of the wire is exposed. After development, the metal can be deposited and then dissolving the resist will 'lift off' the undesired metal.

Every photosensitive material, including our resists, is characterized by what is known as a density–exposure (D–E) curve. This curve plots the density of the remaining resist as a function of the exposure dose. As mentioned above for photographic development, the D–E curve is both resist and developer sensitive. That is, like many photographic films, the development can be 'pushed' to change various properties. The exposure–developer combination has a range of acceptable values, but one can often go outside this range for special effects. Two other factors also are important. One is the sensitivity of the resist; e.g., how much energy is required to initiate the bond breaking (or cross-linking). The second is the molecular weight of the particular resist. In the case of PMMA, one normally uses a molecular weight approaching a million, although good results can be achieved with lighter material, and lines as narrow as 7 nm have been written over a range of molecular weights [25].

The transition from unexposed to fully exposed resist is sensitive to both the resist and the developer. If we talk about the exposure necessary to give 90% of the thickness and the exposure necessary to give 10% of the thickness, we can define a parameter characterizing the D–E curve as

$$\gamma = \frac{1}{ln(E_{10}/E_{90})}. \tag{1.35}$$

Since we want to have as high a contrast as possible, we want this parameter to be as large in magnitude as possible. For a positive resist, we have $E_{90} < E_{10}$, so the above γ is positive. For a negative resist, the inequality will be reversed and we have to flip the ratio to give a positive value for this parameter. It is possible to obtain $\gamma > 10$ with special combinations of developers [81, 82]. High resolution is possible to achieve in a variety of common and uncommon resists [83]. In our discussion above of lifting off undesired metal, it is clear that high resolution is required. Otherwise, one will not achieve a complete break from the metal in the exposed grove and the metal on top of the unexposed resist. Lack of this break will prevent a clean liftoff of the metal. Using multi-level resists, in which the lower levels of resist have less resolution and the top layer provides the thin opening, will allow better liftoff of the metal and better resolution [84].

Other forms of electron-beam lithography have been developed over the years which do not require the use of a photoresist. Erasable electrostatic lithography uses the electron beam to deposit charge on a non-conducting surface [85]. This charge repels electrons and can be used to define a confining potential for electrons in the GaAs 2DEG. An STM can also be used to oxidize the surface of metals [86] and semiconductors [87] to provide a depletion of the electrons under the oxide, which will also form a confining potential.

1.7.2 Etching

Etching involves the removal of material by a chemical process [79]. This can be achieved either by a liquid etch, known as wet etching, or by gaseous etching, known as dry etching, or often as reactive-ion etching. Liquid etches tend to be isotropic, which means that they etch equally as fast in all directions. For our nanostructures, the process of mesa isolation, mentioned above, is most effectively achieved by a wet etch as a great deal of material needs to be removed. As wet etching is a chemical process, the exact chemistry used will depend upon the material that is being etched. The etch rates can depend upon the details of the material as well as properties such as the dopant concentration.

If we are etching grooves or shallow trenches, or transferring a thin line opening in the resist to the underlying material, anisotropic etching is highly preferable. With proper control of the etching conditions, anisotropic etching is readily achieved with dry etching. Here, there are a number of good etching gases that have relatively similar etch rates for a wide variety of materials.

It is important to understand that etching can be used to define nanostructures. We recall that the surface of many semiconductors is pinned by defects so that the Fermi level at the surface is in the band gap, often near the center of the gap. For example, the Fermi level at the GaAs surface is usually pinned about 0.8–0.9 eV below the conduction band regardless of whether the surface is a free surface or a metallized surface. This can work to deplete any electrons in the 2DEG, if the surface is brought sufficiently close to the electron layer. This is the principle used in mesa isolation described earlier; we do not need to etch completely into the GaAs layer. We only need to etch away a sufficient amount of the GaAlAs layer so that the surface depletion reaches the electron layer.

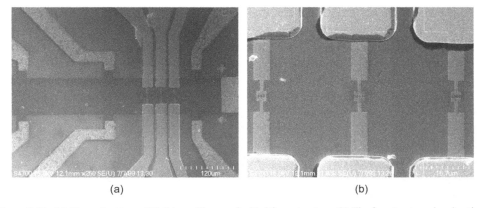

(a) (b)

Figure 1.17. (a) View of patterned Hall bar with an embedded finer structure. (b) The fine structure showing three individual nanostructures, each composed of a triple dot. Reprinted with permission from Chetan Prasad [88].

An example of wet etching is shown in figure 1.17, where an etched hall bar is shown in the GaAs/GaAlAs system [88]. Metal interconnects are also shown, and the mesa is the dark area in the center of the interconnects. Here, an optical positive resist was spun to a depth of 1.5 μm and then soft-baked to drive out the solvent. This was exposed using an optical exposure system and mask. After exposure, the resist was hard-baked to remove any residual moisture that may have been in the system, and then the material was etched in a 1:1:150 solution of H_2O_2:H_2SO_4:H_2O, which was calibrated to remove about 100 nm min^{-1}. After the etch, the resist mask was removed and the mesa measured to make sure it was the desired height. This resist is anisotropic and gives a beveled edge to the mesa, which helps the metal interconnects run over the mesa edge. The interconnects in the figure were created with a second step of optical lithography and a liftoff process.

1.7.3 Bottom-up fabrication

In the past few decades, there has been a growing interest in processes which build the interesting structure from the so-called ground upward. We may call this explosion of interest as one for the study of self-assembled processes that lead to nanostructures. This area actually has a relatively long history driven for many years by the study of self-assembled semiconductor quantum dots that form at hetero-structure interfaces during molecular beam epitaxy (MBE) [27]. The leading example of this approach is provided by the self-assembly of InAs, or InGaAs, quantum dots that form on a GaAs substrate via the Stransky–Krastinov growth process [27]. In this growth mode, a thin layer, less than a monolayer, of InAs is grown on top of a GaAs substrate. If the layer is thin enough, the strain developed from a lattice mismatch will cause the InAs to agglomerate into small three dimensional quantum dots. Growth of a subsequent layer of GaAs or AlGaAs seals the dots into the interface between the two materials. These dots have been of interest for optical applications, such as lasers and LEDs [89], but others have thought about using the dots for tunneling devices [90]. A more recent use of the self-assembled quantum dot is for the generation of single, and entangled, photons for application in quantum computing [91, 92].

While this self-assembly of quantum dots during MBE growth is very interesting, it is not true self-assembly of nanostructures, since the overall device is still rather large. Perhaps more fundamental is the use of chemical processes to create nano-structures, such as graphene nanoribbons [93]. In this fabrication, the topology of the resulting nanoribbon is set by the initial chemical precursors. The precursor monomer is illustrated in figure 1.18 which will be used for the fabrication of a nanoribbon with 7 carbon atoms stretching between the arm-chair edges. The process begins with chemical monomers, shown in the figure, which is 10,10′-dibromo-9,9′-bianthryl monomers. Here, the 10 and 10′ denote the particular carbon atoms (see figure 1.18) on the triple connected benzene rings and the 9 and 9′ denote the positions of the bond that connects the two connected rings. The *dibromo* says that we have two bromine atoms, one at each of the first two points. The connected benzene rings are the anthryl molecules, which are derived from

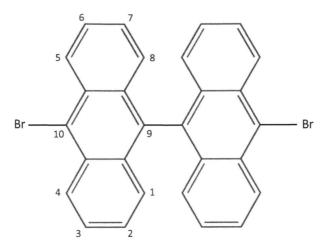

Figure 1.18. Two anthracene molecules are connected and Br atoms added to make the precursor molecule for generating a graphene nanoribbon.

anthracene. The first step is to deposit these molecules on a gold surface, primarily by sublimation. This process removes the bromine atoms from the monomers (known as dehalogenation) which are the building blocks for the final nanoribbon. The structure is then annealed at 200 °C to lead them to diffuse toward one another (known as intermolecular colligation) and form C–C bonds between each monomer and to form polymer chains. Scanning tunneling microscope (STM) images show that these polymers have protrusions that appear alternately on both sides of the chain axis with a periodicity of 0.86 nm [93], and steric hindrance between the H atoms of adjacent anthracene units rotates the H around the σ-bonds leading to opposite tilts of successive anthracene units with respect to the gold surface. Further heating to 400 °C leads to dehydrogenation of these H atoms and the final graphene nanoribbon.

The generation of different numbers of transverse atoms in the final graphene nanoribbon is accomplished with different numbers of benzene rings in the initial monomer. However, the 7 atom width seems to be the easiest due to the presence of the anthracene molecule. For other widths, the precursors are not so easily utilized [94]. Nevertheless, a variety of widths can be fabricated. These nanoribbons can also be easily doped. One method inserts a third anthryl molecule with B atoms at the 9 and 10 sites, by replacement of the C atom at this point, between the two end anthryl units [95]. This produces a p-type graphene nanoribbon. N dopants have also been inserted [96].

The band gap of the resulting nanoribbon is, of course, dependent upon its width. The lateral width governs the effective quantum well in which the electrons of the nanoribbon sit. Hence, this tunes the quantum levels and the resultant energy bands in the direction along the nanoribbon. The band gap for 7 and 13 width nanoribbons (arm-chair edges along the length) have been reported as 2.5 eV and 1.4 eV, respectively [97]. The band gap will also vary with the length of the nanoribbon if the latter is relatively short, or if H endgroups appear on the nanoribbon [98].

This bottom up synthesis of nanoribbons can be used to generated heterojunctions along the length of the nanoribbon [99]. When sections of a graphene nanoribbon have different width, edge (arm-chair versus zig-zag), dopants or termination, these sections can have different electronic topological classes [100]. This can arise because the width leads to different properties as an example. If the nanoribbon has N atoms (in the above definition of width), then for $N = 3p$, where p is an integer) and $N = 3p + 1$, the corresponding nanoribbon has a band gap throughout the width. However, if $N = 3p + 2$, a zero-energy edge state exists along the ribbon and the latter is metallic in behavior. While both 7 and 9 width nanoribbons are both gapped, but they possess different topological states. The 7 nanoribbon is topologically trivial (the $\mathbb{Z}_2$ invariant is 0) while the 9 nanoribbon is topologically nontrivial with $\mathbb{Z}_2 = 1$ [101, 102]. So, when we make a nanoribbon heterostructure 7/9 junctions, there is a change in the $\mathbb{Z}_2$ invariant across the junction and this leads to a one dimensional array of interface states which, if aligned periodically in the superlattice, enable a hierarchy of quantum engineered topological phase discontinuities, which are quite likely to be very useful in the world of topological insulators. These interface states should be half-filled states near midgap, and provide a new tool in engineering the electronic properties of the nanoribbon.

Problems

1. Consider a normal MOS structure on Si at room temperature. Using the classical approach described in the chapter to determine the effective width of the inversion layer, compute the width for densities of 5×10^{11}, 1×10^{12}, 3×10^{12}, 5×10^{12}, 1×10^{13}, all in units of cm^{-2}. Assuming an acceptor density of 5×10^{17}, compute the effective electric field in the oxide. Plot both the effective thickness and the effective electric field as a function of the inversion density on a single plot, using a logarithmic axis for the density.

2. Go to nanoHUB.org and launch the 'bound states calculation lab'. Using the data for the Si MOS structure above, and using the heavy mass for the Si inversion layer, enter the effective electric fields computed in the previous problem. Using the triangular geometry to approximate the inversion layer, determine the effective thickness of each of the first four subbands as a function of the inversion density corresponding to the effective electric field entered in the program.

3. Consider two separate GaAs/AlGaAs quantum wells, A and B, of thickness L and $2L$, respectively. (a) By approximating the AlGaAs barriers of the wells as having infinite height, determine the ratio $E_o{}^A/E_o{}^B$, where are $E_o{}^{A,B}$ are the lowest quantized energy levels in the wells. (b) Now suppose that the quantum wells contain electrons and that they both have the same Fermi energy, $E_F = 3E_o{}^A$ (measured from the bottom of the well) How many quantized subbands will be occupied in each of the wells?

4. Consider a uniformly doped GaAlAs layer, with a composition of $x = 0.25$, grown on top of a GaAs layer. If the Fermi energy is pinned 0.85 eV below the conduction band edge at the surface, what is the minimum thickness of the GaAlAs layer for it to be completely depleted of electrons? Assume the donor ionization energy is 50 meV and the donor density is 10^{18} cm^{-3}.

5. Consider two exposure pixels of an electron-beam lithography system. Each exposure pixel is characterized by a Gaussian beam with a full-width half-maximum of 50 nm. If we assume that each Gaussian has a peak amplitude of unity, compute the exposure (amplitude of the signal) at the mid-point between the centroids of the Gaussians as they are moved from 20 nm center-to-center to 200 nm center-to-center.

6. Consider a 2DEG at the interface of a GaAlAs/GaAs heterostructure at 1.2 K. Assume that the electron density is 4×10^{11} cm^{-2} and the mobility is 500 000 cm^2(Vs)$^{-1}$. (a) Determine the momentum relaxation time from the mobility, and then determine the diffusion coefficient using the Einstein relation (you may need to make some corrections for the Fermi–Dirac distribution function). (b) Estimate the diffusion coefficient from the Fermi energy and Fermi velocity. Is there a difference in the two values? Why?

7. Using the structure and data from problem 7, estimate the elastic mean free path for the electrons. If we assume that the device has a transport length of 1 cm, we can use this for the inelastic mean free path. What is the value of the phase-breaking time in this case?

References

[1] Braun F 1874 *Annal. Phys. Chem.* **153** 556
[2] Turing A 1937 *London Math. Soc., Series II* **42** 230
[3] Child C D 1911 *Phys. Rev., Ser. I* **32** 492
[4] Langmuir I 1913 *Phys. Rev., Ser. I* **2** 450
[5] Atanasoff J V 1984 *Ann. Hist. Comp* **6** 229
[6] Goldstine H H 1972 *The Computer: from Pascal to von Neumann* (Princeton, NJ: Princeton University Press)
[7] Kilby J 1976 *IEEE Trans. Electron. Dev.* **23** 648
[8] Noyce R N 1968 *IEEE Spectr.* **5** 63
[9] Moore G E 1965 *Electronics* **38** 114
[10] Dennard R, Gaensslen F H, Yu H-N, Rideout V L, Bassous E and Leblanc A R 1974 *IEEE Sol.-State Circuits* **9** 256
[11] Evans C 1979 *The Micro Millennium* (New York: Washington Square Press)
[12] Keyes R 1977 *Science* **195** 1230
[13] Ferry D K 2008 *Science* **319** 579
[14] Ferry D K, Gilbert M J and Akis R 2008 *IEEE Trans. Electron. Dev.* **55** 2810
[15] Fang W W, Singh N and Bera L K *et al* 2007 *IEEE Electron Dev. Lett.* **28** 211
[16] Ernst T, Duraffourg L and Dupré C *et al* 2008 *Proc. IEDM Meeting* (Piscataway, NJ: IEEE) 1
[17] Dupré C, Hubert A and Bécu S *et al* 2008 *Proc. IEDM Meeting* (Piscataway, NJ: IEEE) 749

[18] Tachi K, Barraud S and Kakushima K *et al* 2011 *Microelectron. Reliabil.* **51** 885

[19] Huang Y-C, Chiang M-H and Fossum J G 2017 *J. Electron Dev. Soc.* **5** 164

[20] Gaillardon P-E, Amaru L G and Bobba S *et al* 2018 *Phil. Trans. Royal Soc.* A **372** 20130102

[21] Hur J, Lee B-H and Kang M-H *et al* 2016 *IEEE Electron. Dev. Lett.* **57** 541

[22] Veloso A, Matagne P and Simoen E *et al* 2018 *J. Phys. Cond. Matter* **30** 384002

[23] Loubet N, Hook T and Montanini P *et al* 2017 *Symp. VLSI Technol. Dig.* (New York: JSAP) T230

[24] Kelly M 1995 *Low-Dimensional Semiconductors: Materials, Physics, Technology, Devices* (Oxford: Clarendon)

[25] Khoury M and Ferry D K 1996 *J. Vac. Sci. Technol.* B **14** 75

[26] Tao N J, Derose J A and Lindsay S M 1992 *J. Phys. Chem.* **97** 910

[27] Bimberg D, Grundmann M and Ledentsov N N 1999 *Quantum Dot Heterostructures* (Chichester: Wiley)

[28] Ferry D K, Goodnick S M and Bird J P 2009 *Transport in Nanostructures* 2nd edn (Cambridge: Cambridge University Press)

[29] Crommie M F, Lutz C P and Eigler D M 1993 *Science* **262** 218

[30] Mott N F 1970 *Phil. Mag.* **22** 7

[31] Thouless D J 1974 *Phys. Rep.* **13C** 93

[32] Abrahams E, Anderson P W, Licciardello D C and Ramakrishnan T V 1979 *Phys. Rev. Lett.* **42** 673

[33] D'Orio M, Pudalov V M and Semenchinsky S G 1992 *Phys. Rev.* B **46** 15992

[34] Kravchenko S V, Mason W, Furneaux J E and Pudalov V M 1995 *Phys. Rev. Lett.* **75** 910

[35] Pudalov V M, Brunthaler G, Prinz A and Bauer G 1998 *Physica* E **3** 79

[36] Lin J J and Giordano N 1987 *Phys. Rev.* B **35** 1071

[37] Mohanty P, Jariwala E M Q and Webb R A 1997 *Phys. Rev. Lett.* **78** 3366

[38] Price P J 1965 *Fluctuation Phenomena in Solids* ed R E Burgess (New York: Academic) 355

[39] Lillienfeld J E 1926 Method and apparatus for controlling currents *US Patent* 1475175

[40] Atalla M M, Tannenbaum E and Scheibner E J 1959 *Bell Syst. Tech. J.* **38** 749

[41] von Klitzing K, Dorda G and Pepper M 1980 *Phys. Rev. Lett.* **45** 494

[42] Skocpol W J, Mankiewich P M and Howard R E *et al* 1987 *Phys. Rev. Lett.* **58** 2347

[43] Koch J F 1976 *Surf. Sci.* **58** 104

[44] Wilmsen C W 1976 *Thin Sol. Films* **39** 105

[45] Baglee D A, Ferry D K and Wilmsen C W *J. Vac. Sci. Technol.* **17** 1032

[46] Neaman D A 2012 *Semiconductor Physics and Devices* 4th edn (New York: McGraw-Hill)

[47] Tsividis Y 1999 *Operation and Modeling of the MOS Transistor* 2nd edn (New York: McGraw-Hill)

[48] Ando T, Fowler A and Stern F 1982 *Rev. Mod. Phys.* **54** 437

[49] SCHRED 2.0, https://nanohub.org/resources/schred

[50] Ando T 1977 *J. Phys. Soc. Japan* **43** 1616

[51] Goodnick S M, Ferry D K and Wilmsen C W *et al* 1985 *Phys. Rev.* B **32** 8171

[52] Masaki K, Taniguchi K and Hamaguchi C 1992 *Semicond. Sci. Technol.* **7** B573

[53] Ono M, Saito M and Yoshitomi T *et al* 1995 *IEEE Trans. Electron. Dev.* **42** 1822

[54] Chau R, Datta S and Doczy M *et al* 2004 *IEEE Electron Dev. Lett.* **25** 408

[55] Natarajan S, Armstrong M and Bost M *et al* 2008 *IEEE Intern. Electron Dev. Mtg.* (Piscataway, NJ: IEEE)

[56] Susuki Y, Seki M and Okamoto H 1984 *16th Congr. On Solid State Devices and Materials (Tokyo)* **607**

[57] Heiblum M, Nathan M I and Eisenberg M 1985 *Appl. Phys. Lett.* **47** 503

[58] Schubert E F and Ploog K 1965 *Jpn. J. Appl. Phys.* **24** L608

[59] Dingle R, Störmer H L, Gossard A C and Wiegmann W 1978 *Appl. Phys. Lett.* **33** 665

[60] Pfeiffer L, West K W, Störmer H L and Baldwin K W 1989 *Appl. Phys. Lett.* **55** 1888

[61] Lin B J F, Tsui D C, Paalanen M A and Gossard A C 1984 *Appl. Phys. Lett.* **45** 695

[62] Mooney P M 1990 *J. Appl. Phys.* **67** R1

[63] Bound state calculation lab: https://nanohub.org/resources/4875?rev=29

[64] Ambacher O, Smart J and Shealy J R *et al* 1999 *J. Appl. Phys.* **85** 3222

[65] Castro Neto A H, Guinea F, Peres N M R, Novoselov K S and Geim A K 2009 *Rev. Mod. Phys.* **81** 109

[66] Li S-L, Tsukagoshi K, Orgiu E and Samori P 2016 *Chem. Soc. Rev.* **45** 118

[67] Wallace P R 1947 *Phys. Rev.* **71** 622

[68] Onnes H K 1911 *Commun. Phys. Lab. Leiden* **122b** 124c

[69] Cooper L N 1956 *Phys. Rev.* **104** 1189

[70] Bardeen J, Cooper L N and Schrieffer J R 1957 *Phys. Rev.* **106** 162
Bardeen J, Cooper L N and Schrieffer J R 1957 *Phys. Rev.* **108** 1175

[71] Glover R E III and Tinkham M 1957 *Phys. Rev.* **108** 243

[72] Hartree D 1946 *Nature* **157** 527

[73] von Neumann J 1945 *First draft of a report on the EDVAC* (Moore School of Engineering, 1945); archived at https://web.archive.org/web/20130314123032/ http://qss.stanford.edu/~godfrey/vonNeumann/vnedvac.pdf.

[74] Marcovitz A B 2010 *Introduction to Logic Design* 3rd edn (New York: McGraw-Hill)

[75] Guillemin E A 1953 *Introductory Circuit Theory* (New York: Wiley)

[76] Porod W, Grondin R O, Ferry D K and Porod G 1984 *Phys. Rev. Lett.* **52** 232

[77] Schrödinger E 1935 *Naturwiss.* **23** 807, 823, 844
Schrödinger E 1980Trimmer J D *Proc. Am. Phil. Soc.* **124** 323

[78] Einstein A, Podolsky B and Rosen N 1935 *Phys. Rev.* **47** 777

[79] Howard R E, Liao P F and Skocpol W J *et al* 1983 *Science* **221** 117

[80] Rensch D B, Seliger R L and Csanky G *et al* 1979 *J. Vac. Sci. Technol.* **16** 1897

[81] Bernstein G H, Ferry D K and Liu W P 1989 Process of obtaining improved contrast in electron beam lithography *US Patent 4937174*

[82] Bernstein G H and Hill D A 1992 *Superlatt. Microstruc.* **11** 237

[83] Bernstein G H, Liu W P and Khawaha Y N *et al* 1988 *J. Vac. Sci. Technol.* B **6** 2296

[84] Tennant D M, Jackel L D and Howard R E *et al* 1981 *J. Vac. Sci. Technol.* **19** 1304

[85] Crook R, Graham A C and Smith C G *et al* 2003 *Nature* **424** 751

[86] Song H J, Rack M J and Abugharbieh K *et al* 1994 *J. Vac. Sci. Technol.* B **12** 3720

[87] Fuhrer A, Dorn A and Lüscher S *et al* 2002 *Superlatt. Microstruc* **31** 19

[88] Prasad C 2003 private communication

[89] Grundmann M and Bimberg D 1997 *Jpn. J. Appl. Phys.* **36** 4181

[90] Warburton R J, Dürr C S and Karrai K *et al* 1997 *Phys. Rev. Lett.* **79** 5282

[91] Yuan Z, Kardynal B E and Stevenson R M *et al* 2002 *Science* **295** 102

[92] Karrai K, Warburton R J and Schulhauser C *et al* 2004 *Nature* **427** 135

[93] Cai J, Ruffieux P and Jaafar R *et al* 2010 *Nature* **466** 470

[94] Talirz L, Ruffieux P and Fasel R 2016 *Adv. Mater.* **28** 6222

[95] Kawai S, Saito S and Osumi S *et al* 2015 *Nature Commun.* **6** 8098

[96] Bronner C, Stremiau S and Gille M *et al* 2013 *Angew. Chem. Int. Ed.* **52** 4422

[97] Chen Y-C, de Oteyza D G and Pedramrazi Z *et al* 2013 *ACS Nano* **7** 6123

[98] Talirz L, Söde H and Kawai S *et al* 2019 *Chem. Phys. Chem.* **20** 2348

[99] Chen Y-C, Cao T and Chen C *et al* 2015 *Nature Nanotechnol.* **10** 156

[100] Cao T, Zhao F and Louie S G 2017 *Phys. Rev. Lett.* **119** 076401

[101] Rizzo D J, Veber G and Cao T *et al* 2018 *Nature* **560** 204

[102] Gröning O, Wang S and Yao X *et al* 2018 *Nature* **560** 209

IOP Publishing

Transport in Semiconductor Mesoscopic Devices
(Second Edition)

David K Ferry

Chapter 2

Wires and channels

Perhaps the simplest mesoscopic device that can be conceived actually came after some years of work in mesoscopic studies. Hence, we would like to deviate from the historical nature of the field to discuss the topic in a more developmental order. In this chapter we will attempt to lay some foundations for understanding mesoscopic transport based upon this developmental order. We will begin with transport in quasi-one-dimensional (Q1D) structures. Long Q1D structures are nanowires, and these can be either fabricated on a substrate by top-down processing or exist naturally, such as with carbon nanotubes. But, short Q1D structures can also be made, and these are often called quantum point contacts (QPCs), as they are usually fabricated from metallic gates. For, example, a close examination of figure 1.17(b) shows that each of the triple quantum dot structures is created using four QPCs, whose Q1D length varies in the three sets shown. Such a composite structure allows for studying interesting physics, not only of the QPCs, but of the quantum dots themselves in chapters 8 and 9. So, we will begin this chapter with the QPC. Following this, we turn to a discussion of the density of states in nanostructures, and the forms that can arise with different dimensionality. This will be followed by a discussion of ballistic behavior and scattering, as well as a consideration of the role of temperature and magnetic field. Finally, we will introduce two methods of calculating, or simulating, the transport behavior of various mesoscopic devices—the scattering matrix and the recursive Green's function methods.

2.1 The quantum point contact

One of the most important discoveries for the understanding of mesoscopic and nanoelectronic devices has been the observation of one-dimensional conductance quantization. What we mean by one-dimensional is that the transport is not just along a single direction, but is actually characterized as being transport in a narrow one-dimensional channel, not unlike a small pipe. The observed conductance is quantized when the transverse dimensions of this 'pipe' are comparable to the Fermi wavelength, and the conductance is found to increase in steps of $2e^2/h$ as the size of the pipe is slowly increased. Generally, this phenomena is observed in short quantum wires at low temperatures when the electron transport is largely ballistic in nature. We will discuss the reason for this limitation later, but one can simply understand that scattering interferes with the general quantization process.

In figure 2.1(a), we illustrate a mesoscopic structure in which this quantization can be seen, although here we are interested just in the physics of such a structure. The heterostructure in the figure is a layer of $In_{0.53}Ga_{0.47}As$ sandwiched between two layers of $In_{0.52}Al_{0.48}As$, the latter of which has a much wider band gap, so that the electrons in the central layer are confined in a vertical quantum well [1]. Then, two vertical trenches (indicated by the various shades of blue) have been etched through the layers to provide lateral confinement of the electrons as they pass through the narrow region between the two trenches, leaving an opening of about 0.6 µm. The two regions marked 'A' and 'B' are separately contacted and can be biased to provide electrostatic confinement to reduce the size of the opening. In figure 2.1(b), we show a scanning gate microscopy (SGM) image of the transport through the structure. In this image, the effect of moving a biased scanning probe tip on the conductance through the structure (from left to right) is shown. The red regions are the conducting regions, while the blue regions are electrically isolated. By applying a negative voltage to the two gates (the blue regions), we can squeeze the conducting region to a smaller size, eventually cutting off the conduction through the constriction.

The first measurements through a QPC were made by van Wees *et al* [2] and Wharam *et al* [3]. In figure 2.2, we illustrate the quantized conductance through a QPC defined by two electrostatic Schottky barrier gates [4], for a GaAs/AlGaAs heterostructure. We can see the physical structure in the upper inset, with the two split gates (black regions) and source and drain contacts indicated. The overall behavior with gate bias is shown in the lower inset as the raw data, while the main panel shows an expanded view of the discrete steps in conductance. To within the accuracy that can be ascertained from the figure, the steps correspond to plateaus

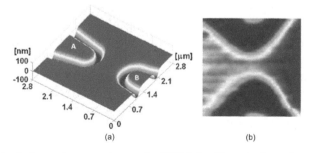

(a)   (b)

Figure 2.1. (a) An atomic force microscopy image of a QPC defined by etched trenches. (b) A SGM image of the transport regions (red) through the structure shown in part (a). (Reprinted from [1] with permission of AIP Publishing LLC.)

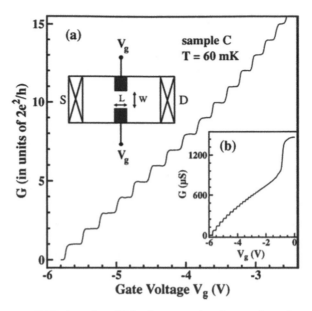

Figure 2.2. Conductance $G(VG)$ through a QPC, after correction for a series resistance of approximately 700 ohm. In the upper inset is a schematic of the sample with the source and drain contacts indicated and the split-gate Schottky barriers in black. The lower inset shows the raw data. (Reprinted with permission from [4]. Copyright 1998 the American Physical Society.)

with a conductance that is an integer multiplier of $2e^2/h$. In this structure, the negative gate voltage controls the electrostatic width W of the opening between the metal gates. As the gate voltage is made more negative, the width is reduced and the conductance is also reduced. The fact that the conductance goes down in steps is a property of the Q1D nature of the channel that passes through the QPC and the nearly ballistic nature of the transport.

The QPCs described above provide the prototypical system for the study of Q1D conductance that can be realized by defining narrow constrictions in a normal quasi-two-dimensional electron gas (Q2DEG). We recall that the Q2DEG exists at the interface between GaAs and GaAlAs, or in the quantum well of the InGaAs/InAlAs system. It has this Q2D behavior as the motion is restricted and quantized in the direction normal to the heterostructure layers. Hence, the electrons can move only in the plane of the heterostructure. The transverse gates provide an additional level of quantization which further restricts the motion of the carriers, limiting them to be nearly in a single direction. Usually these gates define a constriction whose size is less than a micron, but hopefully of a size sufficiently small that no impurities exist in this region, and the transport is ballistic. In the case of electrostatic confinement, such as the split-gate approach, the potential profile has a saddle point structure.

At the center of the QPC, if we move toward the two gates, the potential increases. However, if we move along the channel direction, the potential will decrease. The highest potential in the channel (along the center line) provides a barrier to transport through the QPC. In most cases, it is reasonable to assume that the potential is parabolic in both directions, and this will make the Schrödinger equation solvable exactly. While one might expect that there are reasons for non-parabolic behavior of the potential, in fact the self-consistency that constrains the potential tends to drive it toward a quadratic, parabolic behavior [5]. Thus, we can write the potential approximately as

$$V(x, y) = V_0 - \frac{m^* \omega_{0,x}^2 x^2}{2} + \frac{m^* \omega_{0,y}^2 y^2}{2}. \tag{2.1}$$

where the y-direction is the width of the opening and the x-direction is along the length of the channel indicated in the upper inset to figure 2.2. The quantities $\omega_{0,x}$ and $\omega_{0,y}$ are oscillator frequencies, which are parameters that characterize the strength of the two parabolic potential variations, while V_0 is the height of the saddle potential center point and m* is the effective mass of the channel material. The form of the potential equation (2.1) is separable and allows the Schrödinger equation to be separated into its x- and y-components which can be solved independently.

It is also important to note that in the case of etched trench-isolated gates, as shown in figure 2.1, the confining potential differs from the parabolic form. Instead, the potential is controlled from the side of the channel as opposed to arising from a gate on top of the heterostructure. This lateral potential leads to a confinement closer to a finite quantum well. As the height of the confining potential is finite, the separation of the energy levels will differ from those of an infinite quantum well, and will not have the linear variation that will arise from the parabolic potential of equation (2.1).

The solution to the Schrödinger equation for the x-motion is quite simple. For energies above V_0, the motion is that of a free electron along the channel, while for energies below V_0, the motion is prohibited, although tunneling can occur near the saddle center. As we show in appendix B, solving the one-dimensional Schrödinger equation for the y-direction (the width), yields a series of equally spaced harmonic oscillator levels which correspond to the one-dimensional subbands that arise in the channel, with the corresponding energies

$$E_n(k_x) = V_0 + \left(n + \frac{1}{2}\right)\hbar\omega_{0,y}. \tag{2.2}$$

In a long channel, which has the characteristics of a quantum wire, these subband energies will be more or less uniform along the wire length. In a short QPC, these values exist only at the center of the QPC, and become closer together as one moves away from the center in either direction along the length. Eventually, they merge into the continuum of the Q2DEG when we are well away from the QPC. The gate voltage varies both the height of the saddle point potential V_0, as well as the harmonic parameter $\omega_{0,y}$. From equation (2.2), it is clear that the increase of the harmonic parameter $\omega_{0,y}$ by the gate potential pushes the energy levels upward, and increases the spacing between these levels. The electrons can then move through those channels whose subband energies are less than the Fermi energy in the Q2DEG away from the QPC.

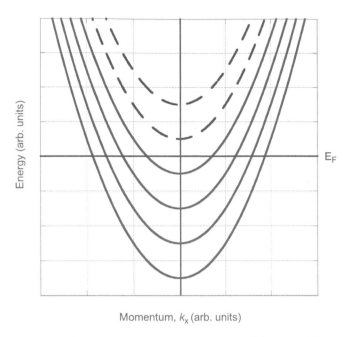

Figure 2.3. Dispersion relation for a quantum wire with a parabolic confining potential, according to equation (2.3).

As remarked, the motion along the length of the QPC is that of a free electron in one dimension. This is the normal quadratic energy dispersion, and modifies equation (2.2) to give

$$E_n(k_x) = V_0 + \left(n + \frac{1}{2}\right)\hbar\omega_{0,y} + \frac{\hbar^2 k_x^2}{2m^*}, \quad n = 0, 1, 2, \ldots. \tag{2.3}$$

Since the energy in the y-direction is quantized, each subband may be considered to be a separate one-dimensional channel with essentially free momentum along the direction of the current flow. In figure 2.3, we illustrate the behavior of these subbands for momentum along the length of the channel. Here, we show four subbands below the bulk Fermi level, and two of the many subbands which have quantized energies above the Fermi level. Thus, there will be four channels flowing through the QPC. As may be inferred from equation (2.3) and figure 2.3, the number of occupied subbands is given by determining the largest value of n in equation (2.3) for which $E_n(k_x = 0)$ is less than the Fermi energy in the region far from the QPC.

For a wire in such a parabolic potential, we can develop an expression for the effective width of the channel. Using equation (2.1) at $x = 0$, and equation (2.2), we can say that

$$E_F = V_0 + \frac{m^*\omega_{0,y}^2}{2}\left(\frac{W}{2}\right)^2. \tag{2.4}$$

Using the value of the Fermi wave vector, in the channel center, defined from

$$E_F - V_0 = \frac{\hbar^2 k_F'^2}{2m^*}, \tag{2.5}$$

we find that the width is related to this value as

$$W = \frac{2\hbar k_F'^2}{m^* \omega_{0,y}}. \tag{2.6}$$

We may use this relation to obtain an expression for the number of occupied channels in the wire, through (note that N here is the number of occupied channels, not the energy level index)

$$E_F' - E_F - V_0 \geqslant \left(N - \frac{1}{2}\right)\hbar\omega_{0,y}, \tag{2.7}$$

and

$$N = int\left[\frac{1}{2} + \frac{E_F'}{\hbar\omega_{0,y}}\right] \sim \frac{E_F'}{\hbar\omega_{0,y}} = \frac{E_F' W}{4} = \frac{\pi W}{2\lambda_F'}. \tag{2.8}$$

Thus, it is clear that the number of modes in the parabolic potential that can propagate through the QPC is related to the number of half-wavelengths of the reduced Fermi wavelength that can be fit into the width of the potential at the Fermi energy.

In the case of the square well potential that arises from the trench-isolated gates of figure 2.1, the relationship is somewhat simpler, and we have

$$N = \frac{k_F W}{\pi} = \frac{2W}{\lambda_F}. \tag{2.9}$$

Thus, a new mode is populated each time the wire width is increased by a half a Fermi wavelength. Note that here the unprimed quantities are used as there is no saddle potential minimum in this case, and the Fermi energy has the same reference point as in the bulk away from the QPC.

Direct experiments for the existence of the modes going through the QPC have been given through experiments using SGM [6, 7], in which a biased scanning microscope tip is scanned over the mesoscopic device, the QPC and its surrounding region in this case, and the conduction modulation caused by the tip measured. The technique works as the negative bias on the tip causes a local reduction in the density, and this is reflected as a change in the conductance through the device. The effect is larger where the local density, or the magnitude squared of the wave function, is larger so that the scanned measurement image gives an indication of the spatial extent of the wave function. We can see the concept of the SGM in the video of figure 2.4. In this video, the propagation through the QPC is shown in the right-hand panel. The conductance is plotted in the left-hand panel. When the probe (indicated by the white shape in the right-hand panel) is at the left, it sits over the

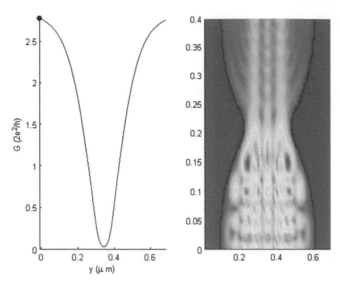

Figure 2.4. The video illustrates the method of operation of the SGM (available at https://iopscience.iop.org/book/978-0-7503-3139-5. (Video by Richard Akis, included with his permission).

potential barrier (red in the right-hand panel) and the conductance measured through the QPC is unaffected by it. However, as the probe moves through the actual channel, the large bias of the tip serves to cut off the transmission through the QPC and the conductance drops to almost zero. Then, as the probe moves onto the potential at the right, the conductance rises back to its normal value. The fact that the conductance is nearly cut off in this video tells us that the probe tip is fairly large. However, it is possible to adjust the distance of the tip from the surface, and the bias applied, such that a spatial resolution of 5 nm, or better, can be achieved with this technique.

Hence, the microscope tip (the movable gate) serves as a probe of the local density and can be used to map this density. Typically, the highest mode through the QPC is measured in this technique. In figure 2.5, we show the results of an experimental study of a QPC, in which the highest mode propagating through the QPC is alternatively set to be the first mode, the second mode and/or the third mode [8]. In the second mode, the peak magnitude of the corresponding eigenfunction has two peaks with non-zero y values, and the transverse momentum constraint is relaxed as the mode leaves the QPC. Hence, we see two diverging beams of density. In the third mode, one of the peaks is at zero transverse momentum, so we see three diverging beams, one of which remains along the axis of the length of the QPC. Also shown in the figure are simulations of the mode propagation away from the QPC. Although there is a difference between the experimental observations and the theoretical predictions, the images are quite close. Presumably, these differences are due to impurities in the heterostructure which are not present in the simulations. We note in the figure that the scanning gate does not enter the QPC proper, presumably due to the use of Schottky split gates on top of the heterostructure. In this situation, the biased gate must be kept away from the Schottky gates to avoid device failure, and

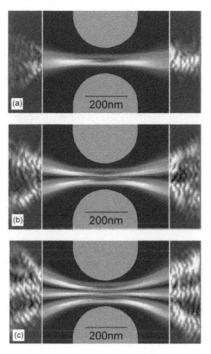

Figure 2.5. Experimental images (outer sections of each panel) and theoretical simulation (inner sections) of the wave functions of electrons passing through a QPC in the first, second, and third modes. (Reprinted [8], with permission from Elsevier, Copyright 2004).

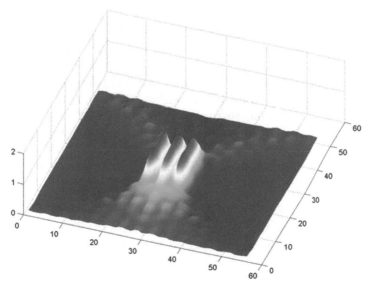

Figure 2.6. The amplitude squared of the third mode wave function within a QPC. The position scales are in nanometers, while the vertical amplitude is arbitrary. The simulation was performed by the author with the scattering matrix wave function approach discussed later in this chapter.

this is why the trench isolated gates, shown in figure 2.1, are used. In figure 2.6, we plot the simulation of the third mode in a square-shaped QPC. The wave function is most well defined within the QPC itself. This simulation was performed using techniques we will discuss near the end of this chapter.

The existence of the modes gives an indication of just why the conductance through the QPC shows the steps that are observed in figure 2.2, but it does not explain why the steps appear as they do, or even why steps are observed at all. What we know so far is that the steps are observed experimentally, but not why they are observed. To understand this, we have to dig deeper into the basics of condensed matter physics and explore both the density of states and the Landauer picture of transport. Let us now turn to these topics.

2.2 The density of states

The density of states is one of the most important quantities for determining the electronic properties of condensed matter systems. Basically, we are asking for the number of states that are available at a given energy in the system. Thus, the density of states is the number of states per unit energy per unit volume, and is a function of the energy itself. In three dimensions, the density of states for a semiconductor, or a free-electron metal, is easily calculated from the quantized solutions of the Schrödinger equation, and leads to a variation as the square root of the energy. But, this depends upon the dimensionality of the system, so that the variations will vary as we move from bulk to two-dimensional to one-dimensional systems. Moreover, it depends upon the nature of the energy bands, with the traditional bulk square-root result only arising for parabolic energy bands.

To understand the properties of the density of states in nanostructures, we begin by determining the properties of the wave functions themselves and how this sets an important quantization on the momentum states that can exist in the material. The problem is to solve the Schrödinger equation in the desired material. The most important initial step, and one that really defines the exact density of states that we find, is the assumption of periodic boundary conditions on the wave function. In fact, the electrons are constrained to stay within the volume of the crystal, so that there are potential barriers that surround the crystal (these are the work function potential required to be overcome to remove an electron from the crystal). The reader would be correct in noting that other boundary conditions can be used, but the best approach is to use the periodic boundary conditions with a length L in each direction, for reasons that will become clearer below:

$$\psi(x, y, z) = \psi(x + L, y, z)$$
$$\psi(x, y, z) = \psi(x, y + L, z) \tag{2.10}$$
$$\psi(x, y, z) = \psi(x, y, z + L).$$

Now, the use of periodic boundary conditions leads to the correct counting of states; that is, there will be two states (of opposite spin) for each bonding electron in the basic crystal lattice. For example, in Si or GaAs, the crystal lattice is the fcc lattice which has four lattice sites per unit cell. But, there is a basis of two atoms at each lattice site: either two Si atoms or one Ga and one As atom, so that there are eight atoms per cell. The periodicity required for the atomic lattice thus gives $4/a^3$, where a is the edge of the fcc cell, states, each of which can hold two electrons of opposite spin. Hence, there are $8/a^3$ total states. Each atom of the basis has four bonding electrons, so this gives a total of 8 electrons per lattice site, which just exactly fills all the states in the band—the valence band, or bonding band in this case. So, these tetrahedrally coordinated semiconductors are insulators at very low temperatures, which is exactly what is observed. Using different boundary conditions would not give this experimentally confirmed result.

Because of the atomic periodicity of the crystalline potential, we find that the momentum wave numbers are not continuous, but are discretized due to the boundary conditions. They form a series of discrete points in the three-dimensional momentum space. Each point in this space corresponds to a particular momentum state which may be occupied by two electrons of opposite spin, according to the Pauli exclusion principle. The traditional approach is to use a cubic lattice of edge a, appropriate to most metals, so that the length $L = Na$, where N is the number of atoms in the given direction of the crystal. Then, we find that the quantization of the momentum, leads to

$$k_x = \frac{2\pi n_x}{L}, \quad n_x = 0, \pm 1, \pm 2, \ldots \frac{N}{2}$$

$$k_y = \frac{2\pi n_y}{L}, \quad n_y = 0, \pm 1, \pm 2, \ldots \frac{N}{2} \qquad (2.11)$$

$$k_z = \frac{2\pi n_z}{L}, \quad n_z = 0, \pm 1, \pm 2, \ldots \frac{N}{2}.$$

Of course, n_x, n_y, n_z can be larger than the limits indicated above, but these larger values are not independent. If we examine our electron waves as having the property

$$\psi \sim e^{ik \cdot r}, \qquad (2.12)$$

then we see that the phase rolls over (past 2π) when the larger numbers are used. So the end result is that the independent values of momentum are precisely equal to the number of atoms in the lattice (not including the basis).

2.2.1 Three dimensions

In three dimensions, the filling of the lowest energy states at low temperature gives rise to the formation of the Fermi sphere, whose properties may be easily computed. In an N electron metal, $N/2$ distinct wave vectors are required to hold these electrons. The volume in momentum space for each wave vector is given by our conditions of equation (2.11) to be $(2\pi/L)^3$.

Hence, our $N/2$ states give rise to the requirement that

$$\frac{N}{2} = \frac{\text{Volume of sphere}}{\text{Volume per state}} = \frac{4\pi}{3} k_F^3 \frac{L^3}{8\pi^3}. \qquad (2.13)$$

We can now solve for the Fermi wave number as

$$k_F = \left(3\pi^2 \frac{N}{L^3}\right)^{1/3} = (3\pi^2 n)^{1/3}, \tag{2.14}$$

where n is the number of electrons per unit volume. In a simple metal, these are all free electrons, and they will fill the band up to the half-way point. That is, one-half of the available states are filled for this simple metal.

At zero temperature, all the states are filled up to the Fermi energy. For our free-electron metal, this value can be found to be

$$E_F = \frac{\hbar^2 k_F^2}{2m} = \frac{\hbar^2}{2m}(3\pi^2 n)^{2/3}. \tag{2.15}$$

This can be inverted to find the density as a function of the Fermi energy, as

$$n = \frac{1}{3\pi^2}\left(\frac{2mE_F}{\hbar^2}\right)^{3/2}. \tag{2.16}$$

This last expression tells us exactly how many electrons fill all the states up to the Fermi energy. There is no requirement that this be the Fermi energy. If we replace the Fermi energy by the energy itself, then equation (2.16) tells us how many electrons can be held in the states up to that energy. Thus, we can find the density of states by a simple derivative, as

$$\rho_3(E) = \frac{dn}{dE} = \frac{1}{2\pi^2}\left(\frac{2m}{\hbar^2}\right)^{3/2} E^{1/2}. \tag{2.17}$$

Now, we see that the density of states is proportional to the square root of the energy in three dimensions.

When we move to semiconductors, or materials other than the simple free-electron metals, there are few changes unless the parabolic energy dependence is modified. This parabolic dependence was invoked in the move to equation (2.15), so is crucial to the development we have used. An additional modification is the replacement of the free-electron mass by the effective mass appropriate to the band and semiconductor in use. The effective mass is introduced to equate the wave momentum to the quasi-particle momentum so as to have consistency (at some level) between the quantum and semiclassical approaches [9].

2.2.2 Two dimensions

Exactly the same approach can be used in two dimensions. In this case, The Fermi energy describes a circle in the two-dimensional momentum space, and the number of states needed to accommodate $N/2$ distinct wave vectors is given as

$$\frac{N}{2} = \frac{\text{Area of the circle}}{\text{Area per state}} = \pi k_F^2 \frac{L^2}{4\pi^2}. \tag{2.18}$$

We can now solve for the Fermi wave vector as

$$k_F = \left(2\pi\frac{N}{L^2}\right)^{1/2} = (2\pi n_s)^{1/2}, \tag{2.19}$$

where n_s is the number of electrons per unit area, or sheet density. At zero temperature, all the states are filled up to the Fermi energy. For our free-electron metal, this value can be found to be

$$E_F = \frac{\hbar^2 k_F^2}{2m} = \frac{\hbar^2}{2m}(2\pi n_s). \tag{2.20}$$

This can be inverted to find the density as a function of the Fermi energy, as

$$n_s = \frac{mE_F}{\pi\hbar^2}. \tag{2.21}$$

This last expression tells us exactly how many electrons fill all the states up to the Fermi energy. There is no requirement that this be the Fermi energy. If we replace the Fermi energy by the energy itself, then equation (2.21) tells us how many electrons can be held in the states up to that energy. Thus, we can find the density of states by a simple derivative, as

$$\rho_2(E) = \frac{dn}{dE} = \frac{m}{\pi\hbar^2}. \tag{2.22}$$

Here, the density of states is independent of energy.

Now, the above discussion is for normal semiconductors or metals which have a band gap. In recent years new materials have been discovered that are characterized by linear bands [10]. One of these is graphene, which is discussed in some detail in chapter 4. Another case is topological edge states which possess linear bands as well [11]. The graphene lattice and Brillouin zone is shown in figure 4.1 and the band structure is shown in figure 4.2. The important point here is that the conduction and valence band touch at two points in the primitive Brillouin zone unit cell. These two points are denoted as K and K'. Near these two points, the energy bands are linear bands given approximately by

$$E = \pm\hbar v_F k, \tag{2.23}$$

where v_F is the slope of the linear band and is denoted the Fermi velocity. The upper sign in equation (2.23) is for electrons and the lower sign is for holes. Here,

$$v_F = \frac{3\gamma_0 a}{2\hbar}, \tag{2.24}$$

where γ_0 is the nearest neighbor coupling energy in a tight-binding computation of the energy bands [12]. In addition to being linear, the bands are chiral, in that the wave function for the positive slope band has positive helicity and the negative slope band has negative helicity. The helicity arises from the pseudo-spin imparted to the wave function by the atomic contributions to the wave functions; e.g., the unit cell of the lattice has two atoms in it, which are usually denoted the A and B atoms.

Usually, one recognizes that the linear bands are Dirac-like, and the Fermi velocity corresponds to an effective 'speed of light.'

An important peculiarity of graphene is this Dirac band structure, which in the Dirac picture corresponds to particles with zero rest mass. That is, the effective mass at the Dirac point (where the two bands cross and $E = 0$) is exactly zero. Away from the Dirac point, however, the particles obtain a dynamic mass which may easily be found by equating the equivalence of crystal momentum to quasi-particle momentum [9], as

$$m^* = \frac{\hbar k}{v_F} = \frac{\hbar}{v_F}\sqrt{\pi n_s}, \tag{2.25}$$

a result that has been confirmed by cyclotron resonance studies [13, 14]. In a manner similar to above, we find that equation (2.19) is still valid as it is a normal property in momentum space. Then, we can use equation (2.23), using only the positive sign, to get

$$E_F = \hbar v_F (2\pi n_s)^{1/2}, \tag{2.26}$$

which leads to

$$n_s = \frac{1}{2\pi}\left(\frac{E_F}{\hbar v_F}\right)^2, \tag{2.27}$$

and

$$\rho_2(E) = \frac{E}{\pi(\hbar v_F)^2}. \tag{2.28}$$

This is a quite different density of states (per unit area) than that for normal 2D or 3D materials, and leads to some interesting behavior, that will be discussed in chapter 4.

2.2.3 One dimension

Similarly, if we consider an artificial system of free electrons with allowed motion in one dimension and with only a one-dimensional wave function, such as a nanowire, the same arguments above lead to the result

$$\rho_1(E) = \left(\frac{m}{2\pi\hbar^2}\right)^{1/2}\frac{1}{\sqrt{E}}, \tag{2.29}$$

and the density of states varies as the inverse square root of the energy.

2.2.4 Multiple subbands

While we have discussed the systems of interest as two-dimensional and one-dimensional, the GaAs/GaAlAs heterostructure system, and others of interest, are more typically fully three-dimensional in character. However, the confinement near the interface produces the Q2D character. What we notice in figure 2.2 is the fact that there are many steps in the QPC conductance, which correspond to many modes propagating through the structure. In our solution of the harmonic oscillator in appendix B, we also find many subbands, which correctly reflect the experimental observations. Now, the question is how do these subbands modify the density of states found above.

The important view is that each subband corresponds to a channel of transport through the QPC. The electrons are free to move in any direction parallel to the heterostructure interface, so this motion appears to be two-dimensional systems. Each of the subbands produces a similar Q2D layer of electrons. For low energies (relative to the eigen-energy of the lowest subband), only the first subband is occupied, and the density of states is the constant value given in equation (2.22). However, when the Fermi energy is increased such that it exceeds the eigen-energy of the second subband, then the electrons will be associated with both of the two subbands, with the result that the net density of states is double at this energy. And, this multiplicative effect continues as the Fermi energy is raised further. The net density of states is thus a series (in two dimensions) of steps, the height of which is equal and given by equation (2.22). A similar behavior occurs in one dimension, where the electron is free to move along the wire. In figure 2.7, we illustrate this additivity of the density of states in two and one dimension. In panel (a) we show the additivity of the density of states for three subbands, each of which arises for a different eigen-energy. The lowest contributor (blue) arises from the ground state, and provides the first step in the total density of states. The next contribution comes from the second subband (red), and provides a second step in the total density of states. Similarly, the third contribution comes from the third subband (green) and provides a third step in the total density of states. An important point is that for a given Fermi energy, the kinetic energy of the carriers in the different subbands will be different. In the lowest subband, the kinetic energy is measured from the corresponding lowest eigen-energy. In the second subband, the kinetic energy is measured from the corresponding second eigen-energy. And, this behavior repeats

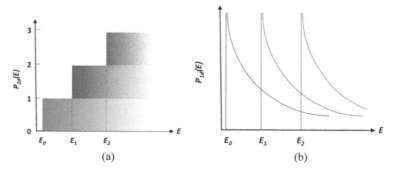

Figure 2.7. (a) Additive density of states for three subbands (blue, red, and green) in two dimensions, in units of $m^*/\pi\hbar^2$. (b) Subband density of states for three subbands (blue, red, green) in one dimension in arbitrary units. The characteristic shape of the fall of each density of states is due to the square root singularity in equation (2.29).

as we march up through the ladder of subbands. Of course, this can greatly complicate the interpretation of transport measurements when many subbands are occupied. For this reason, most experiments are designed so that only a single subband is occupied by electrons, and this of course is the lowest subband.

In panel (b) of this latter figure, we plot the equivalent ladder of contributions to the density of states for a Q1D structure. Here, we do not add them together, but display them independently to show how they are replicas of one another. Thus, the shape of the second subband is identical to that for the first subband, but is shifted to higher energy according to the two eigen-energies. The total density of states will thus show a saw tooth shape on a rising background as these individual contributions are added together, as one can see by adding the three curves of figure 2.7(b) together.

Because these quasi-low-dimensional systems have these multiple subbands, we need to account for this behavior in the corresponding total density of states. Hence, for the Q2D system, equation (2.22) becomes

$$\rho_2(E) = \frac{m}{\pi \hbar^2} \sum_{E_n < E} \Theta(E - E_n). \tag{2.30}$$

Similarly, for a Q1D system equation (2.30) becomes

$$\rho_1(E) = \left(\frac{m}{2\pi \hbar^2}\right)^{1/2} \sum_{E_n < E} \frac{1}{\sqrt{E - E_n}} \Theta(E - E_n). \tag{2.31}$$

In these last two equations, $\Theta(x)$ is the Heaviside step function, equal to unity when $x > 0$, and equal to zero otherwise.

In some nanowires, complications in the density of states can arise due to confinement in two directions. If the two transverse directions have the same dimension, then the quantization has degenerate eigenvalues. For example, if we consider hard wall boundary conditions for a square nanowire, then the lowest eigenstate has $n_x = n_y = 1$, and this state is unique (singly degenerate). The second eigenstate, however, is doubly degenerate, arising for $n_x = 2$ and $n_y = 1$, or for $n_x = 1$ and $n_y = 2$. This subband density of states will be twice as large as that for the first subband. The third subband is again singly degenerate, but much higher degeneracies can arise for higher subbands. Thus, it is not unusual to have the chain of subband contributions to equation (2.30) be non-uniform among the subbands. We will see these multi-subband effects in a number of experiments later. Now, however, it is time to bring all these quantized states together to understand how the steps occur in figure 2.2.

2.3 The Landauer formula

Many years ago, Rolf Landauer presented an approach to transport, and the calculation of conductance, that was dramatically different from the microscopic kinetic theory based on the Boltzmann equation that had been utilized previously (and is still heavily utilized in macroscopic conductors) [15, 16]. He suggested that one could compute the conductance of low-dimensional systems simply by computing the transmission of a mode from an input reservoir to a similar mode in an output reservoir. The transmission of this probability from one mode to the other was then very similar to the computation of a tunneling probability, except that there was no requirement that the process be one of tunneling. The only real constraint was that of lateral confinement so that the two reservoirs could be discussed in terms of their transverse modes. The key property of these two reservoirs was that they were in equilibrium with any applied potentials. That is, the electrons in the reservoirs were to be described by their intrinsic Fermi–Dirac distributions with any applied potentials appearing only as a shift of the relative energies (which would shift one Fermi level relative to the other). While he originally considered that the transport was ballistic, this is not required. Rather, the requirement is that we can assign a definitive mode to the electron when it is in either of the two reservoirs, which means that if scattering is present, it must be described specifically as a transfer of the electron from one internal mode to another.

In figure 2.8, we give a schematic view of the mesoscopic device, in which the central constriction is connected to a pair of reservoirs which are maintained in equilibrium. The bias is applied to the right reservoir so that the left Fermi energy can be used as the reference level for the applied potential (and indeed for the energies throughout the structure). The right contact now emits carriers into the constriction with energies up to the local Fermi level plus the applied bias, $E_F + eV$ (note that the energy eV will be negative for a positive voltage). The left contact emits electrons into the constriction with energies only up to E_F. In the following discussion, we will assume that the applied voltage is quite small, although this also is not a stringent requirement.

By making the initial assumption that no scattering takes place within the constriction, we can make a definitive association between the energy of the carriers and their direction of propagation. That is, electrons injected into the constriction from the left reservoir have a positive momentum and travel from left to right in the figure. On the other hand, electrons injected into the constriction from the right reservoir have a negative momentum and thus travel from the right to the left. We sketch this bit in figure 2.9, where we again remind ourselves that positive voltage applied to the right reservoir lowers the energy of that reservoir. In figure 2.9, we schematically plot the dispersion relation for these electrons. The momentum in the figure essentially corresponds to the motion of the electrons through the constriction.

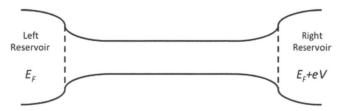

Figure 2.8. One can treat the region between the two reservoirs, which may be a quantum wire or a QPC as a ballistic constriction, although the ballistic requirement is not necessary. The bias is assumed to be applied to the right reservoir, so that the left reservoir provides the reference level for the energy.

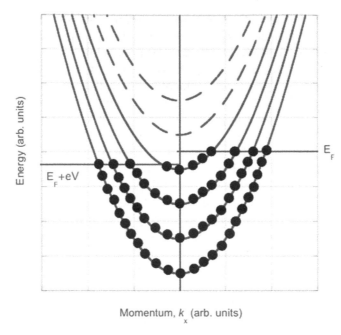

Figure 2.9. Electrons populate the allowed bands differently for an applied bias (here a negative voltage applied to the cathode). It can be seen that more electrons are moving to the left than are moving to the right, giving a current flow in the device.

Those with positive momentum come from the left reservoir and travel to the right, while those with negative momentum come from the right reservoir and travel to the left. Because of the applied bias, there are more electrons traveling to the left than are traveling to the right and this gives us a net current through the constriction. As discussed previously, the dispersion in momentum actually consists of discrete points according to equation (2.11). The imbalance between left-going and right-going electrons means that the constriction is in a condition of non-equilibrium, which is required to support a net current.

In order to compute the current through the constriction, we need to first evaluate the charge that is occupied in each channel in the energy range between E_F and $E_F + eV$. As we remarked above, if the applied bias is small, then we can estimate the excess charge as

$$\delta Q \approx \frac{e^2 V}{2} \rho_{1d}(E_F) = e^2 V \sqrt{\frac{m^*}{2\pi^2 \hbar^2 E_F}}.$$ (2.32)

It should be noted that we used only one-half of the density of states given by equation (2.29) since we are only interested in those electrons that are traveling to the left (those with positive momentum). Also, because of our definition of the density of states, equation (2.32) is the charge per unit length. In order to compute the current, we need only multiply equation (2.32) by the group velocity of the carriers at the Fermi energy, which is

$$v_g = v_F = \frac{1}{\hbar} \frac{dE}{sk} \Big|_{E=E_F} = \frac{\hbar k_F}{m^*} = \sqrt{\frac{2E_F}{m^*}}, \tag{2.33}$$

and this leads to the current

$$I_c = \delta Q v_g = e^2 V \sqrt{\frac{m^*}{2\pi^2 \hbar^2 E_F}} \sqrt{\frac{2E_F}{m^*}} = \frac{2e^2}{h} V. \tag{2.34}$$

A very interesting effect occurred in this development of the current, and that is the energy has dropped out of the equation. Hence, the current is completely independent of the energy, and this is true for each Q1D channel that flows through the constriction. The result (2.34) thus has no dependence on the channel index; the current in each channel is identical. This result has been called the equipartition of the current, which is a unique event for one-dimensional channels. Hence the current flow is divided equally among the available channels. Since the current in each channel is equal, the total current carried in the constriction is simply the number of such channels that are occupied and the current per channel. Hence,

$$I_{\text{total}} = N I_c = N \frac{2e^2}{h} V. \tag{2.35}$$

From this last expression, we obtain the conductance through the constriction as

$$G = \frac{I_{\text{total}}}{V} = N \frac{2e^2}{h}, \tag{2.36}$$

and this last expression is usually called the Landauer formula. Thus, the conductance of the quantum wire (our constriction) is quantized in units of $2e^2/h$ ($\sim$77.28 μS) with a resulting magnitude that depends only upon the number of occupied channels. Finally, the results of figure 2.2 become clear. The steps in the conductance occur as individual channels are occupied or emptied (depending upon a rising or decreasing potential), and the steps occur due to this important behavior in one-dimensional channels and the cancellation between the energy terms in the density of states and in the group velocity. The factor of 2 in equation (2.36) arises from the density of states and is the form for zero magnetic field where we can assume that the electrons are spin degenerate. Thus, in the experiment with split gates on the surface of the heterostructure, as the gate voltage is made more positive, the width of the QPC is increased (and the saddle potential is decreased), which leads to an increase in the number of channels conducting through the constriction. As each new channel is occupied, the conductance jumps upward by $2e^2/h$, and this increase is the same for each and every channel. To clearly see the steps in the conductance requires high mobility materials so that the transport is almost ballistic. As we will see below, the presence of significant scattering will wash out the steps by introducing transitions in which electrons jump from one channel to another and this will upset the balance in the channels.

One method in which the contribution of the separate spin subbands can be observed is to apply a large magnetic field to a material with significant spin splitting. The magnetic field lifts the spin degeneracy due to the Zeeman shift in the electron energy, thus splitting the spin-up and the spin-down electrons. Now, each subband splits into two distinct subbands, each of which has an opposite spin and now contributes only e^2/h to the total conductance. The same behavior will be observed as the gate voltage is varied, although there will be twice as many steps, each of which will be only one-half the previous height [17]. We will return to the role of the magnetic field in a later section where we see that the spin does not always follow this simple idea, particularly in the single channel case.

In the above, we did not consider the probability that an electron in one mode in one reservoir might actually wind up in a different mode in the other reservoir. This is not reflected in the formula (2.35). Nor, does this latter equation offer the possibility of partial transmission of a mode, which is clear in the rounding of the steps in figure 2.2, and will become even clearer in the discussion of the next section. The transitions from one mode to another will become even more important when we allow scattering within the channel region. Consequently, it is more intuitive, if we adapt equation (2.35) to the actual experimental situation in which partial transmission of a mode is possible. To do this, we rewrite equation (2.35) as

$$I_{\text{total}} = \frac{2e^2}{h} V T(n), \tag{2.37}$$

where $T(n)$ is the transmission of the n channels in one reservoir, which we define as

$$T(n) = \sum_{j=1}^{n} T_{mn}, \tag{2.38}$$

with the caveat that $m \leqslant n$. This caveat is based upon the assumption that we are dealing with electrons which enter the structure through the left reservoir and exit through the right reservoir. Thus, the entering electrons occupy modes up to the nth, which may be partially occupied. They will exit through the same number of modes in the right reservoir unless there is scattering, in which there may be a different number occupied in the case of partial occupation of modes. Of course, then the number of entering modes is an integer, such as when the conductance is on one of the steps, then we may say that $T(n) = N$, and we recover equation (2.36). In equation (2.38), the term T_{nm} is the probability that an electron entering through mode n exits through mode m. We will use this notation again below, and several more times through this book.

2.3.1 Temperature dependence

So far, we have not talked about the transition from the reservoirs to the channel region. In figure 2.8, this transition is shown as a relatively smooth one with no sharp edges, and the steps in figure 2.2 are relatively distinct and well formed. But, this is not always the case. The important point is that the transition should be smooth enough that no reflections are generated at the transitions. While waves can be

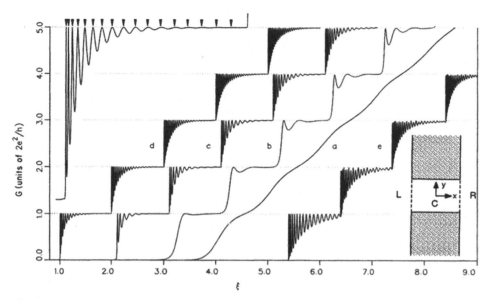

Figure 2.10. Simulations of a QPC at $T = 0$ and with pure ballistic transport. Curves 'a', 'b', 'c', 'd' are for zero potential in the channel and $2d/W = 0, 1, 5, 10$. Curve e is expanded for a saddle potential of $V_0 = 2.5\Delta$ and $2d/W = 10$. The first plateau of curve 'd' is enlarged at the upper left. The horizontal scale is expanded for this latter curve with the other curve offset to the right by 1.1 units. (Reprinted with permission [18]. Copyright 1989 the American Physical Society.)

turned around, this must be a smooth process so as not to mix modes at these transitions. Such a transition is often called an adiabatic transition, and when this is achieved, nice curves are obtained as in figure 2.2. While the transition is shown as sharp in the upper inset to this latter figure, split gate Schottky barriers usually do have relatively smooth transitions due to the slow variations of the electrostatic barriers.

In figure 2.10, we show a simulation of a QPC with sharp corners and pure ballistic transport at $T = 0$ [18]. The model is shown in the lower right inset, and includes an infinite potential for $|y| > W/2$, where W is the width of the opening and has a channel of length $2d$. In the channel, the potential is taken as V_0 so as to simulate the saddle potential. The major reason for the steps and the rounding on the transitions lies with the properties of the Fermi–Dirac distribution function. At zero temperatures, the electrons fill the states exactly up to the Fermi energy. As the temperature is raised, however, electrons begin to be thermally excited above the Fermi energy, leaving empty states just below the Fermi energy. This, of course, leads to the well-known form of the Fermi–Dirac distribution

$$f_{FD}(E) = \frac{1}{1 + \exp\left(\frac{E - E_F}{k_B T}\right)}. \tag{2.39}$$

If one computes the energy derivative of the Fermi–Dirac distribution, then one finds a negative peaked function, similar in form to a Lorentzian curve (but, of course, different from this). The full-width of this peak at half-height is $3.5k_BT$. So, as the temperature rises, the transition from a value of unity to zero in equation (2.39) extends over a greater and greater energy range. For reference, at 4.2 K, the full-width at half-maximum is just over 1.25 meV. The energy levels that give rise to the steps in the conductance will have to be spaced considerably wider than this if the steps are to be observed in an experiment, as obviously they are in figure 2.2.

In the figure, the Fermi energy and the saddle potential are normalized to the quantity $\Delta = \hbar^2/8m^*W^2$. The Fermi energy is parameterized in the figure as the quantity $\xi = (E_F - V_0)/\Delta$. The most striking feature in this figure is the oscillations that appear at the start of each step. In the most dramatic case, they range in amplitude from the bottom to the top of the step; e.g., a maximum of $2e^2/h$. But, they become damped out as one progresses past the threshold energy for the step. These oscillations are most distinct on the first step, where there is only a single longitudinal mode propagating through the channel. Then, they become more complex as more modes propagate and interfere within the channel. The source of the oscillations is longitudinal resonances in the transmission through the constriction. With the sharp boundaries between the constriction and the two reservoirs, there is a distinct mismatch in the wave functions for the allowed modes in the two regions. One can see such resonances in transmission over a tunneling barrier, and they are often referred to as 'over the barrier resonances'. This is exactly the source for these oscillations, since there is a wave mismatch at the transitions, this sets up interferences much like in thin dielectric multi-layers. The transmission rises to unity when the length of the constriction is such that the momentum wave vector is a multiple of $\pi/4d$ (remember that the length is $2d$). Clearly, as the length of constriction is raised, the required momentum for resonance is reduced and many more oscillations are observed. We will see below that raising the temperature and introducing scattering both work to wash out these oscillations. But, they are seen in some experiments.

In figure 2.11, we show this temperature behavior for the simulations given in figure 2.10. Here, the quantity μ is the chemical potential, which is effectively our Fermi energy. One can see that, for a temperature of only a half percent of the Fermi energy, the oscillations on the third plateau are completely gone, and only a smooth step remains. Remarkably, the oscillations completely disappear over a temperature increase of only a factor of five. As the value of the Fermi energy in the figure is 10 meV, the two non-zero temperatures are 0.1 K and 0.5 K. This reinforces the view that the steps can be washed out at temperatures of just a few kelvin since the subband spacing in split gate QPCs may be only a few meV. In figure 2.12, we give data for a split gate Schottky defined QPC in GaAs/AlGaAs [19]. Here, the role of the temperature can be clearly observed, as the steps become quite rounded as the temperature is raised.

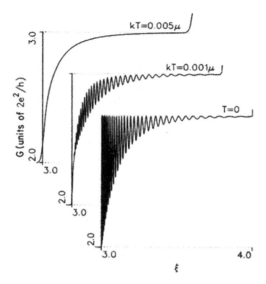

Figure 2.11. Effect of temperature on the simulated step for the third plateau of figure 2.10. Here, μ is 10 meV, so the center curve is for a temperature of 0.1 K. (Reprinted with permission from [18]. Copyright 1989 the American Physical Society.)

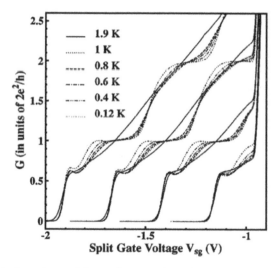

Figure 2.12. Temperature dependence of the conductance through a QPC. The different sets of temperature traces are for a different bias at the mid-point of the QPC corresponding to central gate potentials of 0, −1.2, −1.8, and −2.4 V (left to right). The observed structure below the first plateau will be discussed later. (Reprinted with permission from [19]. Copyright 2000 the American Physical Society.)

2.3.2 Scattering and energy relaxation

Although we have related the density of states for our quantum wires to their ideal forms in various dimensions, these are really only approximations to the real density of states. Certainly these forms are valid at zero temperature and pure ballistic transport, but we have already talked about the broadening of the step transitions due to non-zero temperature. In the presence of scattering, there is a finite lifetime in any quantum state and this causes a broadening of the density of states. For example, the step function that arises at the edge of each step can be considered to be the integral over a delta function which resides at the step position. That is, the step function in equations (2.30) and (2.31) can be written as

$$\Theta(E - E_n) = \int_{-\infty}^{E} \delta(E' - E_n) dE'. \tag{2.40}$$

When the energy is greater than the subband eigen-energy, then the integral encloses the delta function and has the value of unity. Otherwise, the integral gives zero, and these two results define the limiting cases of the step function.

In the presence of scattering the broadening changes the shape of the delta function giving it a finite height and width in energy. At low temperatures, and near equilibrium, the shape of the broadened function is nearly a Lorentzian line [20]. The scattering process is treated via perturbation theory, and this leads to a self-energy whose imaginary part is given by the total scattering rate, and this leads to

$$\Theta(E - E_F) \rightarrow A(E - E_F) = \frac{\hbar}{\pi\tau} \int_{-\infty}^{E} \frac{dE'}{(E' - E_F)^2 + \hbar^2/\tau^2}. \tag{2.41}$$

In the case of very strong scattering, there can actually be a shift in the eigen-energy at which the step occurs, much like a heavily damped LC circuit in which this damping can cause a shift of the resonance. But, the shift is seldom observed as such strong damping would already have washed out any chance to observe the steps.

In our derivation of the Landauer formula and the conductance quantization, use was made of the equipartition of current so that each occupied channel carried the same current. This was a reasonable approximation as long as there was no scattering between the different subbands in the channel. However, if inter-subband scattering occurs, we can no longer make this approximation, and we cannot continue to associate a definite direction of momentum with a given energy level. Hence, this will also contribute to a reduction of the quantization properties of the conductance. So, scattering not only smooths the onset of the step, it also provides a smoothing by introducing scattering between the various subbands in the channel.

It might have occurred to some readers that, when we have purely ballistic transport, the conductance should have been infinite rather than the finite value given by the Landauer relation (2.36). In fact, the transition between the reservoir and the channel has a non-zero resistance which is a contact resistance between the one-dimensional channel and the reservoirs which give access to it. Normally, the reservoirs are very wide regions which are approximately Q2D, and the contact resistance arises from a mode mismatch between these latter regions and the channel. This occurs even when the transition is very smooth and adiabatic. This contact resistance goes directly to the question of just where the voltage drop across

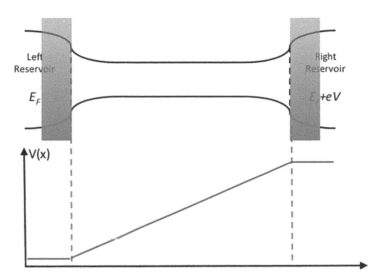

Figure 2.13. The constriction of figure 2.8 is redrawn here with a linear potential drop (bottom portion) corresponding to a constant electric field.

the constriction actually occurs. We have assumed that the voltage exists between the two reservoirs, but can we be more specific about this. In a normal resistive conductor we would expect that this voltage should be dropped uniformly along the length of the constriction, just like a normal resistor, but this still requires unusual behavior in the contacts.

Since there is no scattering within the constriction, the voltage drop across it must correspond to a constant electric field, which can be quite low as we will see. In reality, the fact that the voltage is not dropped smoothly across the device means that charge must accumulate at the contacts, which leads to the variations in the voltage. As a result, the field in the center of the constriction is quite small, and the voltage drops are forced to occur in the transition regions. Suppose we take the case of a linear voltage drop across the constriction so that there is a constant electric field in this region. Then the potential drop will look like that in the bottom panel of figure 2.13. But notice that, at the two points where the dashed line has been extended from the transition region, the electric field is discontinuous. Such a discontinuity in the electric field requires that a sheet of excess charge be present at this discontinuity. In fact, the charge at the left must be positive charge, while the charge at the right must be negative charge. So, we must have a depletion of electrons at the right and an accumulation of electrons at the left. If electrons are the major charge carrier, then the left hand reservoir is the cathode, through which the electrons enter the structure. The accumulation of charge at the transition point contributes to the contact resistance.

There is still an additional problem with the potential as drawn in this figure. The linear potential drop leads to a sizable electric field, which will accelerate the electrons. We can see this in figure 2.14, in which we plot the electron energy instead of the voltage. If the ballistic electron is injected from the left contact, it will gain

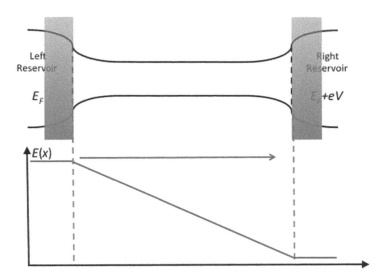

Figure 2.14. In this figure, the energy E(x), rather than the voltage, is plotted through the structure. An electron entering from the cathode at the left, and traveling ballistically, will gain energy as it moves through the channel, exiting with an additional energy defined by the applied voltage.

energy from the electric field. That is, the ballistic electron travels at constant energy, and the potential energy at the cathode is converted to kinetic energy as it travels. By the time it reaches the anode at the right-hand side, it will have gained kinetic energy given by eV. A more important aspect is introduced by Kirchoff's current law. As the electrons are accelerated by the electric field, their increase in velocity requires that fewer electrons are present as we move along the channel. That is, Kirchoff's current law requires the current to be uniform throughout the structure. Since the cross-sectional area does not change, this requires the product of density and velocity to be a constant value. As the velocity rises, the density must decrease. But, in a semiconductor, the density is usually set by the doping (as we discussed in the previous chapter in our discussion of heterostructures). As the number of carriers drops, this creates an additional space charge in the channel, which in turn requires that the potential be nonlinear; e.g., the potential drop cannot be a linear one. The result of this argument is that we cannot have the linear potential drop shown in figures 2.13 and 2.14 if the carriers are moving via ballistic transport. The linear potential drop only can occur if there is sufficient scattering to assure that the carriers move with a near-equilibrium energy. Our conclusion then can only be that the electric field must essentially be quite near zero in the channel, if the carriers are to move via ballistic transport.

As a result, the potential drop must divide between the cathode and the anode transition regions, as shown in figure 2.15. While we have drawn the two voltage drops as nearly the same, there is no real requirement that this be the case. In fact, it is usually assumed in the transport world that most of the discontinuity is related to the cathode. But now, ballistic transport can occur through the constriction without the carriers gaining excess energy from the applied bias. At the same time, we note

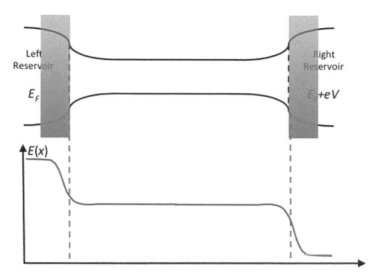

Figure 2.15. Potential drop required for ballistic transport through the constriction.

that the potential drops in the transition regions now require that dipole charge densities exist at each transition. The potential drop can only occur through the existence of such dipole charge densities. Originally, in figure 2.13, the dipole was split between the two transition regions. Now, each transition region has its own dipole charge density, and this leads to the contact resistance.

These results give rise to an important modification of the Landauer formula. In equation (2.36), the transmission from one contact to another is defined in the reservoirs, which means to the left of the cathode transition region and to the right of the anode transition region. What if we actually do measurements in the channel itself? If these are two-terminal measurements, in which the same contacts are used to source the current and to measure the voltage, we will obtain an answer, but the voltage drop has to occur in these contacts, which have merely replaced the normal ones. On the other hand, if we use four-terminal measurements, then we will find that the Landauer formula becomes modified. We will address this modification in a later section when we deal with the multi-terminal Landauer formula. But, the modification arises from the presence of these contact resistances. We will return to these contact resistances below.

2.3.3 Contact resistance and scaled CMOS

When a large level of carrier scattering is present in the channel discussed above, then it becomes quite likely that many, if not most, of the individual mode transmissions will deviate from unity due to the greatly broadened probability for mode occupancy. This can change the nature of the Landauer formula. In order to simplify the argument, let us first assume that only a few modes are occupied so that each mode has the same partial transmission $T_{pc} < 1$. Then, we can invert the conductance to talk about the resistance of the device as

$$\frac{1}{G} = \frac{h}{2e^2}\frac{1}{NT_{pc}} = \frac{h}{2e^2 N} + \frac{h}{2e^2}\frac{1 - T_{pc}}{NT_{pc}} \equiv \frac{1}{G_C} + \frac{1}{G_D}. \qquad (2.42)$$

We can see now how the resistance can be divided into a contact resistance, which can never vanish, and a device resistance which can go away in the absence of scattering. The contact resistance is connected with the manner in which the actual voltage drop is proportioned across the charge regions in the transitions between the reservoirs and the channel.

While we initially developed the Landauer formula with a treatment of transport in which all the energy relaxation occurred in the reservoirs (true ballistic transport), this was not really required. The Landauer formula provides a rigorous framework for the analysis of nanostructures so long as current is conserved and the reservoirs can be analyzed in terms of modes. This allows us to determine transmission coefficients, as described above, for transport between these modes from, e.g. the Schrödinger equation, or from a classical approach such as the Duke tunneling formula [21]. As long as the scattering can be described as taking a particle from one channel to another, so that particle conservation accompanies current conservation, then any level of scattering can be incorporated within the Landauer formula. This does not mean that energy cannot be transferred to the crystal lattice by the scattering process. We need only particle and current conservation for the validity of the Landauer formula. Further equations are required if we desire to examine energy flow within the nanostructure.

In recent years, it has been popular to try to stretch the Landauer formula to more common every-day devices, such as the scaled CMOS transistor [22]. It turns out that this is not a particularly useful or straightforward approach, but one can gain some insight into the behavior of the MOSFET with this approach. In the transmission approach, one often jumps to the conclusion that the potential will have the linear drop shown in figure 2.13 and that the transport is ballistic. For reasons discussed in connection with this figure in the previous paragraphs, such an approach is incorrect. The linear potential drop can only occur with considerable scattering to keep the energy of the carriers from rising in the electric field. Rather than start with the transmission itself, let us look at the common equation for the MOSFET and see how we can connect with contact effects and the Landauer formula. This will shed light on just how the behavior of the common MOSFET is not that different from our channel discussed in the earlier parts of this chapter.

The common formula for the MOSFET within the gradual channel approximation is given by an expression for the drain current in terms of the various potentials applied to the device, as

$$\begin{aligned} I_D &= \frac{W\mu C_{ox}}{L}\left(V_G - V_T - \frac{V_D}{2}\right)V_D \\ &= \frac{W\mu C_{ox}}{2L}[(V_G - V_T)^2 - (V_G - V_T - V_D)^2], \end{aligned} \qquad (2.43)$$

where W and L are the gate width and length (along the channel), C_{ox} is the gate capacitance per unit area, μ is the electron mobility, V_T is the threshold voltage where charge begins to accumulate in the inversion layer, and V_G and V_D are the gate and drain bias voltages (the source is taken to be grounded). In writing the equation the way in which it is shown in the second line of equation (2.43), we can immediately make contact with the Landauer formula. The first term in the square brackets represents current entering the device from the source (left) electrode, while the second term represents current entering the device from the drain (right) electrode. Thus, the source and drain are our reservoirs, and we connect with particle flow exactly as in the Landauer formula.

In thermal equilibrium, with no bias applied to the gate and drain electrodes, there is no current flow through the device, since the two terms cancel each other. This cancellation remains the case even with gate voltage applied, but with the drain voltage set to zero. The gate voltage changes the properties of the channel, but does not upset the detailed balance of particle flow through the structure. However, when a small drain bias is applied, the second term in the square brackets is reduced relative to the first term, a small current begins to flow. As the drain biased is raised, the current increases. Once the second term is reduced to zero (it is not allowed to go negative, as this violates the conditions under which it was derived), the current saturates through the device at the normal value

$$I_{D,sat} = \frac{W\mu C_{ox}}{2L}(V_G - V_T)^2. \tag{2.44}$$

When the drain bias is applied, the drain region is lowered in energy relative to the source region, just as shown in figures 2.13–2.15. Saturation occurs when this energy difference is sufficiently large that electrons in the drain can no longer surmount the barrier to reach the source. No matter whether the transport is ballistic or not (there is lots of scattering), the energy regions in the source and drain will be the same. What will be different between these two cases is how the energy drops between the source and drain.

Let us now take (2.44) and write it in terms of an 'equilibrium' current that depends upon the gate voltage, as

$$I_{eq}^+ = I_{D,sat} = \frac{W\mu C_{ox}}{2L}(V_G - V_T)^2. \tag{2.45}$$

Then, we can rewrite the second line of (2.43) as

$$I_D = I_{eq}^+\left[1 - \left(1 - \frac{V_D}{V_G - V_T}\right)^2\right]. \tag{2.46}$$

For small values of the drain voltage, we can approximate write this result in terms of a resistance (likely seen as a contact resistance)

$$R_C = \frac{V_D}{I_D} \sim \frac{V_G - V_T}{2I_{eq}^+}. \tag{2.47}$$

Even for a very short device with perfect transmission, we cannot get around this minimum resistance. This resistance characterizes the transition region between the pure source and the channel (just as in the Landauer approach) and appears regardless of whether the transport is described by the mobility or is ballistic. In some sense, this resistance describes a barrier between the source and the channel [23]. The current flux of equation (2.45) is due to carriers from the source with sufficient kinetic energy to overcome this barrier between the source and the channel. It is important to note in equation (2.45) the flux is actually metered by the gate potential, so that this is a gate-controlled barrier. Once the carriers supplant this barrier, they are going to flow downhill (down the potential drop) to the drain. They will flow to the drain regardless of the nature of the transport. The only effect that ballistic transport can play is to change the shape of the potential landscape between the source and the drain. In this sense, scattering actually affects this potential landscape and serves to isolate the drain from the source. In a ballistic device, this isolation does not occur and the source–channel barrier can be affected by the drain voltage, an effect known as *drain-induced barrier lowering*.

2.4 Beyond the simple theory for the QPC

In the previous few sections, it has been shown that the understanding of the conductance steps in the QPC, or in a longer nanowire, fit nicely with our understanding of the role played by the density of states within the Landauer formula. This has led to a beautiful example of condensed matter theory being demonstrated by the experiments on these devices. But, there is always a chance to go beyond the simple theory to gain new insight into the devices, or to discover new things which may not be so well understood.

2.4.1 High bias transport

In deriving the above expressions for the conductance quantization, we made the assumption that the voltage applied across the device was quite small, in which case the voltage variation across the device is essentially unaffected by the applied bias. This supported the idea that transport in the small QPCs is essentially ballistic in nature, so that the actual voltage was dropped mainly across the transition regions between the reservoirs and the channel. However, if we apply a larger bias, we can induce significant band bending in the QPC itself, which of course must be accompanied by various non-equilibrium charge distributions within the device. But, these effects can lead to new and different experimental observations of the quantization phenomena in the observed conductance. For example, the usual plateaus that appear at integers of $2e^2/h$ evolve with voltage to half-integer values. We can see this in figure 2.16, which gives the results for a very high quality QPC in the GaAs/AlAs system. The Si dopants are placed in a thin GaAs quantum well sandwiched between two AlAs layers, and the 2DEG is located a distance of 160 nm

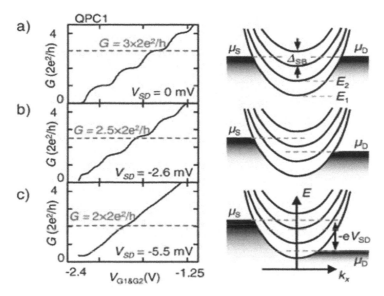

Figure 2.16. The pinch-off trace, and the schematic view of the potential, is shown for the QPC. (a) The linear response for a vanishingly small source–drain bias across the QPC. (b) Response for a larger bias (negative voltage on source side) where the plateaus have evolved to near the half-integer values. (c) A higher bias such that the plateaus have evolved further and returned to their integer values. (Reprinted with permission from [24]. Copyright 2011 IOP Publishing and Deutsche Physikalische Gesellschaft).

below the surface. The Si donors lead to an electron density of 3.5×10^{11} cm^{-2} with a mobility of $1-2 \times 10^7$ cm^2 Vs^{-1} [24]. In the figure, the conductance through the QPC and a schematic diagram of how the voltage is dropped across it, are shown for convenience. In panel (a), the normal results for a very small applied bias are shown. Here, the steps have their normal plateaus at integer values of $2e^2/h$. In panel (b), however, a bias of 2.6 mV has been applied to the drain (note that the figure gives the values at the source relative to the drain). With this bias, the plateaus have moved basically to the half-integer values, appearing at 0.5, 1.5, 2.5, and so on. There remains a small feature at 1.0 though. From the schematic diagram on the right, it can be seen that the Fermi energies in the two reservoirs are such that a subband minimum resides between them. Hence, the characteristic of this half-integer plateau arises when one reservoir injects into one fewer subbands than the other. Finally, in panel (c) the bias has been increased further, and the integer plateaus have returned. But, now there are two subband minima between the two Fermi levels. At still higher values of bias, the plateaus sometimes become washed out due presumably to heating of the electrons in the QPC [25].

When the source–drain bias is applied, this changes the Fermi potential in the center of the QPC, which shifts the gate voltages at which the transitions occur. Tracking the plateaus as one varies the source–drain bias allows a method of spectroscopy in which we may analyze the subband spacing in the device. This is illustrated in figure 2.17, where the transconductance is plotted as the two voltages are varied [24]. The plot uses the source–drain voltage for the vertical axis and the

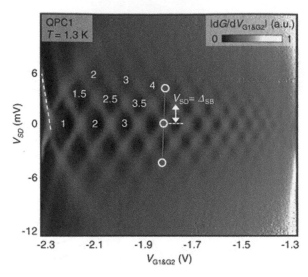

Figure 2.17. The transconductance $G_{TC} = dG/dV_G$ for one QPC (the two split gates are biased equally). The source–drain voltage is plotted vertically and the gate voltage horizontally, with the transconductance color coded according to the scale at the upper right corner. Plateaus in conductance (small transconductance) appear as dark areas, and the various ones are labels accordingly. (Reprinted with permission from [24]. Copyright 2011 IOP Publishing and Deutsche Physikalische Gesellschaft.)

gate voltage in the horizontal axis, with the transconductance color coded. Bright yellow is the largest value and black is the smallest value. Hence, the plateaus appear as dark regions between the brighter 'lines' which indicate the transitions between the plateaus. It can be easily seen how the integer plateaus gradually change with bias into the half-integer ones, and then change again into integer ones with increasing source–drain bias. Observation of these returning integer plateaus can be difficult to see, and the very high mobility in these samples makes this possible. This high mobility manifests itself as a reduction in the back-scattering of various modes at the transitions and leads to enhanced ability to do spectroscopy in this nonlinear transport regime.

Also indicated in the figure is how one can obtain analytical results. Three white circles are indicated at the transition between the fourth and fifth plateaus at different values of source–drain bias. We recall from the discussion above, that in the nonlinear regime, the two Fermi levels (left and right) span subband bottoms. On the dashed line between the circles, the observed change as the source–drain bias is increased obviously measures the distance between the subbands. That is, as one moves across the plateau from low bias to the bias corresponding to the next transition, one is clearly coupling exactly two new subbands between the Fermi energies. The transition occurs as the subband crosses one Fermi level, so clearly one-half of this applied bias corresponds to the subband spacing, as indicated in the figure. This suggests that, at this bias, the subband spacing is about 2.25 meV. But, one can go further, as each of the transitions allows us to study the subband spacing as a function of the two bias voltages. Thus, one can see just how the eigen-energies of the harmonic oscillator vary with gate voltage and with source–drain bias voltage. Thus, for the devices discussed here, the authors have found that the subband spacing varies from almost 5 meV down to about 1 meV when the QPC is sufficiently open such that ten modes are propagating through it [24]. It is clear that, with good quality material, one can fully characterize both the device under study and the models used to explain the results.

2.4.2 Below the first plateau

It may be clearly noted in figure 2.12 that there is a feature that appears below the first plateau, near the value of $0.7(2e^2/h)$. This additional shoulder in the conductance has been seen almost since the first measurements of the quantized conductance through a QPC. This feature cannot be explained by the simple theory, since it is not correlated with any common behavior, such as the number of modes. Moreover, this quasi-plateau does not seem to have any universal behavior. Some researchers have suggested that the 0.7 plateau, as it is usually referred to, merges with the first plateau as the temperature is lowered and an anomaly at zero source–drain bias emerges, which they identify with a Kondo-like feature (associated with charging a localized state in most cases) [26]. On the other hand, there appears to be no movement at all of the plateau with temperature in figure 2.12, nor is such behavior seen in many other structures which show conductance quantization, such as cleaved-edge overgrowth wires [27]. Many of the observations have been brought together by Pepper and Bird [28] in a Special Issue of the *Journal of Physics: Condensed Matter* which collects the work of a great many authors. More recently, Micolich has provided an extensive review of experimental studies of the 0.7 plateau [29]. The conclusion which one might draw from these works is that the effect may not be a single effect, but may be the appearance of several different, but similar, effects which depend upon the details of the actual device upon which the experiments are being conducted. The discussion here will argue with such a conclusion, but will present a few of the more universally accepted ideas.

The idea that the 0.7 plateau is, in at least some manner, connected with the electron spin is supported by observations that the plateau evolves to one at $0.5G_0$, where $G_0 = 2e^2/h$, as an applied magnetic field is raised to values to show full spin splitting [30]. In general, there is only a weak variation with the electron density in the reservoirs [4]. As remarked above, there is some consideration that the 0.7 plateau is related to the Kondo effect [31]. The Kondo effect is often seen in quantum dots (discussed later) [26]. Such a result is somewhat surprising, since the Kondo effect is well known to involve the interaction of conduction electrons with a localized spin (or magnetic moment). In quantum dots, in which the defining QPCs are pinched off so that only single-electron tunneling can contribute to the transport, the idea of a localized spin within the dot is easy to connect with a single localized electron in the highest occupied state of the dot; e.g., a total odd number of electrons in the dot. For this to occur in a QPC suggests that there must be a localized state that forms in the QPC below the first plateau, and this has been suggested to form self-consistently by some simulations [32, 33]. Others, however, have suggested that the localized state could produce the 0.7 plateau without the need for a Kondo effect. But, if the localized state contains a single spin state, as has been suggested [34, 35], it seems reasonable that Kondo physics could certainly play a part.

As mentioned, simulations in the linear spin-density approximation have shown that a localized state can form in the QPC below the first plateau [33, 35]. These features can be associated with the formation of a spin-dependent energy barrier that arises once one spin is trapped in the localized state. This barrier prevents the other spin from transporting through the QPC. These barriers rise and fall as a function of the local density, and with the local potential. By varying the density, the 0.7 plateau can be induced to evolve into the fully formed first step or to drop near a value of 0.5 [32]. In addition, other features can be observed, such as an anomaly near $0.25G_0$ that has been observed experimentally [26, 36, 37]. But, what is clear from these simulations is that the localized state is often not a simple minimum in the potential. Rather it is much more complicated, perhaps due to the existence of Friedel oscillations which extend from the minimum of the saddle potential far into the reservoir regions [38]. Moreover, the local potential tends to have a double minimum, which leads to three peaks. The amplitude of these peaks varies not only with the gate voltage, but also with the local density in the QPC (along the wire for example). In our own simulation, a model potential proposed by Timp [39] was used in place of the pure harmonic oscillator saddle potential discussed earlier. This confining potential had the form

$$V(x, y) = F\left(\frac{2x - L}{2z}, \frac{2y + W}{2z}\right) - F\left(\frac{2x + L}{2z}, \frac{2y + W}{2z}\right)$$
$$+ F\left(\frac{2x - L}{2z}, \frac{-2y + W}{2z}\right) - F\left(\frac{2x + L}{2z}, \frac{-2y + W}{2z}\right), \quad (2.48)$$

where

$$F(u, v) = \frac{eV_G}{2\pi}\left[\frac{\pi}{2} - \tan^{-1}(u) - \tan^{-1}(u) + \tan^{-1}\left(\frac{uv}{\sqrt{1 + u^2 + v^2}}\right)\right]. \quad (2.49)$$

and z is the distance of the Q2DEG from the surface of the heterostructure. W and L are the lithographic dimensions of the QPC at the surface of the heterostructure, so are taken from an experimental structure (140 nm and 350 nm, respectively, in the simulations discussed here). However, the actual local potential is solved self-consistently with the density in the linear spin-density approximation [34, 35]. In figure 2.18, we show the conductance plots for two values of the quantum wire (average) line density, in order to illustrate the role played by this quantity. In panel (a), the density is 2.01×10^6 cm^{-1} ($E_F = 13.5$ meV), and the first plateau has almost disappeared in favor of the 0.7 plateau. In panel (b), the density is 1.96×10^6 cm^{-1} ($E_F = 13.4$ meV), and the first plateau is well formed, while the 0.7 plateau has dropped closer to 0.5. In both cases, the feature near 0.2 is distinct. In these curves, the solid curve is the total conductance, while the dotted curve is the spin-down conductance and the dashed curve is the spin-up conductance. The unusual feature

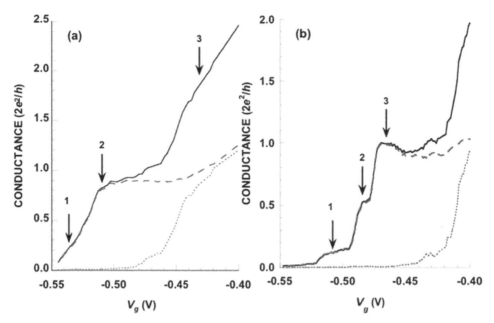

Figure 2.18. Conductance curves at two different line densities in the QPC: (a) 2.01×10^6 cm^{-1} and (b) 1.96×10^6 cm^{-1}. The solid lines are total conductance while the dotted and dashed curves are the spin-down and spin-up conductances, respectively, as discussed in the text. Data taken from [35].

here is that the spin-up modes are almost completely blockaded until the second plateau begins to appear. The entire first plateau, with the sub-structure, is entirely due to the spin up modes. Hence, there appear to be two spin-up modes propagating to form the first plateau.

As remarked above, the local potential seems to have three peaks in the region of the center of the QPC. We can see this in figure 2.19, where we plot the local potential profiles for the two cases shown in figure 2.18. The two panels refer to the same two panels of the conductance plots. Curves (1) are the potential profiles for the conductance near the feature at $0.2G_0$, while curves (2) correspond to the feature near $0.7G_0$. The set of curves (3) correspond to other plateaus as indicated by the arrows. Curves (1) and (3) have been offset for clarity. This three peak structure in the potential was also reported by Hirose *et al* [38].

One can probe these anomalies with the SGM discussed previously. We have used an InGaAs quantum well clad with InAlAs to form the heterostructure, and then formed in-plane, trench isolated gates such as those shown in figure 2.1. In this way, the scanning probe tip can be brought into the heart of the QPC without fear of hitting surface metallic gates. In figure 2.20, we show an SGM image obtained with the QPC in the 0.7 plateau [40]. Amazingly, essentially the same figure is obtained in the region $0.2–0.3 G_0$ [1], which suggests a common origin for these two features, just as found from the simulations mentioned earlier. It is noteworthy that the peaks in the figure are resistance peaks; e.g., the red color denotes high resistance. These resistance peaks correspond to the potential peaks that are observed in the

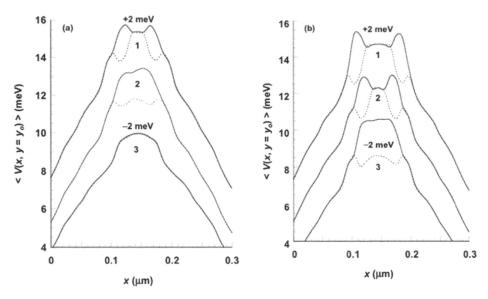

Figure 2.19. Potential profiles for the conductance plots of figure 2.18. Panels (a) and (b) correspond to the respective panels in this earlier figure. In each panel, curves (1), (2) and (3) correspond to the features identified by the arrows in figure 2.18.

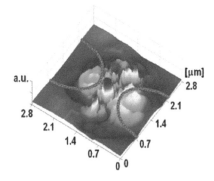

Figure 2.20. SGM image of a QPC at the 0.7 plateau. Very similar images are also obtained near the 0.25 plateau that is sometimes observed. The position of the trench isolation for the gates is indicated by the dotted curves.

simulations. These suggest that the conductance through these quasi-plateaus is quite probably via tunneling as one might guess. But, it seems to be clear that the peaks below the first plateau are quite probably connected to the spin of the electron, and a complete understanding of what is really occurring is still in question.

As mentioned earlier, some investigators do find a simple single localized minimum in the potential [32]. More recent simulations by Meir have also shown that a single localized state can exist which allows only a single spin state to propagate through the QPC [41]. Thus, even the nature of the localized state (or states) seems to vary with the properties of the particular device under study, and even with the local carrier density in the device. Thus, it is not surprising that there is so much variation in the experimental details, and there remains a significant likelihood that the structure below the first plateau is certainly a many-body effect, and may well be not a single effect but many possible ones which all have about the same likelihood of occurring.

Perhaps, it is best summarized by results from a recent paper, in which one or more localized states are observed to appear, and these are felt to arise from the Friedel oscillations in the electron charge density within the QPC [42]. These latter authors also report that there are spins associated with the localized states, as discussed above, and the Kondo effect, associated with transport through these localized spin states, contributes to the formation of the resulting many-body state. Of course, the self-consistent potential is going to be crucial to the formation of the localized state(s), and so some authors may not see the role of the Kondo effect, as has been discussed [29].

2.5 Simulating the channel: the scattering matrix

There are a great many approaches to solve the Schrödinger equation in a finite-sized system. When the system is large, it is often computationally more useful to use a recursive approach, in which the solution is generated by propagating a known contact solution through the structure with a slice by slice recursion. That is, the properties of a single slice are determined by those of the preceding slices starting from one reservoir or the other. The best known approach is the mode matching in which the wave function and its derivative are matched at each interfacial boundary between slices within the system. Unfortunately, this approach is known to be unstable for more than a few slices. An important variation of this approach is to switch the interface matching into the scattering matrix [43, 44], an adaptation borrowed from earlier microwave theory. We have used this approach for many years to study quantum dots [45]. A second approach is to use Green's functions, derived from the Schrödinger equation, in a recursive approach, and this will be discussed in the next section. Both of these approaches actually use a Green's function in the recursion through the incorporation of Dyson's equation. In each case, the transmission through the entire structure is computed, and the conductance evaluated using the Landauer formula (2.37) and (2.38), with

$$T_{nm} = |t_{nm}|^2, \tag{2.50}$$

where t_{nm} is the complex transmission of the wave function through the device, from mode n in the left reservoir to mode m in the right reservoir. In fact, a correction needs to be made to account for the fact that we are dealing with semiconductors and not metals. Hence, the mode velocities are not just the Fermi velocity. The velocities in each mode may be different, as they are measured from the subband minima corresponding to the modes. Thus, we correct equation (2.50) as

$$T_{nm} = \frac{v_n}{v_m} |t_{nm}|^2, \tag{2.51}$$

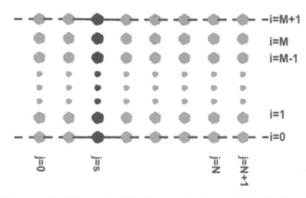

Figure 2.21. A two-dimensional grid around which the recursion formula is built. One slice is indicated by the darker shade of blue.

where the v are the mode velocities. In appendix A, we discuss the basic principles of discretizing the Schrödinger equation and the Poisson's equation and setting up slices for both approaches that we will discuss.

As discussed above, the normal approach to matching the wave function and its derivative at each slice of the structure (across the interface) is generally unstable when a large number of such interfaces are present. Stability can be restored by modifying the recursion to one based upon the scattering matrix, which has long been a staple of microwave systems and entered quantum mechanics through the Lippmann–Schwinger equation [46]. The strength of this approach lies in the fact that modal solutions to the Lippmann–Schwinger equation maintain their orthogonality through the scattering process [46]. In mesoscopic structures and nanostructures, the connection to microwave theory becomes stronger as the transport becomes dominated by the modes introduced by the lateral confinement. Even in the non-recursive mode, the use of scattering states provides a viable method of building up an orthogonal ensemble, even with weighting of the states by e.g., a Fermi–Dirac distribution [47]. This is particularly useful in providing the initial conditions for the wave function in a particular semiconductor device [48]. In the recursive approach, it reaches new levels of viability when combined with the modal structure in small devices [43]. In the end, this approach allows one to determine the transmission from one end to the other and evaluate the conductance via the Landauer formula.

The approach is based upon the discretization scheme for a uniform array of points spaced by a (see appendix A). We show such an array in figure 2.21, in which the array has a slice width in the transverse direction of $M + 1$ points and a length of $N + 1$ slices. A useful fact is that if we just limit our size as shown, the solution automatically assumes that the wave function will satisfy

$$\psi(0, j) = \psi(M + 1, j). \tag{2.52}$$

As a result of this, we do not concern ourselves with the first and last points of the array, and can write the slice wave function as

$$\psi_s = \begin{bmatrix} \psi(M, s) \\ \vdots \\ \vdots \\ \psi(2, s) \\ \psi(1, s) \end{bmatrix}. \tag{2.53}$$

Note here, that we have defined i in the y-direction and j in the x-direction, which is the current flow direction. Note also that the single subscript wave function is the *slice* wave function, as opposed to the *site* wave function which has two variables. Then, the Schrödinger equation can be rewritten as a matrix equation relating the slice vectors

$$H_{0,j}\psi_j - t\psi_{j+1} - t\psi_{j-1} = EI\psi_j, \tag{2.54}$$

where t is the diagonal matrix whose elements are the hopping energy

$$E_t = \frac{\hbar^2}{2m^*a^2} \tag{2.55}$$

(and not to be confused with the transmission matrix elements, which have subscripts). In equation (2.54), the Hamiltonian for the isolated slice is written as

$$H_{0,j} = \begin{bmatrix} (V_{Mj} - 4E_t) & E_t & 0 & & & \\ E_t & (V_{M-1j} - 4E_t) & . & & 0 & \\ . & . & . & & & \\ & & & . & . & 0 \\ 0 & & & . & (V_{2j} - 4E_t) & E_t \\ & & & . & E_t & (V_{1j} - 4E_t) \end{bmatrix}, \tag{2.56}$$

where the upper right block and lower left block are identically zero, and I is an identity matrix. We note that the individual elements of the slice wave function are the corresponding site wave functions. Generally, the transverse wave functions for the width confinement are found from the corresponding slice Hamiltonian, e.g., equation (2.56) with $j = -1$. This process produces the site wave functions and the slice wave function, which is doubly degenerate with the ability to propagate to the right and to the left. These wave functions may also be found through solving the equation

$$det[T_0 - \lambda I] = 0, \tag{2.57}$$

where T_0 is a matrix whose rank is twice that of the Hamiltonian (2.56) and is given as

$$T_0 = \begin{bmatrix} U_+ & U_- \\ \lambda_+ U_+ & \lambda_- U_- \end{bmatrix}. \tag{2.58}$$

Here, $U_\pm$ are matrices whose rank is equal to that of the Hamiltonian and whose columns give the slice wave functions in site representation. These are the modes of the transverse confinement and the positive and negative signs correspond to positive (right-directed) propagation and negative (left-directed) propagation. Similarly, the $\lambda_{\pm}$ are the corresponding eigenvalues. These matrices give the site-to-mode transformations of the various wave functions. There is a corresponding Fermi energy (the simulation is at zero temperature at this point), so that modes whose eigenvalues are below the Fermi energy will be propagating, while those whose eigenvalues lie above the Fermi energy are evanescent. Once, we get slice 0 in the mode representation, we can proceed with the recursion.

To carry out the recursion, we have to put everything into the scattering matrix form. The site to mode transformation creates slice wave functions (2.53) in which each element corresponds to a mode in the system, as suggested above. These modes can be either propagation or evanescent. To continue, we create two new matrices, which are the 'scattering matrices' C_1^s and C_2^s at the general site $j = s$. The recursion relation for the are now expressed as

$$\begin{bmatrix} C_1^{s+1} & C_2^{s+1} \\ 0 & 1 \end{bmatrix} = \begin{bmatrix} 0 & I \\ -I & t^{-1}(H_{0,s} - EI) \end{bmatrix} \begin{bmatrix} C_1^s & C_2^s \\ 0 & 1 \end{bmatrix} \begin{bmatrix} I & 0 \\ P_{1s} & P_{2s} \end{bmatrix}, \quad (2.59)$$

where the first matrix on the right-hand side is denoted as T_s, and H_{0s} is the slice Hamiltonian (2.56). For the zero temperature formulation, C_1^s is set to a unit matrix, although for a finite temperature simulation, the elements will be given by the Fermi–Dirac distribution for the energy of the corresponding eigenvalue. Similar values are given to C_2^s *at the right-hand end of the system*. Now, equation (2.59) is required to be satisfied, and this allows us to thus determine the propagation matrices P_{1s} and P_{2s} through the recursions

$$\begin{aligned} C_1^{s+1} &= P_{1s} = -P_{2s}(-I)C_1^s \\ C_2^{s+1} &= P_{2s} = (-IC_2^s + t^{-1}(H_{0,s} - EI))^{-1}. \end{aligned} \quad (2.60)$$

The unit matrix (the $-I$ term) is actually the 2,1 element of the T_s matrix which will be different if there is a magnetic field applied, as it arises from $t^{-1}t$ and will be different when time reversal symmetry is broken by the magnetic field.

At the end of the sample, the waves are transformed back into the mode representation, and the transmission used to determine the conductance. Also at the end of the structure, one determines the wave function at the output in terms of the mode number and its values at the various sites. Thus, we determine at site i and mode m the contribution to the wave function as $\psi_j(i, m)$. This can now be back propagated to find the equivalent wave functions at each slice via

$$\psi_j(i, m) = P_1^j + P_2^j \psi_{j+1}(i, m), \quad (2.61)$$

so obviously this recursion now proceeds from right to left. The velocities for each mode are found from the imaginary part of the eigenvalue for that mode. At the same time, the density can be determined at each site by the square magnitude of the wave function at that site (initially, the normalization is to unity, but this can be understood to be in terms of the density in the initial contact). A fuller description of the approach can be found in Akis *et al* [45].

When we use a vanishingly small bias between the two reservoirs (between the source and the drain), then we need only compute the transmission and then use the Landauer formula. However, when a real, non-zero level of bias is applied, the situation becomes more difficult, on several levels. First, we have to compute both a source-derived flow and a drain-derived backflow, just as in equation (2.44). Both flows must be computed to determine the total wave function and the density at each grid point since the non-zero bias will drive the system out of equilibrium. This implies that we will then need to use this density to solve self-consistently for the potential within the device, as the value of the potential at each grid point goes into the Hamiltonian equation (2.56). Now, we have an iterated loop that solves the quantum transport simulation and then solves for the self-consistent potential (see appendix A), and then once again for each until convergence is obtained. The scattering matrix approach is well suited to this process, as the density is a natural output of the simulation through equation (2.61). Secondly, we have to account for the temperature at each end of the device, as we can no longer assume that each mode has unity amplitude in the reservoir. Rather, the amplitude is determined from the Fermi–Dirac distribution at the temperature of the reservoir. The drain reservoir can be at a higher temperature, as this is ultimately where the last of the excess carrier energy is deposited. This approach has been extended to three dimensions and nanoscale MOSFETs at room temperature with real scattering [49]. Again, the scattering matrix approach is well suited to this task, as the Fermi–Dirac functions are applied in the reservoirs and not within the active region of the device.

By its very form, it is quite simple to modify the recursion method to include scattering by an on-site self-energy that is added to the diagonal elements of equation (2.54) and represents the interactions with scattering centers. Generally, the self-energy has both real and imaginary parts and it is the latter that are of interest for dissipative scattering. In semiconductors, the scattering is generally quite weak, and is traditionally treated by first-order, time-dependent perturbation theory, which yields the common Fermi golden rule for scattering rates. With such weak scattering, the real part of the self-energy can generally be ignored for the phonon interactions, and the real part that arises from the carrier-carrier interactions is incorporated into the solutions of Poisson's equation. In the many-body formulations of the self-energy, the latter is usually expressed as a two site function [20]

$$\Sigma(r_1, r_2), \tag{2.62}$$

where the two positions are usually three-dimensional vectors. Since we are using transverse modes in the recursion, we rewrite this expression as

$$\Sigma(i, j; i', j'; x_1, x_2). \tag{2.63}$$

As usual, the i and j (and their primed versions) are the transverse site and mode, respectively, and correspondingly the x's represent positions along the nanowire. We can then introduce a center-of-mass transformation [50]

$$X = \frac{x_1 + x_2}{2}, \quad \xi = x_1 - x_2, \tag{2.64}$$

and then Fourier transform on the difference variable to give

$$\Sigma(i, j; i', j'; X, k_x) = \frac{1}{2\pi} \int d\xi e^{ik_x\xi} \Sigma(i, j; i', j'; X, \xi). \tag{2.65}$$

The center-of-mass position X remains in the problem as the mode structure may change as one moves along the channel. At this point, the left-hand side of equation (2.65) the self-energy computed by the normal scattering rates, such as is done in quantum wells and quantum wires [51, 52]. As mentioned above, since scattering in semiconductors is relatively weak, it is sufficient to compute these using Fermi's golden rule, which is an evaluation of the bare self-energy in equation (2.65), rather than incorporating more elaborate many-body effects. Since the recursion is in the site representation, we have to reverse the Fourier transform in equation (2.65) to get the x- axis variation, and do a mode-to-site unitary transformation to get the self-energy in the form necessary for the recursion. We thus proceed by using the Fermi golden rule expression for each scattering process of interest and generating a real space self-energy from it. The imaginary part of the self-energy is related to the scattering rate via

$$\mathrm{Im}\{\Sigma(i, j; i' j'; X, k_x)\} \equiv \hbar\left(\frac{1}{\tau}\right)_{i,j}^{i'j'}. \tag{2.66}$$

It is the latter scattering rate which we calculate, which is a function of the x-directed momentum (which is related, in turn, to the energy of the carrier) in a cross-section of the device, which can be thought of locally as a quantum wire. This scattering rate must be converted to the site representation with a unitary transformation. At site s,l,η, the correction due to scattering that gets added to the local potential $V_{s,l,\eta}$ is given by

$$\Gamma(s, l, \eta) = Im\{\Sigma\} = U_s^\dagger\left(\frac{\hbar}{\tau}\right)_{i,j}^{i'j'} U_s, \tag{2.67}$$

where U_s is the mode-to-site transformation matrix discussed above. The actual scattering rates are computed in the normal manner [49, 53].

This methodology, based upon the scattering matrix recursion technique, has been applied to a number of both ballistic and scattering approaches to gated silicon quantum wire structures with excellent results [54]. In these approaches, it was possible to directly determine the diffusive to ballistic crossover length.

2.6 Simulating the channel: recursive Green's functions

A second approach is to use Green's functions, derived from the Schrödinger equation, in a recursive approach [55–57]. Again, we have used this technique successfully in the past [58]. With the recursive Green's function, the contact area is assumed to be a semi-infinite metallic wire of a given width. The modes and eigenvalues are then computed for this strip for a given Fermi energy. To begin, the region is discretized as in figure 2.21, and there is a width of the contact given as $W = Ma$, where a is the grid spacing. Thus, the transverse eigenfunctions are those of an infinite potential well, given as

$$\varphi_r(y) = \sqrt{\frac{2}{W}} \sin\left(\frac{r\pi y}{W}\right), \quad r = 1, 2, 3.... \tag{2.68}$$

Here, the index r corresponds to the transverse mode in the confined wire contact. This wave function is evaluated at each grid point i in the transverse direction to evaluate the wave function at the sites in the zero slice. Once we are given the Fermi energy, then we can compute the longitudinal wave number from a knowledge of whether or not the eigenenergy is greater or smaller than the Fermi energy. We first compute the quantity

$$\xi = 2 - \cos\left(\frac{r\pi}{M+1}\right) - \frac{E_F}{E_t}, \tag{2.69}$$

where E_t is the hopping energy (2.55). If $\xi \leqslant 1$, then the wave is propagating, and we write the longitudinal wave function in the contact as

$$\begin{aligned} \psi_r(x) &\sim \sin[\cos^{-1}(\xi)] \\ k &= \frac{\xi}{a}, \quad x \leqslant 0. \end{aligned} \tag{2.70}$$

This wave function needs to be properly normalized, of course, but represents the assumption that the contact leads to a vanishing of the wave function at $x = 0$ in the absence of the active channel. If the eigenenergy of the wave function is greater than the Fermi energy, then we are dealing with an evanescent wave, and the momentum in equation (2.70) is replaced by

$$\lambda = \frac{1}{a}[\xi - \sqrt{\xi^2 - 1}]. \tag{2.71}$$

The recursion begins by establishing a self-energy correction which acts upon the first slice, but is based upon the left contact and its connection to the first slice. That is, we define the left self-energy by the action

$$\Sigma_L = H_{10}G_{00}H_{01}, \quad \Gamma_L = 2Im\{\Sigma_L\}. \tag{2.72}$$

Here, Σ_L and Γ_L are diagonal matrices, G_{00}, H_{10}, and H_{01} are $M \times M$ matrices. The matrix G_{00} is the on-site Green's function given by

$$G_{00} = (E_F I - H_0 + iI\eta)^{-1}, \tag{2.73}$$

H_0 is the site Hamiltonian (2.56), here for the zeroth slice where the potential is zero, and a small damping factor η has been added to assure convergence of the matrix inversion process. The use of a positive damping factor assures us that the

Green's functions are the retarded (causal) Green's functions. In the above equation, I is a unit matrix as before. The two matrices H_{10} and H_{01} are inter-slice connection matrices and are adjoints to each other. For example, H_{10} contains a hopping energy on the diagonal connecting a site i along one row with its neighbor on the row to the right. Equation (2.73) is just used to determine the self-energy coupling between the contact and the channel, and the actual zeroth slice Green's function is computed from

$$G_{00} = (E_F I - H_0 + \Sigma_L + iI\eta)^{-1}. \tag{2.74}$$

Now, we propagate along the active channel with a recursion that computes first the slice Green's function and then the connecting Green's functions. For slice $j \geqslant 1$, we can compute the slice Green's function as

$$G_{j,j} = (E_F I - H_j - H_{j,j-1}G_{j-1,j-1}H_{j-1,j})^{-1}. \tag{2.75}$$

Note that we have added a comma to avoid confusion in the subscripts of the Green's functions. In this expression, the slice Hamiltonian H_j is just equation (2.56) and includes the local site potential at each grid point, while the two nonlocal Green's functions are just the diagonal hopping terms described previously. These are equal to one another except when we add a magnetic field, which will be discussed below. In addition to equation (2.75), we also construct the two propagating Green's functions

$$G_{0,j} = G_{0,j-1}H_{j-1,j}G_{j,j}, \tag{2.76}$$

and

$$G_{j,0} = G_{j,j}H_{j,j-1}G_{j-1,0}. \tag{2.77}$$

At the right contact, we have to connect to the second lead, which we assume occurs for $j = L$. Then, the right contact Green's functions become

$$\begin{aligned} G_{L+1,L+1} &= (E_F I - H_R + \Sigma_R - H_{L,L+1}G_{L,L}H_{L+1,L})^{-1} \\ G_{0,L+1} &= G_{0,L}H_{L,L+1}G_{L+1,L+1} \\ G_{L+1,0} &= G_{L+1,L+1}H_{L+1,L}G_{L,0}. \end{aligned} \tag{2.78}$$

Here, the right self-energy has been computed in exactly the same manner as in equation (2.72), but with the right contact Hamiltonian. This approach allows one to actually use leads that have different sizes, neither of which actually needs to be the same as the active strip. The transmission through the total system is now found from (there are several variants of this formula available in the literature, for example, [59])

$$T = Tr\{\Gamma_L G_{0,L+1}\Gamma_R G_{L+1,0}\}. \tag{2.79}$$

A more extensive derivation of the approach is given in [60]. A magnetic field can be easily added to the simulation.

As previously, the recursive Green's function approach discussed here is well suited to vanishingly small bias applied between the two reservoirs. Temperature can be included by computing the transmission for a range of energies around the Fermi energy and then weighting each transmission by the appropriate value of the Fermi–Dirac distribution. But, this is an approximation that cannot be extended to larger values of the applied bias. Nor can the recursive Green's function approach discussed here be used in this latter case. The approach here is based upon equilibrium Green's functions. Even weighting each value of transmission as mentioned is an approximation, as reasonable temperatures really require use of the thermal, or Matsubara, Green's functions [20]. And, if we go to larger values of bias applied between the two reservoirs, then the density is driven out of equilibrium, and one has to go to the more complicated non-equilibrium, or real time Green's functions, which can also yield the non-equilibrium distribution function within the device [51, 61, 62]. This is required to determine the local density, as it does not follow directly from the calculation as in the case of the previous section. Progress in this area is best described by OMEN, a powerful simulation package for quantum device simulation, which can use the real-time Green's functions [63].

The recursive Green's function has been used extensively to study conductance fluctuations, weak localization, and strong localization. We discuss weak localization in chapter 3, strong localization in chapter 5, and conductance fluctuations in chapter 9. But, we can introduce the idea here by looking at the Hamiltonian (2.56), and considering the potential to be a random potential, or to have a random component. The transverse confinement introduces a quantization in the transverse eigenvalues, some lying below the Fermi energy and others lying above the Fermi energy as non-propagating evanescent modes. Let us assume that the random amplitudes of the site potential lie between, e.g., $-V_0/2$ and $V_0/2$, with no uniform potential present. Now, if V_0 is smaller than the Fermi energy, a few modes near this value will see the random potential, and there will be interference between those modes that see the potential. This interference can change dramatically for small changes in the Fermi energy, or a magnetic field, and this can lead to conductance fluctuations. That is, the conductance changes as the interference changes by a small amount. This is the conductance fluctuations discussed in chapter 9, although they are strongly related to the weak localization effect. The latter arises when the interference leads to back-scattering of some modes. Technically, when there is back-scattering, without a magnetic field, one can consider a particle can traverse the back-scattering trajectory in either of two time-reversed directions, and this leads to interference between the time-reversed paths and, in turn, to a small decrease in the conductance. This will be discussed further in chapter 3. Finally, when the potential amplitude gets larger, lower energy modes are in fact cut off, as they cannot propagate through the random potential. Hence, the number of propagating modes decreases, and we have strong localization. All of these behaviors can be seen in a typical nano-ribbon and are easily studied with the recursive Green's function, as well as the scattering-matrix approach [64]. We will illustrate these effects with further examples in the following chapters.

Problems

1. A Poisson solver that may be used to calculate the energy bands of different heterostructures may be downloaded at: http://www.nd.edu/~gsnider/. Consider a heterostructure comprised of (starting from the top layer): a 5 nm thick GaAs cap layer; 40 nm of undoped $Al_xGa_{1-x}As$ ($x = 0.33$); 10 nm of $Al_xGa_{1-x}As$ ($x = 0.33$) doped with Si at a concentration of 1.5×10^{18} cm^{-3}; 20 nm of undoped $Al_xGa_{1-x}As$ ($x = 0.33$), 100 nm of undoped GaAs; and a GaAs substrate with an unintentional p-type doping of 5×10^{14} cm^{-3}.

   (a) The Poisson solver is a self-consistent program that computes the band structure by solving two important equations simultaneously. What are these equations?

   (b) Plot the calculated energy bands and ground-state electron wave function for the heterostructure at 1 K. You may assume full ionization of the donors.

   (c) Explain quantitatively the reasons for the different energy variations in each of the layers of the heterostructure.

   (d) What is the number of two-dimensional subbands that are occupied at this temperature? Explain this result by using the value of the electron density determined from the program.

   (e) Determine the minimum doping density that may be used to realize a 2DEG in the heterostructure. Plot the energy bands for this doping condition and the ground state electron wave function.

2. A rectangular quantum wire with cross section 15×15 nm^2 is realized in GaAs.

   (a) Write an expression for the subband threshold energies of the quantum wire.

   (b) Write a general expression for the electron dispersion in the quantum wire.

   (c) In a table, list the quantized energies and their degeneracies (neglecting spin) for the first 15 energetically distinct subbands.

   (d) Plot the density of states of the wire over an energy range that includes the first fifteen subband energies.

3. In problem 2, we solved for the density of states in a rectangular GaAs quantum wire with cross section 15×15 nm^2. For this same wire: (a) Write an expression for the electron density (per unit length) as a function of the Fermi energy of the wire. (b) Plot the variation of the electron density as a function of energy for a range that corresponds to filling the first five distinct energy levels, taking proper account of level degeneracies.

4. When a voltage (V_{sd}) is applied across a QPC, it is typical to assume that the quasi-Fermi level on one side of the barrier is raised by $\alpha e V_{sd}$ while that on the other drops by $(1 - \alpha)e V_{sd}$, where α is a phenomenological parameter that, in a truly symmetrical structure, should be equal to ½. If we consider a device in which only the lowest subband contributes to transport, then the current flow through the QPC may be written as:

$$I_{sd} = \frac{2e}{h}\left[\int_L T(E)dE - \int_R T(E)dE\right],$$

where $T(E)$ is the energy-dependent transmission coefficient of the lowest subband and L and R denote the left and right reservoirs, respectively. If we assume low temperatures, we can treat the transmission as a step function, $T(E) = \theta(E - E_1)$, where E_1 is the threshold energy for the lowest subband. (a) Write this integral with limits appropriate to determine the current. (b) Use this information to obtain an expression for the current flowing through the QPC when the source–drain voltage is such that both reservoirs populate the lowest subband, and when it populates the subband from just the higher-energy reservoir.

5. At non zero temperature, the conductance through a nanodevice is determined by weighting the transmission at each energy by the value of the (negative of the) derivative of the Fermi–Dirac distribution. This leads to the fact that the conductance takes place near the Fermi level. Compute the derivative of the Fermi–Dirac function and evaluate the full-width at half-maximum for this function.

6. Consider the potential profile shown in figure 2.14. let us assume that the total potential drop across the device is 10 mV. Assume that the two transitions can be approximated by an arctangent function with a full-width between the 10% and 90% values of the transition of 2 nm. Determine the dipole charge that exists at each transition, and the corresponding density profile and value.

7. In treating the QPC with the harmonic oscillator approximation, the different energy levels are all equally spaced at a given value of the curvature, which is determined by the gate potential. Using the data shown in figure 2.17, estimate the energy level spacing as a function of the gate voltage (horizontal axis) in this device. Include at least the first nine transitions above the first plateau.

8. Plot the potential profile given in equation (2.48) for a QPC of width $W = 100$ nm and length $L = 200$ nm. You may assume that the 2DEG lies 180 nm below the surface and the bias on the metal gates is −5 V.

9. Consider a MOSFET with a W/L ratio of 50, a mobility of 500 cm^2 Vs^{-1}, a gate SiO$_2$ dielectric thickness of 2 nm, and a threshold voltage of 1 V. Plot the 'contact resistance' between the source and channel as a function of the gate voltage in the range $0 \leqslant V_G \leqslant 5$ V.

10. Plot the first three harmonic oscillator wave functions as a function of their normalized spatial variable.

Appendix A Coupled quantum and Poisson problems

In many real world cases, such as the inversion layer of the Si MOSFET, the shape of the potential depends upon the actual charge density in the inversion layer. Moreover, the actual shape of the potential must be computed from the charge density via Poisson's equation. Then, this potential must be used in the Schrödinger equation to find new values of the energy eigenvalues and the corresponding wave functions. This loop must be iterated until convergence is obtained. Here, we will discuss the approach first and then talk about some available software that can handle the problem for us. To begin with, let us write down the Schrödinger equation as

$$-\frac{\hbar^2}{2m}\nabla^2\psi + V\psi = E\psi, \tag{A.1}$$

while Poisson's equation is

$$\nabla^2 V = -\frac{\rho}{\varepsilon}. \tag{A.2}$$

Here, V is the potential that appears at each site, ρ is the charge density, and ε is the dielectric permittivity. The wave function ψ corresponds to the eigenenergy E. The charge density is related to the wave function, as the latter provides the probability that leads to the charge carrier distribution, as

$$\rho = -\sum_i |\psi_i|^2 \tag{A.3}$$

for electrons, and the sum runs over all occupied modes at a site.

It is clear from equations (A.1)–(A.3) that one can iterate easily from wave function to density to potential to wave function, etc and continue to convergence. To proceed one has to adopt a gridding scheme, such as that of figure 2.21. Here, all of the various quantities discussed above are evaluated on the grid points (complications arise if something like the electric field, which is the gradient of the potential is wanted, as it has to be evaluated midway between the grid points). If some arbitrary function is known at a series of equally spaced points, one can use a Taylor series to expand the function about these points and evaluate it in the regions between the known points. Suppose we wish to know the solution of the second-order differential equations above (we begin with only one dimension) in a region between 0 and L. We can divide this into $N + 1$ segments of length a. Thus, there are N interior points and the two end points of $x = 0$ and $x = L$. These end points represent boundary values that will be imposed upon the solutions as necessary. But, we assume that these segments are all of equal length, and each point is separated from its neighbor by the distance a. This allows us to develop a finite-difference scheme for the numerical evaluation of the function. If we first expand the function in a Taylor series about the points on either side of x_0, we obtain

$$f(x_0 + a) = f(x_0) + a\frac{\partial f}{\partial x}\bigg|_{x_0} + \frac{a^2}{2}\frac{\partial^2 f}{\partial x^2}\bigg|_{x_0} + \ldots$$

$$f(x_0 - a) = f(x_0) - a\frac{\partial f}{\partial x}\bigg|_{x_0} + \frac{a^2}{2}\frac{\partial^2 f}{\partial x^2}\bigg|_{x_0} + \ldots. \tag{A.4}$$

If we add these two equations together, we can write the second derivative as

$$\left.\frac{\partial^2 f}{\partial x^2}\right|_{x_0} \approx \frac{f(x_0 + a) + f(x_0 - a) - 2f(x_0)}{a^2}. \tag{A.5}$$

Using the properties of our grid, we can write x_0 as ja, as indicated in figure 2.21. Hence, we can rewrite equation (A.5) in grid indices as

$$\left.\frac{\partial^2 f}{\partial x^2}\right|_j \approx \frac{f_{j+1} + f_{j-1} - 2f_j}{a^2}. \tag{A.6}$$

We can now represent the one-dimensional Poisson's equation as

$$\frac{1}{a^2}[S][V] = -\left[\frac{\rho}{\varepsilon}\right] + [B], \tag{A.7}$$

where V, ρ/ε, and B are column matrices of length N. The matrix B contains the boundary conditions, which are the defined voltages at sites 0 and $N + 1$, which appear at the first and last elements of B, the other elements being 0. The matrix S is a square tri-diagonal matrix of rank N, whose elements are all zero except for the diagonal and first off-diagonal elements. The diagonal elements are all -2, while the off-diagonal elements are 1. Inversion of the S matrix used to be difficult due to memory and speed limitations, so approximate iterative schemes were developed for the purpose. With modern machines, this matrix can be inverted just once, and then solving equation (A.7) is simple matrix multiplication.

The Schrödinger equation can be similarly cast into a simple matrix form for the one-dimensional situation. This becomes an eigenvalue equation with the form

$$[S][\psi] = 0. \tag{A.8}$$

Once more, S is a tri-diagonal matrix, whose elements are given by

$$S_{j,j} = V_j + \frac{\hbar^2}{ma^2} - E, \quad S_{j,j\pm1} = -\frac{\hbar^2}{2ma^2}. \tag{A.9}$$

The last term on the right equation is the hopping energy (2.55) and gets its name as the energy connecting two adjacent sites. Solving equation (A.8) involves finding the determinant of S which gives an equation whose solutions are the eigenvalues. Generally, modern techniques diagonalize S which then leaves the eigenvalues on the main diagonal, and also generate a transformation matrix whose columns are the site representation for the eigen functions.

But, as mentioned above, the time consuming aspect of this is the iteration between Poisson's equation and the Schrödinger equation that is required to reach convergence of the charge density and the potential consistent with the statistics of the particular problem. At non-zero temperatures, each eigenstate is occupied according to the Fermi–Dirac probability for its energy value and the position of the Fermi energy. Then, each eigenstate contributes to the density according to its spatial variation. This provides the spatial variation for the total density which drives the Poisson equation solution. Once the potential is determined, this drives the solution to the Schrödinger equation. And, of course, the total density determines the position of the Fermi energy. This can be an unstable process if it is not put together carefully. And, as remarked, several people have developed simulation packages which solve this problem successfully. One such package, mentioned earlier is SCHRED 2.0 [65], available at NanoHUB.org, a site supported by the National Science Foundation as a resource for nanoscience and nanotechnology. SCHRED solves the above two equations self-consistently for a typical MOS structure. A typical set of input parameters is shown in figure A1. The tool assumes that the surface normal is one of the three major coordinate axes, and the default is for the z-direction. The three sets of valleys then have their major axes along the k_x, k_y, and k_z directions, with the latter the default. One difference is that the tool numbers the energy levels as 1, 2, ... instead of the 0, 1, ... that is common in the literature for silicon (and used above). One drawback of this tool is that the

Figure A1. A screen shot of the parameter panel for SCHRED 2.0, illustrating the ease of setting up the simulation. The simulator is available at NanoHub.org.

minimum temperature for which it will simulate the system is 50 K, so it will not be useful at low temperatures. In figure A2, we show a screen shot of the two lowest energy wave functions for the conditions simulated in figure 1.5.

Now, when we go to two dimensions, such as for the gridding of the nanowire in figure 2.21, we have to extend the approach to both x and y derivatives. We can immediately extend equation (A.6) to

$$\left.\frac{\partial^2 f}{\partial x^2}\right|_{i,j} \approx \frac{f_{i,j+1} + f_{i,j-1} - 2f_{i,j}}{a^2} + \frac{f_{i+1,j} + f_{i-1,j} - 2f_{i,j}}{a^2}. \tag{A.10}$$

This result has been found using the notation shown in figure A3. The problem is now that we need a fourth rank tensor for the expansion matrix. To solve this, we note that figure 2.21 can be considered to be a stack of one-dimensional wires along the j direction, with the wires denoted by the index i. This suggests to create a two-dimensional matrix of rank MN. This matrix will be block tri-diagonal, with each block being the matrix S from equation (A.9) corresponding to wire i. The off-

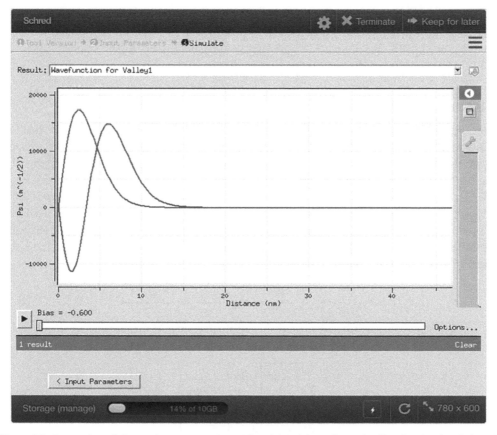

Figure A2. Screen shot of the two lowest energy wave functions of the z-directed valleys for the simulation of figure 1.5.

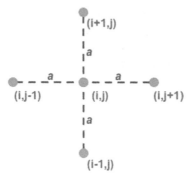

Figure A3. A five-point discretization for the Schrödinger equation uses a grid in two dimensions.

diagonal blocks with be diagonal matrices whose elements are the hopping energies from equation (A.9). One important change is that

$$\frac{\hbar^2}{ma^2} \to \frac{2\hbar^2}{ma^2} \tag{A.11}$$

in the diagonal blocks. This arises from the factor of 4 in equation (A.1) for the on-site term, rather than the factor of 2 in equation (A.6); each on-site point has four nearest neighbors now instead of just two nearest neighbors in the one-dimensional case.

Appendix B The harmonic oscillator

The Hamiltonian that results from the potential of equation (2.4) is doubly quadratic; that is, it contains one term quadratic in the momentum operator, arising from the kinetic energy term, and one term quadratic in the position, from the quadratic position in the potential. We will reduce the Hamiltonian by introducing a set of operators that combine position and momentum in a way that leads to a simpler approach to the problem. We suppose that a proper combination of the operators will provide a set of new operators that correspond to movement through the ladder of energy levels (which are yet to be found). We begin by writing the total Hamiltonian for the time-independent Schrödinger equation as

$$H = -\frac{\hbar^2}{2m^*}\frac{\partial^2}{\partial y^2} + \frac{m^*\omega_{0,y}^2 y^2}{2} = \frac{p^2}{2m^*} + \frac{m^*\omega_{0,y}^2 y^2}{2}. \tag{B.1}$$

To proceed, let us introduce two complex operators

$$a = \sqrt{\frac{m^*\omega_{0,y}}{2\hbar}}\left(y + i\frac{p}{m^*\omega_{0,y}}\right)$$

$$a^\dagger = \sqrt{\frac{m^*\omega_{0,y}}{2\hbar}}\left(y - i\frac{p}{m^*\omega_{0,y}}\right). \tag{B.2}$$

With these operators, we find that the Hamiltonian (B.1) can be rewritten as

$$H = \hbar\omega_{0,y}\left(a^\dagger a + \frac{1}{2}\right).$$ (B.3)

The momentum and position operators satisfy a basic commutator relationship in quantum mechanics, since they do not commute with each other. The new operators in equation (B.2) also will satisfy such a commutator relationship because they do not commute with one another either. Let us consider the two possible products of these operators:

$$a^\dagger a = \frac{1}{\hbar\omega_{0,y}}\left(\frac{p^2}{2m^*} + \frac{m^*\omega_{0,y}^2 y^2}{2}\right) - \frac{1}{2}$$

$$aa^\dagger = \frac{1}{\hbar\omega_{0,y}}\left(\frac{p^2}{2m^*} + \frac{m^*\omega_{0,y}^2 y^2}{2}\right) + \frac{1}{2},$$ (B.4)

so that

$$[a, a^\dagger] = aa^\dagger - a^\dagger a = 1.$$ (B.5)

It is also observed here that the operator products themselves are not operators, but are what are called c-numbers, or simple numbers, which may be complex in some cases. This will be important in the following operations by which we determine the properties of these operators.

Since the product $a^\dagger a$ is a c-number, multiplying a function by it simply scales the amplitude of that function. Hence, if we multiply the nth wave function by it, we can write the corresponding eigenvalue equation

$$a^\dagger a \psi_n = \lambda_n \psi_n,$$ (B.6)

where λ_n is the eigenvalue, and we must still determine the value for this quantity. It is easy to show that this value must be a positive quantity. Let us multiply by the adjoint of this wave function and integrate, while using the properties of adjoint operators, to yield

$$\int \psi^\dagger a^\dagger a \psi dy = \int (a\psi)^* a\psi dy \geqslant 0.$$ (B.7)

The adjoint operator is moved to the conjugate term due to the properties of adjoint, or Hermitian conjugate, operators.

Let us take one of the eigen functions and operate on it with the two operators, and then apply equation (B.5). This leads to

$$a^\dagger a(a^\dagger \psi_n) = a^\dagger(aa^\dagger + 1)\psi_n = (\lambda_n + 1)a^\dagger \psi_n$$
$$a^\dagger a(a\psi_n) = (aa^\dagger + 1)a\psi_n = (\lambda_n + 1)a\psi_n.$$ (B.8)

Thus, we see that operation upon an eigenfunction by the adjoint operator produces an eigenvalue equation in which the eigenvalue has been increased, or decreased, by unity. The last form gives us some insight, as there must always be a lowest eigenvalue and its corresponding eigen function. If we apply the operator a to this lowest eigen function, we find

$$a\psi_0 = 0, \tag{B.9}$$

since there can be no wave function below the lowest one. Thus, comparison of equations (B.8) and (B.9) tells us that $\lambda_0 = 0$, and

$$\lambda_n = n. \tag{B.10}$$

We can now rewrite (B.3) as

$$H = \hbar\omega_{0,y}\left(n + \frac{1}{2}\right). \tag{B.11}$$

The conclusions we can draw from the above arguments is that the adjoint operator $a^\dagger$ raises the energy of the harmonic oscillator by one unit of energy $\hbar\omega_{0,y}$ and alters the wave function to that of the next level up. Similarly, the operator a lowers the energy of the harmonic oscillator by one unit of energy $\hbar\omega_{0,y}$ and alters the wave function to that of the next level down. For this reason, the two operators are termed the *raising* and *lowering* operators. With their action, the harmonic operator absorbs or emits a quantum of energy, respectively.

We can now use equation (B.9) to find the wave function for the ground state. If we expand the operator in equation (B.9), we obtain

$$a\psi_0 = \sqrt{\frac{m^*\omega_{0,y}}{2\hbar}}\left(y + i\frac{p}{m^*\omega_{0,y}}\right)\psi_0 = 0, \tag{B.12}$$

which leads immediately to

$$\psi_0 = C_0\exp\left(-\frac{m^*\omega_{0,y}^2 y^2}{2\hbar}\right). \tag{B.13}$$

This may be recognized as the lowest order Hermite polynomial with its exponential weighting function. The constant C_0 can be determined by standard normalization to be

$$C_0 = \left(\frac{m^*\omega_{0,y}}{\pi\hbar}\right)^{1/4}. \tag{B.14}$$

We can now determine any of the wave functions at the various levels simply by repeated application of the raising operator, as

$$\psi_n = (a^\dagger)^n\psi_0. \tag{B.15}$$

We now turn to the expectation values for the operators themselves, which will also give us the full normalization of the various eigenstate wave functions. The chain generation in equation (B.15) does not automatically guarantee proper normalization of the wave functions, so that we have to find the correct normalization for all higher lying states. From the basic properties expressed in equations (B.5) and (B.6), we know that for unnormalized wave functions such that

$$\int \psi_n^\dagger \psi_n dy = C_n^2,$$ (B.16)

we can write (we take the coefficients as real)

$$\int \psi_n^\dagger a^\dagger \psi_k dy = \int \psi_n^\dagger \psi_{k+1} dy = \alpha_n^2 \delta_{n,k+1} C_n^2$$

$$\int \psi_n^\dagger a \psi_k dy = \int \psi_n^\dagger \psi_{k-1} dy = \beta_n^2 \delta_{n,k-1} C_n^2.$$ (B.17)

We still have to determine the constants α_n and β_n, which in fact are related to the normalization of the eigen functions (the two eigen functions in the above expectation values have different normalizations). For this, we use

$$\int (a\psi_n)(a\psi_n) dy = \int \psi_n a^\dagger a \psi_n dy = nC_n^2 = C_{n-1}^2.$$ (B.18)

This now leads us to be able to say

$$C_{n-1} = \sqrt{n}\, C_n.$$ (B.19)

We can now use this property to give the following results

$$\int \psi_k a^\dagger \psi_n dy = \int \psi_k \psi_{n+1} dy = \sqrt{n+1}\, \delta_{k,n+1}$$

$$\int \psi_k a \psi_k dy = \int \psi_k \psi_{n-1} dy = \sqrt{n}\, \delta_{k,n-1}.$$ (B.20)

We can now finally write the exact normalization for the general wave function as

$$\psi_n = \frac{1}{\sqrt{(n+1)!}} (a^\dagger)^n \psi_0.$$ (B.21)

References

[1] Aoki N, da Cunha C R, Akis R and Ferry D K 2005 *Appl. Phys. Lett.* **87** 223501
[2] van Wees B J, van Houten H, Beenakker W J, Williamson J G, Kouwenhoven L P, van der Marel D and Foxton C T 1988 *Phys. Rev. Lett.* **60** 848
[3] Wharam D A, Thornton T J, Newbury R, Pepper M, Ahmed H, Frost J E F, Hasko D G, Peacock D C, Ritchie D A and Jones G A C 1988 *J. Phys. C: Sol. State Phys.* **21** L209
[4] Thomas K J, Nichols J T, Appleyard N, Simmons M Y, Pepper M, Mace D R, Tribe W R and Ritchie D A 1998 *Phys. Rev. B* **58** 4846

[5] Laux S E, Frank D J and Stern F 1988 *Surf. Sci.* **196** 101

[6] Topinka M A, LeRoy B J, Shaw S E J, Heller E J, Westervelt R M and Maranowski A C G 2000 *Science* **289** 2323

[7] Crook R, Smith C G, Simmons M Y and Ritchie D A 2000 *J. Phys.: Condens. Matter* **121** L735

[8] Westervelt R M, Topinka M A, LeRoy B J, Bleszynski A C, Aidala K, Shaw S E J, Heller E J, Maranowski K D and Gossard A C 2004 *Physica* E **24** 63

[9] Ferry D K 2019 *Semiconductors: Bonds and Bands* 2nd edn (Bristol: IOP Publishing) section 2.6

[10] Castro Neto A H, Guinea F, Peres N M R, Novoselov K S and Geim A K 2009 *Rev. Mod. Phys.* **81** 109

[11] Raghu S and Haldane F D M 2008 *Phys. Rev.* A **78** 033834

[12] Wallace P R 1947 *Phys. Rev.* **71** 622

[13] Novoselov K S, Geim A, Morozov S V, Jiang D, Katsnelson M I, Grigorieva I V, Dubonos S V and Firsov A A 2005 *Nature* **438** 197

[14] Bostwick A, Ohta T, Seyller T, Horn K and Rotenberg E 2006 *Nat. Phys.* **3** 37

[15] Landauer R 1957 *IBM J. Res. Develop.* **1** 223

[16] Landauer R 1970 *Phil. Mag.* **21** 863

[17] Thomas K J, Nicholls J T, Simmons M Y, Pepper M, Mace D R and Ritchie D A 1996 *Phys. Rev. Lett.* **77** 135

[18] Kirczenow G 1989 *Phys. Rev.* B **39** 10452

[19] Thomas K J, Nicholls J T, Pepper M, Tribe W R, Simmons M Y and Ritchie D A 2000 *Phys. Rev.* B **61** R13365

[20] Fetter A L and Walecka J D 1971 *Quantum Theory of Many-Particle Systems* (New York: McGraw-Hill)

[21] Duke C B 1969 *Tunneling in Solids* (New York: Academic)

[22] See, e.g.,; Datta S, Assad F and Lundstrom M S 1998 *Superlatt. Microstruc.* **23** 771

[23] Kroemer H 1968 *IEEE Trans. Electron. Dev.* **15** 819

[24] Rössler C, Baer S, de Wiljes E, Ardelt P-L, Ihn T, Ensslin K, Reichl C and Wegscheider W 2011 *New J. Phys.* **13** 113006

[25] Patel N K, Nicholls J T, Martin-Moreno L, Pepper M, Frost J E F, Ritchie D A and Jones G A C 1991 *Phys. Rev.* B **44** 13549

[26] Cronenwett S M, Lynch H J, Goldhaber-Gordon D, Kouwenhoven L P, Marcus C M, Hirose K, Wingreen N S and Umansky V 2002 *Phys. Rev. Lett.* **88** 226805

[27] de Picciotto R, Baldwin K W, Pfeiffer L N and West K W 2008 *J. Phys.: Condens. Matter* **20** 164204

[28] Pepper M and Bird J P 2008 *J. Phys.: Condens. Matter* **20** 160301

[29] Micolich A 2013 *J. Phys.: Condens. Matter* **23** 443201

[30] Thomas K J, Nicholls J T, Simmons M Y, Pepper M, Mace D R and Ritchie D A 1996 *Phys. Rev. Lett.* **77** 135

[31] Goldhaber-Gordon D, Shtrikman H, Abush-Magder D, Meirav U and Kastner M A 1998 *Nature* **391** 156

[32] Wang C-K and Berggren K-F 1998 *Phys. Rev.* B **57** 4552

[33] Meir Y, Hirose K and Wingreen N S 2002 *Phys. Rev. Lett.* **89** 196802

[34] Akis R and Ferry D K 2006 *Proceedings of the 14th International Conference on Nonequilibrium Carrier Dynamics in Semiconductors* (Berlin: Springer) pp 163–7

[35] Akis R and Ferry D K 2008 *J. Phys.: Condens. Matter* **20** 164201

[36] Kristensen A *et al* 2000 *Phys. Rev.* B **62** 10950

[37] Shailos A, Ashok A, Bird J P, Akis R, Ferry D K, Goodnick S M, Lilly M P, Reno J L and Simmons J A 2006 *J. Phys.: Condens. Matter* **18** 1715

[38] Hirose K, Meir Y and Wingreen N S 2003 *Phys. Rev. Lett.* **90** 026804

[39] Timp G 1992 *Semiconductors and Semimetals* vol 35 (New York: Academic) pp 113–90

[40] Aoki N, da Cunha C R, Akis R, Ferry D K and Ochiai Y 2014 *J. Phys.: Condens. Matter* **26** 193202

[41] Meir Y 2008 *J. Phys.: Condens. Matter* **20** 164208

[42] Igbal M J *et al* 2013 *Nature* **501** 79

[43] Ando T 1991 *Phys. Rev.* B **44** 8017

[44] Usuki U, Saito M, Takatsu M, Kiehl R A and Yokoyama N 1995 *Phys. Rev.* B **52** 771

[45] Akis R, Ferry D K and Bird J P 1996 *Phys. Rev.* B **54** 17705

[46] Merzbacher E 1970 *Quantum Mechanics* 2nd edn (New York: Wiley)

[47] Kriman A M, Kluksdahl N C and Ferry D K 1987 *Phys. Rev.* B **36** 5953

[48] Kluksdahl N C, Kriman A M and Ferry D K 1989 *Phys. Rev.* B **39** 7720

[49] Ferry D K, Akis R, Gilbert M J and Ramey S 2016 *Nanoscale Silicon Devices* ed S Oda and D K Ferry (Boca Raton, FL: CRC Press) pp 1–36

[50] Kadanoff L P and Baym G 1962 *Quantum Statistical Mechanics* (Reading, MA: Benjamin Cummins)

[51] Ferry D K, Goodnick S M and Bird J P 2009 *Transport in Nanostructures* 2nd edn (Cambridge: Cambridge University Press)

[52] Kotylar R, Obradovic B, Matagne P, Stettler M and Giles M D 2004 *Appl. Phys. Lett.* **84** 5270

[53] Gilbert M J, Akis R and Ferry D K 2005 *J. Appl. Phys.* **98** 094303

[54] Akis R, Ferry D K and Gilbert M J 2009 *J. Comp. Electron.* **8** 78

[55] Lee P A and Fisher D S 1981 *Phys. Rev. Lett.* **47** 882

[56] Thouless D J and Kirkpatrick S 1981 *J. Phys. C: Sol. State Phys.* **14** 235

[57] MacKinnon A 1985 *Z. Phys.* B **39** 385

[58] Takagaki Y and Ferry D K 1993 *Phys. Rev.* B **47** 9913

[59] Meir Y and Wingreen N S 1992 *Phys. Rev. Lett.* **68** 2512

[60] Shepard K 1991 *Phys. Rev.* B **44** 9088

[61] Lake R and Datta S 1992 *Phys. Rev.* B **45** 6670

[62] Ferry D K 2018 *An Introduction to Quantum Transport in Semiconductors* (Singapore: Pan Stanford)

[63] https://engineering.purdue.edu/gekcogrp/software-projects/omen/

[64] Liu B, Akis R and Ferry D K 2013 *J. Comp. Electron.* **25** 395802

[65] SCHRED 2.0, https://nanohub.org/resources/schred

IOP Publishing

Transport in Semiconductor Mesoscopic Devices
(Second Edition)

David K Ferry

Chapter 3

The Aharonov–Bohm effect

One of the earliest observations of quantum effects in nanostructures, or mesoscopic devices, was the experimental observation of phase interference in the conductance through metallic [1] or etched semiconductor rings [2, 3]. In these rings, electron waves pass from an input port, around the two halves of the ring and interfere at the output port, presumably due to the Aharonov–Bohm (AB) effect [4]. Interference between differing waves can occur over distances on the order of the coherence length of the carrier wave, and the latter distance is generally different from the inelastic mean free path for quasi-ballistic carriers (those with weak scattering). The inelastic mean free path is related to the energy-relaxation length $l_e = v\tau_e$, where τ_e is the energy-relaxation time and v is a characteristic velocity (typically the Fermi velocity in degenerate material and the thermal velocity in non-degenerate material). The inelastic mean free path can be quite long, on the order of several tens of microns for electrons at low temperatures in the inversion channel of a high-electron-mobility transistor in GaAs/AlGaAs. On the other hand, the coherence length is often defined for weakly disordered systems by the diffusion constant (as we discuss below). But, this can be misleading as it has meaning in materials with very little disorder, where it is often determined by the relevant electron–electron scattering mean free path. At first thought, one is usually aware that two interacting electrons do not dissipate either energy or momentum, as both are conserved. However, this does not mean that each of the two electrons cannot exchange energy or momentum, and it is this change that breaks up the phase coherence. Hence, electron–electron scattering is a major source of phase decoherence.

We can illustrate the idea of phase interference by considering two waves (or one single wave which is split into two parts which propagate over different paths) given by the general form $\psi_i = A_i e^{i\varphi_i}$. When the two waves are combined, at some point different from the origin, the probability amplitude varies as

$$P = |\psi_1 + \psi_2|^2 = |A_1|^2 + |A_2|^2 + 4A_1^* A_2 \cos(\varphi_2 - \varphi_1). \tag{3.1}$$

The probability can therefore range from the sum of the two amplitudes to the differences of the two amplitudes, depending on how the phases of the two waves are related. In most cases, it is not important to retain any information about the phase in device problems because the coherence length is much smaller than any device length scale and because ensemble averaging averages over the phase interference factor so that it smooths completely away in macroscopic effects. This ensemble averaging requires that a large number of such small phase coherent regions are combined stochastically. In small structures this does not occur, and many observed quantum interference effects are direct results of the lack of ensemble averaging [5].

In this chapter, we start with a discussion of the AB effect, and discuss its observation in some various systems. We will then discuss the role of the phase-breaking length and its determination. Following this, we discuss a related effect in which interference effects are seen via another process which produces a different dependence upon the magnetic field. In appendix C, we discuss the role of gauge in mesoscopic systems, and those not familiar with the vagaries of this quantity might start with the appendix before proceeding to the following section.

3.1 Simple gauge theory of the AB effect

A particularly remarkable illustration of the importance of the phase is the magnetic AB effect [4]. It was almost immediately verified using a biprism (which causes the electrons to take two different paths) in a transmission electron microscope (TEM) [6]. Yet, in spite of this first experiment and the familiarity of the effect in today's semiconductor world, there were strong debates about whether the effect was even real or not for nearly the first quarter of a century after the original paper. The debate was largely settled when the first observation of it was made in electron holography (electrons within a TEM) [7]. A few years later, the effect was observed in small metal rings [1, 8], and in semiconductor etched rings [2, 9]. From the beginning, though, it was inferred that the AB effect was a pure quantum phenomenon, in which it appears through interference caused by varying the phase of the wave function. However, it is more proper to think of it as a general wave mechanical effect which has its analog in classical wave mechanics [10]. For example, the effect has been studied in water surface waves [11, 12], and via light scattering from a hydrodynamic 'vortex' [13]. We consider it a quantum phenomenon in semiconductors solely because we are treating our electrons as quantum mechanical waves. But, our treatment in this section is in terms of the fields and the gauge, and is not particularly limited to quantum mechanics.

The basic structure of the experiment is illustrated in figure 3.1. The semiconductor structure is formed on an AlGaAs/GaAs heterostructure. A Q1D conducting channel is fabricated on the surface of a semiconductor by using electron-beam lithography to deposit a NiCr pattern by liftoff, and then using this pattern as a mask for reactive-ion etching away parts of the heterostructure to leave electrons in the ring structure [3]. The waveguide is sufficiently small so that only one

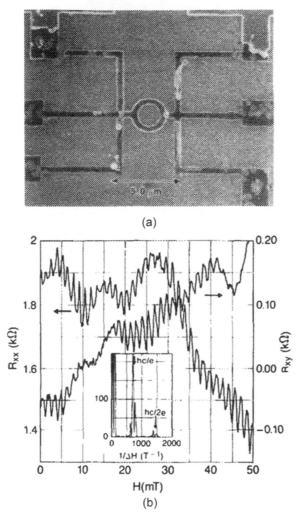

Figure 3.1. (a) Micrograph of a conducting semiconductor ring etched into an AlGaAs/GaAs heterostructure. (b) The magnetoresistance and Hall resistance measured for the ring of panel (a). The inset shows the Fourier transform of the resistance. Reprinted with permission from [3]. Copyright 1988 American Vacuum Society.

or a few electron modes are possible. The incident electrons, from the left of the ring in figure 3.1(a), have their wave split at the entrance to the ring. The waves propagate around the two halves of the ring to recombine (and interfere) at the exit port. The overall transmission through the structure, from the left electrodes to the right electrodes, depends upon the relative size of the ring circumference in comparison to the electron wavelength. If the size of the ring is small compared to the inelastic mean free path, the transmission depends on the phase of the two fractional paths. In the AB effect, a magnetic field is passed through the annulus of the ring, and this magnetic field will modulate the phase interference at the exit port. There are two types of measurements, which are labeled R_{xx} and R_{xy} in figure 3.1(b).

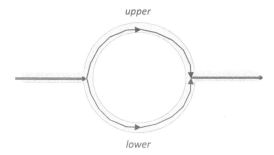

Figure 3.2. A symbolic representation of the two paths electrons can take through the ring.

The R_{xx} measurement is made directly across the ring and is a direct measure of the resistance of the ring as measured by the voltage drop across it [3] The other measurement is a transverse measurement, which in some sense is a nonlocal one as it does not measure the voltage drop across the ring, but the effect the ring voltage has on the rest of the circuit.

We understand the measured behavior from the assumption that the magnetic field passes vertically through the ring. The vector potential for a magnetic field passing through the annulus of the ring is azimuthal, so that electrons passing through either side of the ring will travel either parallel or anti-parallel to the vector potential, and this difference produces the phase modulation, as indicated by the ring in figure 3.2. The vector potential will be considered to be directed counterclockwise around the ring. (We adopt cylindrical coordinates, with the magnetic field directed in the z-direction and the vector potential in the θ-direction.) The phase of the electron in the presence of the vector potential is given by the Peierls' substitution (see appendix C), in which the normal momentum vector k is replaced by $(p - eA)/\hbar$ and

$$\phi = \phi_0 + \frac{1}{\hbar}(\boldsymbol{p} - e\boldsymbol{A}) \cdot \mathbf{r}, \tag{3.2}$$

so that the exit phases for the upper and lower arms of the ring can be expressed as

$$
\begin{aligned}
\phi &= \phi_0 + \int_\pi^0 \left(\mathbf{k} + \frac{e}{\hbar}\mathbf{A}\right) \cdot \mathbf{a}_\vartheta r d\varphi \\
\phi &= \phi_0 - \int_\pi^0 \left(\mathbf{k} - \frac{e}{\hbar}\mathbf{A}\right) \cdot \mathbf{a}_\vartheta r d\varphi
\end{aligned}
\tag{3.3}
$$

and the net phase difference is just

$$\delta\phi = \frac{e}{\hbar} \int_0^{2\pi} \mathbf{A} \cdot \mathbf{a}_\vartheta r d\varphi = \frac{e}{\hbar} \int \boldsymbol{B} \cdot \mathbf{a}_z dS = 2\pi \frac{\Phi}{\Phi_0}, \tag{3.4}$$

where $\Phi_0 = h/e$ is the quantum unit of flux and Φ is the flux enclosed in the ring. The phase interference term in equation (3.4) goes through a complete oscillation each time the magnetic field is increased by one flux quantum unit. This produces a modulation in the conductance (resistance) that is periodic in the magnetic field, with a period h/eS, where S is the area of the ring. This periodic oscillation is the AB effect, and in figure 3.1(b) results are shown for such a semiconductor structure. One can see that exactly this 'frequency' appears in the Fourier transform shown as the inset to this figure. More interestingly, the transverse resistance, which is not measured across the ring also shows such an oscillation. This appears to be a nonlocal effect, but is a reflection of the fact that the voltage oscillations across the ring directly affect the rest of the circuit.

Now, we can use these results to analyze the experimental data shown in figure 3.1. The peak of the Fourier transform occurs roughly at 730 T^{-1}, which corresponds to a radius of about 1.0 μm, which is about that shown in figure 3.1(a). We can go a little further, however, and note that the full-width of the Fourier peak is about 100 T^{-1}, from which we can estimate the effective width of the nanowire forming the ring. Unless the nanowire is a single-mode structure, then one can conceive of orbits going around the inner diameter as well as the outer diameter, and anywhere in between. Thus, the width of the Fourier peak is a form of measure of this spread of orbit areas. The range in the Fourier transform thus suggests a width of the nanowire of about 70 nm, which is a bit larger than one might guess from the length scale in the figure. But, we need to remember that the electrical width, which we are calculating, can be different from the physical width due to edge depletion and other effects. The authors themselves state that the electrical widths vary from 60 nm to about three times higher, depending upon the carrier density [3]. The value found here is at the lower end of that range.

We note that the AB effect is a single-electron effect. But, it was recognized early on that this experiment is a variation on the famous two-slit experiment in quantum mechanics. In this regard, the upper branch and the lower branch form the two slits through which the electron can pass. How can one electron decide which path to take? This is the famous problem of quantum mechanics, and is resolved by requiring the electron to be a wave which can flow simultaneously through both slits, or branches. Only then do we obtain the interference at the output of the ring. This understanding was tested by the experiments in TEMs, where a biprism can be used to create the two paths for the electrons [6, 7]. As remarked, however, there was a large debate over whether or not this was a real effect. The observation in condensed matter came later, as most people thought that the scattering that occurs in this material would break up the phase coherence needed for the effect. Yet, it can be found in both metals and semiconductors, which suggests that the phase coherence time can be much longer than the scattering time. One important conclusion to be drawn is that the electron phase is not destroyed by elastic scattering events, as these basically leave the phase undamaged. So, impurity scattering, for example, is not a major source of phase-breaking collisions. Electrons which traverse the ring can undergo tens to thousands of collisions due to impurities, but still clearly show the oscillatory behavior.

3.2 Temperature dependence of the AB effect

An important feature of the AB effect in many structures, particularly metal rings, is that they wash out fairly rapidly with increasing temperature. Since there is no temperature dependence in the basic effect, as shown in equation (3.4), this decay with increasing temperature must arise from the role of scattering and the decoherence that it introduces. In particular, the scattering must break up the phase coherence, so we are concerned with the so-called phase breaking, or phase coherence, time τ_φ. Naturally, this is going to be related to a phase coherence length, which is the distance that the particle travels before the phase is randomized by the inelastic collisions. These events tend to give the amplitude A of the AB oscillations a dependence upon this length via a common formula

$$A(L) = A_0 \exp\left(-\frac{L}{l_\varphi}\right), \tag{3.5}$$

where L is the path length of the electrons, or essentially one-half the circumference of the ring, and l_φ is the phase coherence length. In figure 3.3, we show a typical ring fabricated in an InGaAs/InAlAs heterostructure [14]. The $In_{0.64}Ga_{0.36}As$ quantum well was slightly strained and was 10 nm thick and located 50 nm from the surface. The darker areas in the image, delimited by the white edges, are etched trenches which define the ring structure (lighter gray areas). The electrons are thus forced to move around the ring between the top contacts and the bottom contacts. In figure 3.4, typical data obtained are shown, both in the raw form and in the processed form, in which an oscillatory background has been removed. The Fourier transform is shown in panel (c), where the clear h/e signal is the strongest component of the data. Finally, in figure 3.5, the amplitude of the AB signal is shown as a function of the lattice temperature. Note that the amplitude is plotted on a logarithmic scale, so the linear decay of the amplitude clearly satisfies the exponential form (3.5), provided that the coherence length varies linearly with the temperature, a point we return to below. In the last figure, two different sets of data are shown, which arise from different cool-downs of the sample in the cryogenic

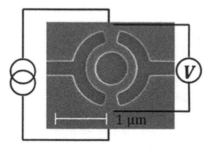

Figure 3.3. An SEM micrograph of a representative AB ring in an InGaAs/InAlAs heterostructure, with a schematic of the measurement set up. The darker areas, outlined by the white edges, are etched trenches by which the ring is formed. The current flows from the top to the bottom as indicated by the source on the left. (Reprinted with permission from [14]. Copyright 2013 IOP Publishing).

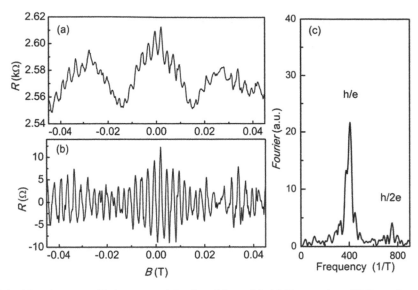

Figure 3.4. AB resistance oscillations around the ring of figure 3.3. (a) The raw data. (b) Data after removing the oscillatory background. (c) The Fourier transform of the data. Reprinted with permission from [14], copyright 2013 by IOP Publishing.

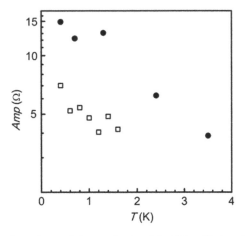

Figure 3.5. The temperature dependence of the amplitude of the AB oscillation, for two different cool-downs of the same sample. Reprinted with permission from [14], copyright 2013 by IOP Publishing.

refrigerator. Note that the amplitude varied by almost a factor of two between these two runs. This is not unusual, as warming the sample up and then re-cooling it corresponds to an annealing cycle. During the warm-up, various impurities can move around, and affect the details of the sample. This is perhaps one significant point about mesoscopic experiments: each cool-down presents a different sample with a different set of characteristics.

Since the transport through the ring is normal scattering limited behavior in a near equilibrium situation, the mobility and the diffusion constant are related. This transport is different from the ballistic or quasi-ballistic only in that many scattering events occur. While the actual transport may well be quasi-ballistic in nature, the phase-breaking events depend upon the scattering that occurs. Hence, the mobility, or the diffusion constant, can be used to define a characteristic phase-breaking length from the inelastic phase-breaking time, via

$$l_\varphi = \sqrt{D\tau_\varphi}, \tag{3.6}$$

where D is the diffusivity of the semiconductor material. One often thinks about this type of transport as diffusive transport, although the majority carriers are the important quantity and their motion is mainly determined by the mobility. But, in equilibrium, the mobility and diffusivity are connected by the Einstein relation, although one needs to use a slightly modified form for the degenerate materials. The modification is a factor that is the ratio of two Fermi integrals, but the exact integrals are determined by the dimensionality of the sample.

In order to see interference effects, it is necessary that the phase coherence length be comparable to, or larger than, the size of the device under study, as can clearly be seen from the form of equation (3.5). It is for this reason that nearly all mesoscopic experiments are carried out at low temperature, where the scattering is much weaker than at room temperature. At non-zero temperatures, an additional source of dephasing arises from the fact that a range of energies, near the Fermi energy, are involved in the transport. As discussed in the previous chapter, the width of energies involved is of order k_BT. We can therefore define a *thermal broadening time* by using the broadening that scattering induced and relating it to the spread in energy as

$$\frac{k_BT\tau_T}{\hbar} \sim 1 \rightarrow \tau_T \approx \frac{\hbar}{k_BT}. \tag{3.7}$$

Now, we can define a thermal diffusion length in analogy to the phase coherence length (3.6) with this new time, as

$$L_T = \sqrt{D\tau_T} = \sqrt{\frac{\hbar D}{k_BT}}. \tag{3.8}$$

This last contribution to decoherence is an unavoidable result of operating at a non-zero temperature, as this decoherence effect would exist even if it were possible to eliminate all the inelastic scattering processes.

The temperature decay of equation (3.5) is said to be a dynamic dephasing, as it is caused by the scattering processes. On the other hand, the thermal broadening is thought of as a static dephasing, and contributes differently to the temperature dependence. In this latter case, one considers each of the different energies as being uncorrelated with one another, so we have to simply consider the number of such energies. The important aspect is when the thermal length is comparable to the sample size L. Thus, we expect that the amplitude will show an additional reduction which varies as $(L_T/L)^{1/2}$. When this is added to equation (3.5), we get the final temperature dependence to vary as

$$\Delta G \sim A_0 \sqrt{\frac{\hbar D}{L^2 k_B T}} exp\left(-\frac{L}{l_\varphi}\right). \tag{3.9}$$

When dynamic dephasing can be ignored ($l_\varphi \gg l_T$), then we expect the prefactor to dominate and lead to a temperature which decays as $1/T^{1/2}$. When the reverse is true, then the exponential term is dominant.

3.3 The AB effect in other structures

In the above two sections, we have pretty much concentrated on standard quantum rings in which the AB effect is the dominant effect in the conductance. But, we remember that the effect is basically quite similar to the double-slit experiment in quantum mechanics, as the passage of the quantum wave through the two slits actually creates a quantum ring topology. Hence, one can conceive of many different structures in which quantum interference can be created with the result that one can see the AB effect in the system response. For example, it has been shown that two particle correlations can also show the AB effect. This has been shown theoretically [15] and experimentally [16, 17]. In this last experiment, an electronic analog to the optical correlation experiment of Brown and Twiss [18] was created. Brown and Twiss created a new type of radio interferometer, in which the signal was based upon the correlation between two independent receivers spaced out along a common baseline for reference. In the experiment, they looked for correlations in the intensity fluctuations that each receiver measured. Oberholzer *et al* [17] created an electronic equivalent to this experiment for a beam of electrons in a Q2DEG. In this latter experiment, a tunable metallic beam splitter was used to partition the electron beam into transmitted and reflected partial beams, and the current fluctuations in these two partial beams were measured. They found that these fluctuations were fully anti-correlated, which demonstrated that fermions tend to exclude one another and lead to anti-bunching. In subsequent work, it was shown that almost any electronic interferometer would exhibit the two-particle effect [19].

As we will see in a later chapter, electrons are confined near the edges of a rectangular sample when a large magnetic field is applied normal to the sample. This is a result of the quantum Hall effect. The motion of the electrons for certain magnetic field is confined to edge states which are channels located at the edges of the sample. We can create an anti-dot within the sample, which is a local region of high potential which excludes electrons from the particular local region. Then, the edge states can be localized around the anti-dot as well, and if tunnel coupling between the various edge states occurs, one will observe the AB effect in this geometry as well [20, 21]. More interestingly, if one puts a quantum dot in the center of the sample in such a manner that most of the edge states cannot propagate through the quantum point contacts (QPCs) which provide the openings to the dot, as shown in figure 3.6(a), then there will be trapped edge states within the dot [22]. As these trapped states essentially form AB loops, their occupation will depend upon the specific magnetic field, and this will lead to observable oscillations in the transmission through the dot. A typical conductance sweep in a magnetic field is shown in figure 3.6(b).

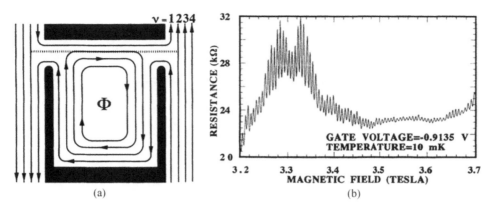

Figure 3.6. (a) Schematic diagram of the edge-state configuration at 4 T for a dot located in a quantum Hall bar. (b) The magnetoresistance trace reveals AB oscillations. (Reprinted with permission from [22]. Copyright 1994 the American Physical Society.)

More recently, the AB effect has been studied in rings formed in InGaAs [23], where the connection with the Berry phase was more extensively studied. The AB effect in biexcitons propagating around a GaAs/AlGaAs quantum ring have also been studied optically [24]. The existence of AB oscillations has also been reported in graphene p–n junctions [25]. Perhaps more interesting is the observation of AB oscillation in gate-all-around topological insulator nanowires of Bi_2Se_3 [26] and Bi_2Te_3 [27].

3.4 Gated AB rings

In the general AB rings discussed above, the magnetic field was tuned to provide the interference and oscillations. But, we note from equation (3.3) that the propagation through each arm of the ring depends upon the wave vector in that arm. This suggests that one could use a local gate to tune the wave vector, which would also lead to oscillatory wave interference as the wave vector is varied. Almost as soon as the effect was observed in semiconductors, this tuning idea was patented [28]. The use of a metallic gate on one arm of an electronic double-slit experiment was demonstrated shortly afterward [29].

We can illustrate how this occurs with a simple description. Consider figure 3.7, in which we have modified the earlier figure to include a gate over part of one arm of the ring. Let us assume that the ring is created in a GaAs/AlGaAs heterostructure, in which the reservoirs are a 2DEG with a density of 4×10^{11} cm^{-2}. This corresponds to a Fermi energy of about 14.3 meV, and a Fermi wave vector of about 1.6×10^6 cm^{-1}. We assume that the wire that composes the ring is about 50 nm wide electrically, so this produces only about 2.3 meV quantization energy in the ring, which lowers the Fermi momentum in the ring to about 1.5×10^6 cm^{-1}. Now, let us assume that the ring is 1 μm in diameter. Thus, in the absence of a magnetic field, an electron in each half of the ring (without gate voltage applied) accumulates about 236 radians of phase (or about 75π). Of course, the electrons emerge at the exit slit in phase with each other. Now, let us ask just how much do we need to reduce the density in order

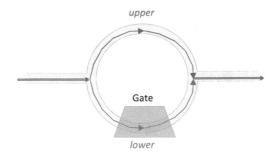

Figure 3.7. We have gated one arm of the ring in order to use the gate to control the wave vector and cause interference.

to shift the phase by π in one arm so that there is destructive interference at the output port. Let us assume that the angular spread of the gate covers about $\pi/3$ of the ring (about 60°, or one-third of the arm). Then, the electrical length of the gate is $\pi r/3$, where r is 0.5 µm. Hence, we need

$$\Delta k \cdot \frac{2\pi}{3} = \pi \rightarrow \Delta k = 6 \times 10^6 \, \text{m}^{-1} \tag{3.10}$$

which is a very small reduction in the wave number (about 4%). So, we should be able to obtain an entire series of oscillations for a modest variation in the gate voltage. Of course, we have assumed ideal conditions here, and the width of the ring as well as scattering will degrade the signal substantially, just as for the magnetic tuning of the rings. In actual fact, there has been little work on studying the role of gating on such rings, although several theoretical analyses have been performed [30–32]. Such experiments as there are have not shown strong voltage-dependent oscillations [29, 33].

Typical of the experiments are those of Krafft *et al* [33], for which the device is shown in figure 3.8(a). Here the ring is fabricated in an AlGaAs/GaAs heterostructure, with the Q2DEG located about 60 nm below the surface. Measurements show that the electron density is about $3.7 \times 10^{11} \, \text{cm}^{-2}$ at the 50 mK temperature of the dilution refrigerator. What can be seen in the figure is that the ring is defined through the use of two gates, A and B, which are isolated from the ring itself via in-plane trenches which are etched into the heterostructure with wet chemical etching. The transport measurements were performed with four terminal connections, and the various leads can be seen in the figure. Typically, current was passed between terminals 1 and 4, while terminals 2 and 3 were used to measure the voltage (resistance, as a constant current of 10 nA was passed through the system). The ring structure shows clear, if weak, oscillations as a function of the applied magnetic field, as shown in figure 3.8(b). From the inset to this latter figure, one can see that there is a strong single peak in the Fourier transform of the oscillations, with the peak corresponding to a period of about 3.1 mT, which gives a ring radius of 650 nm. This corresponds fairly well with the length key in figure 3.8(a), although the wide width of the ring compared to its radius leads to significant broadening of

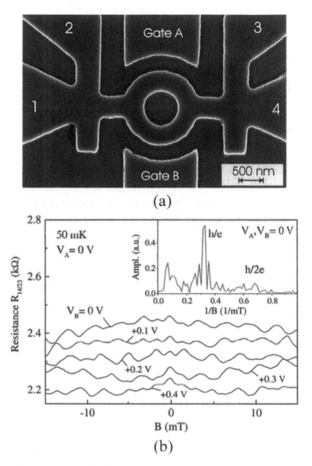

Figure 3.8. (a) Electron-beam micrograph of the ring structure and the in-plane gates A and B. (b) Resistance of the ring structure of (a) as a function of the magnetic field, with the voltage applied to gate B varied as a parameter. The inset shows the Fourier transform for zero applied gate voltage. (Reprinted with permission from [33]. Copyright 2001 Elsevier.).

the Fourier peak, for reasons discussed in section 3.1. In figure 3.8, the voltage applied to gate B is varied as a parameter. As this voltage is raised to a positive value, there are observable changes in the various curves, which can be interpreted as a phase shift of the h/e oscillation. At 0.2 V, the peaks at $B = 0$ are diminished, while those near ± 1.5 mT increase in amplitude, and the authors suggest this arises from a phase shift of π [33]. It is possible to see a gradual shift in a number of the peaks as the voltage is raised, which supports the idea of the tuning of the oscillation phase by changing the wave momentum in the arm with the biased gate. But, the gate tuning effects do not produce as striking a resistance oscillation as that due to the magnetic field, and this seems to be generally true in such measurements.

More recently, the interaction between the AB effect in a gated ring and the Coulomb blockage (discussed in chapter 8) has been discussed [34]. Of more interest is the use of a scanning gate microscope (as discussed in chapter 2) to provide the gate bias on the AB ring [35]. The use of the scanning gate allows one to taylor the potential landscape and observe interference fringes in the transport through the ring as a function of voltage and magnetic field. As a result, one can create a ring with a specific number of modes in each arm of the AB ring. Studies of the AB effect in semiconductors still appear regularly [36].

3.5 The electrostatic AB effect

The magnetic AB effect is often called the vector AB effect due to it arising from the vector potential, as in equation (3.3). There is another AB effect, which is often called the electrostatic, or scalar, AB effect. In the scalar AB effect, the interference between two propagating electrons, or electron waves, is modulated by the scalar potential generated by an electric field that does not itself exist in the regions where the electrons propagate. While the electric field does not exist in the region where the waves are propagating, the scalar potential does, which is in complete analogy to the vector form [4]. It is important to point out that this scalar form is not the same as the gate modulation we discussed in the last section. In the last section, the electric field was in the region in which the waves were propagating and merely modulated the properties of the waves rather than the phase directly. The electrostatic AB effect has become of interest for semiconductors, since it has been suggested as a source for novel device concepts [37].

In the scalar AB effect, the presence of the scalar potential adds to the energy of the particle or wave, and thus can contribute to a phase shift as

$$\Delta\varphi = \frac{-eVt}{\hbar} \sim \frac{eVL}{\hbar v_F} = -\frac{eVm^*L}{\hbar^2 k_F},$$ (3.11)

where V is the voltage and L is an effective length. If the electrons, or waves, are confined in a one-dimensional wire, the density and the Fermi wave vector are related by $k_F = \pi n/2$. We can use this in equation (3.11) to give the phase shift as a function of the applied voltage, as

$$\Delta\varphi = -\frac{2eV_a m^*L}{\hbar^2 n\pi}.$$ (3.12)

If we assume GaAs with a wire density of 5×10^7 cm^{-1}, and a length of 2 μm, then we would find a phase shift of about 220 rad V^{-1}, which is significant.

But, reality must be called into play here. According to a recent article, the scalar AB effect has never been experimentally confirmed [38]. As we pointed out in the introduction to this chapter, there was considerable debate about the reality of the AB effect when it was first put forward, and this debate lasted for almost a quarter of a century. Walstad [39] has re-examined in particular the electrostatic AB effect. He concludes that the electrostatic AB effect rests upon a theoretical error on the part of the original authors, and that this effect just does not exist, and claims of such are often mistaken observations of the magnetic AB effect [40]. So, apparently, the debate has not died down, and there is still significant doubt about such an effect.

Nevertheless, there are still scientists that give interesting predictions for observations of the electrostatic AB effect [41–43].

3.6 The AAS effect

Most of the treatment above has dealt with semiconducting rings which are nearly ballistic in nature. But, there is another effect which can occur in such rings when they are heavily disordered. In the normal AB effect, the waves go around each side of the ring and interfere at the output port, giving an oscillatory transmission. In a strongly disordered ring, it is possible for the two trajectories to deflect past the output port and continue on around the ring, where they eventually interfere at the input port, giving an oscillatory reflection signal whose periodicity is $h/2e$, since the two trajectories enclose twice as much area. The effect is known as the AAS effect for the three who developed the theory—Altshuler, Aronov, and Spivak [44]. They developed the theory based upon dirty metals; e.g., Fermi liquids with very high electron densities, but it has been seen in other systems as well. Shortly after the theoretical work, the AAS oscillations were seen, apparently for the first time, in a Mg cylinder, in which the current flowed from one side to the other [45], although it was quickly measured in gold and copper rings [46].

Interestingly enough, in small gold rings, both the AB oscillations and the AAS oscillations can be seen simultaneously [47]. In these rings, the $h/2e$ oscillations are typically seen near zero magnetic field, while the h/e oscillations appear at a higher magnetic field. This arises because the AAS oscillations are strongly damped at high magnetic fields. We noted above that the two trajectories both went completely around the ring, making a full circuit. In this situation, these two trajectories can be time-reversed paths, as they are mirror images (across the center line of the ring). In that sense, the AAS effect is really the continued interference between these two time-reversed paths as they interact everywhere around the ring. The magnetic field, however, breaks time-reversal symmetry, and so one expects that these oscillations will be heavily damped as the amplitude of the magnetic field increases. Hence, at higher magnetic field, only the AB oscillations remain. This allows one to distinguish between the AAS effect and any second harmonic signal in the Fourier transform of the normal AB effect.

It has also been shown that the AAS effect can also be seen in diffusive systems such as semiconductors [48, 49]. It has been seen in an AlGaAs/GaAs heterostructure in which InAs quantum dots were embedded near the interface to provide correlated disorder in the system [50]. More recently, however, both oscillations have been seen in InAs/AlGaSb and InSb/InAlSb heterostructures [51]. Similarly, both types of oscillation have been seen in heterostructures of AlGaN/GaN [52].

It is instructive to look a little closer at a typical good quality semiconductor heterostructure. In the work of Lillianfeld *et al* [51], a 15 nm InAs quantum well was located under a 20 nm $Al_{0.2}Ga_{0.8}Sb$ cap layer and is unintentionally doped. At low temperature, the electron density was found to be about 8.5×10^{11} cm^{-2}. The sample composed a 7×7 array of rings whose average radius was 350 nm and had a ring width of 130 nm. Measurements were made at temperatures in the range 0.4–10 K.

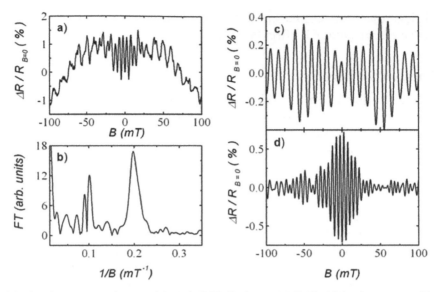

Figure 3.9. (a) The magnetoresistance of the InAs/AlGaSb rings at 0.4 K. The higher frequency oscillations at low B are the AAS oscillations with period $h/2e$. At higher values of B, the AB oscillations dominate for $|B| >$ 50 mT. (b) The Fourier transform of the data for the range $|B| <$ 100 mT, so that the AAS period is the strong signal at about 0.2 mT. Digital filtering of the signal isolates the (c) h/e and (d) $h/2e$ components of (a). (Reprinted with permission from [51]. Copyright 2010 Elsevier).

As previously, both types of oscillations were found to occur, with the AAS oscillations existing at low values of the magnetic field and going away at higher magnetic field. Typical data are shown in figure 3.9 for the InAs rings. Digital filtering of the magnetoresistance signal allows one to separate the AAS component from the normal AB component, with the former lying mainly below $|B| <$ 50 mT. These two different components are shown in panels (c) and (d) of the figure. It may clearly be seen that the AAS signal is heavily damped above a few tens of mT, and that the AB oscillations dominate over the larger magnetic field range.

3.7 Weak localization

In the previous section, the AAS effect arose from two trajectories moving around a ring in opposite directions such that they interfered with one another all around the ring. This resulted in an enhanced back-scattering which was oscillatory in the magnetic field for small magnetic fields. Now, this effect does not require the ring. Disorder can lead to a set of localized scatterers which produce exactly the same back-scattering. In the bulk, this process is known as weak localization. The idea is illustrated in figure 3.10, where we show a set of scatterers and two time-reversed trajectories in which the particles scatter from each site and eventually return to their original direction as a back-scattering. Since we cannot know which direction the particle scatters around the 'ring', both directions are possible and represent the time-reversed paths. The presence of weak localization results in a reduction in the actual conductivity of the sample, that is, weak localization is a negative

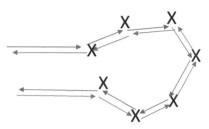

Figure 3.10. The red and blue paths are the time-reversed set of trajectories which scatter from the set of scatterers shown here. They interfere along their entire path, producing weak localization from the net back scattering.

contribution to the conductivity due to the back scattering effect. This is usually a small but significant effect if there is sufficient disorder in the system.

The weak localization correction is damped by the application of a magnetic field. This follows as the magnetic field breaks time-reversal symmetry, and causes the two reverse paths to diverge, hence reducing the quantum interference. This divergence arises from the Lorentz force, which produces a force normal to both the magnetic field and the local velocity of the particle. Since the two paths have opposite velocities, the Lorentz force on the two will have opposite directions, hence causing the paths to diverge and breaking up the interference effect. We will see below that a critical magnetic field can be defined as that at which the weak localization correction has been reduced to one-half of its peak value, and this is called the correlation magnetic field. It will be argued that this value of critical magnetic field is just enough to enclose one flux quanta (h/e) within the area of the phase-coherent loop.

3.7.1 A semiclassical approach to the conductance change

The conductivity can be generally related to the current–current correlation function, a result that arises from the Kubo formula, but it also is dependent upon an approach involving the retarded Langevin equations [53]. In general, the conductivity can be expressed simply as

$$\sigma = \frac{ne^2}{m^*\langle v^2(0)\rangle} \int_0^\infty \langle v(t)v(0)\rangle dt \sim \frac{ne^2}{m^*} \int_0^\infty e^{-t/\tau}dt = \frac{ne^2\tau}{m^*}, \tag{3.13}$$

where it has been assumed that the velocity decays with a simple exponential, and the conductivity is measured at the Fermi surface (although this is not strictly required for the definitions used in this equation). The exponential in equation (3.13) is related to the probability that a particle diffuses, without scattering, for a time τ, which is taken as the mean time between collisions. In one dimension, we can use the facts that $D = v_F^2\tau$ (or more generally, $D = v_F^2\tau/d$, where d is the dimensionality of the system), $l_e = v_F\tau$, and $n_1 = k_F/\pi$ to write equation (3.13) as

$$\sigma_1 = \frac{e^2}{\pi\hbar} 2D \int_0^\infty \frac{1}{2l_e} e^{-t/\tau} dt. \tag{3.14}$$

In two dimensions, we use the fact that $D = v_F^2\tau/2$ and $n_2 = k_F^2/2\pi$ to give

$$\sigma_2 = \frac{e^2}{\pi\hbar} 2D \int_0^\infty \frac{k_F}{2l_e} e^{-t/\tau} dt. \tag{3.15}$$

We note that the integrand now has the units L^{-d} in both cases. It is naturally expected that this scaling will continue to higher dimensions. However, the quantity inside the integral is no longer just the simple probability that a particle has escaped scattering. Instead, it now has a prefactor that arises from critical lengths in the problem. To continue, one could convert each of these to a conductance by multiplying by L^{d-2}. This dimensionality couples with the diffusion constant and time integration to produce the proper units of conductance. To be consistent with the remaining discussion, however, this will not be done.

In weak localization, we seek the correlation function that is related to the probability of return to the initial position. We define the integral analogously to the above as a correlator, specifically known as the particle–particle correlator. Hence, we define

$$C(\tau) = \int_0^\infty W(t)dt \sim \int_0^\infty \frac{1}{2L^d} e^{-t/\tau} dt, \tag{3.16}$$

where $W(\tau)$ is the time-dependent correlation function describing this return, and the overall expression has exact similarities to the above equations (in fact, the correlator in the first two equations is just the integral). What is of importance in the case of weak localization is that we are not interested in the drift time of the free carriers. Rather, we are interested in the diffusive transport of the carriers and in their probability of return to the original position. Thus, we will calculate the weak localization by replacing C or W by the appropriate quantity defined by a diffusion equation for the strongly scattering regime. Hence, we must find the expression for the correlation function in a different manner than simply the exponential decay due to scattering.

Following this train of thought, we can define the weak localization correction factor through the probability that a particle diffuses some distance and returns to the original position. If we define this latter probability as $W(t)$, where t is the time required to diffuse around the loop, we can then define the conductivity correction in analogy with equation (3.15) to be

$$\Delta\sigma = -\frac{2e^2}{h} 2D \int_0^\infty W(t)dt, \quad x(t) \to x(0), \tag{3.17}$$

and the prefactor of the integral is precisely that occurring in the first two equations of this section. (The negative sign is chosen because the phase interference reduces the conductance.) In fact, most particles will not return to the original position. Only a small fraction will do so, and the correction to the conductance is in general small. It is just that small fraction of particles that actually does undergo back-scattering (and reversal of momentum) after several scattering events that is of interest. In the above equations, the correlation function describes the decay of 'knowledge' of the initial state. Here, however, we use the 'probability' that particles can diffuse for a time t and return to the initial position while retaining some 'knowledge' of that initial condition (and, more precisely, the phase of the particle at that initial position). Only in this case can there be interference between the initial wave and the returning wave, where the 'knowledge' is by necessity defined as the retention of phase coherence in the quantum sense. While we have defined $W(t)$ as a probability, it is not a true probability since it has the units L^{-d}, characteristic of the conductivity in some dimension (DWt is dimensionless). In going over to the proper conductance, this dimensionality is correctly treated, and in choosing a properly normalized probability function, no further problems will arise. This discussion has begun with the semiclassical case, but now we are seeking a quantum mechanical memory term, exemplifying the problems in connecting the classical world to the quantum mechanical world. In fact, Alt'shuler, Aronov, and Spivak derived precisely this correlation function in a full Green's function approach in deriving the effect named for them [44]. We will not go to that extreme, as the essence can be illustrated quite simply.

It has been assumed so far that the transport of the carriers is diffusive, that is, that the motion moves between a great many scattering centers so that the net drift is one characterized well by Brownian motion. By this, we assume that quantum effects cause the interference that leads to equation (3.17), and that the motion may be described by classical motion. This means that $kl_e < 1$, where k is the carrier's wave vector (usually the Fermi wave vector) and l_e is the mean free path between collisions, which is normally the elastic mean free path (which is usually shorter than the inelastic mean free path). This means that the probability function will be Gaussian (characteristic of diffusion), and this is relatively easily established by the fact that $W(t)$ should satisfy the diffusion equation for motion away from a point source (at time $t = 0$), since the transport is diffusive. This means that [54]

$$\left(\frac{\partial}{\partial t} - D\nabla^2\right)W(t) = \delta(t)\delta(r), \tag{3.18}$$

which has the general solution

$$W(r, t) = \frac{1}{(4\pi Dt)^{d/2}}\exp\left(-\frac{r^2}{4Dt}\right). \tag{3.19}$$

In fact, this solution is for unconstrained motion (motion that arises in an infinite d-dimensional system). If the system is bounded, as in a two-dimensional quantum well or in a quantum wire, then the modal solution must be found. At this point we will not worry about this. Our interest is in the probability of return, so we set $r = 0$. There is one more factor that has been omitted so far, and that is the likelihood that the particle can diffuse through these multiple collisions without losing phase memory. Thus, we must add this simple probability, which is an exponential. This leads us to the probability of return after a time t, without loss of phase, being

$$W(r, t) = \frac{1}{(4\pi Dt)^{d/2}}\exp\left(-\frac{t}{\tau_\varphi}\right),\qquad (3.20)$$

where the phase-breaking time τ_φ has been introduced to characterize the phase-breaking process. We note at this point that the dimensionality of $W(t)$ is L^{-d} (Dt has the dimensions of L^2), which is the dimensionality of the integrand for the conductivity, not the conductance. Thus, this fits in with the discussion above.

One further modification of this simple semiclassical treatment has been suggested by Beenakker and van Houten [55]. This has to do with the fact that we do not expect to find these diffusive effects in ballistic transport regimes. Thus, it can be expected that on the short-time basis, these effects go away. Here, 'short time' is appropriate in that collisions must occur before diffusive transport can take place. If there are no collisions, there is little chance for the particle to be back-scattered and to return to the original position. Thus, these authors suggest modifying equation (3.20) to account for this process. This gives the new form for the probability of return to be

$$W(r, t) = \frac{1}{(4\pi Dt)^{d/2}}\exp\left(-\frac{t}{\tau_\varphi}\right)(1 - e^{-t/\tau}).\qquad (3.21)$$

At this point, we have slipped in the only quantum mechanics in the current approach. This quantum mechanics is connected with the phase of the electrons and is described by the phenomenological phase relaxation time τ_φ. We have not actually carried out a quantum mechanical calculation, yet we have introduced all the necessary phase interference through this phenomenological term. The actual quantum calculations are buried at this point, but it is important to recognize where they have entered in the discussion.

The dimensionality correction to the probability of return will go away if we work with the total conductance, rather than the conductivity. However, as above, we will not make this change. We now use equation (3.21) in equation (3.17). This gives the conductivity corrections for weak localization to be

$$\Delta\sigma = -\frac{e^2}{\pi\hbar}\begin{cases}\dfrac{1}{2\pi l_\varphi}\left(\sqrt{1 + \tau_\varphi/\tau} - 1\right), & d = 3 \\[2ex] \dfrac{1}{2\pi}ln\left(1 + \dfrac{\tau_\varphi}{\tau}\right), & d = 2 \\[2ex] l_\varphi\left(1 - \sqrt{\dfrac{\tau}{\tau_\varphi + \tau}}\right), & d = 1\end{cases}.\qquad (3.22)$$

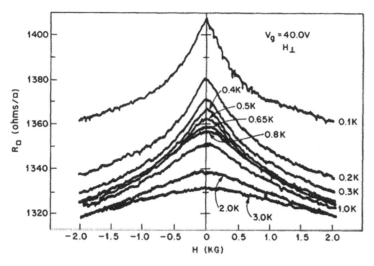

Figure 3.11. Low-field magnetoresistance for a Si (111) MOSFET in a perpendicular magnetic field for various temperatures. For the younger readers, 1 kG = 0.1 T. (Reprinted with permission from [44]. Copyright 1982 the American Physical Society).

It is clear that the important length in this diffusive regime is the phase coherence length $l_\varphi = \sqrt{D\tau_\varphi}$. To be sure, this result is the simplest one that can be obtained within reasonable constraints. Nevertheless, the results are quite useful and show that the weak localization reduction of conductance is relatively universal in its amplitude, but also that it has an adjustment depending upon the ratio of the important time scales in the transport problem. Nevertheless, the quantum mechanics is buried in the ad hoc introduction of the phase coherence time τ_φ. Without this introduction, none of the above formulas would be meaningful.

In figure 3.11, we show data taken from studies of the weak localization in a Si MOSFET structure [56]. In this case, the surface is a (111) plane, and the electron density was about 4.5×10^{12} cm^{-2}. The measurements were performed in a dilution refrigerator, and several curves at different substrate temperatures are shown in the figure. It is clear that there is a distinct peak structure which occurs at $B = 0$, and which decays away rather quickly in magnetic field. As the temperature rises, the phase-breaking time obviously gets shorter and the enhancement peak at $B = 0$ is reduced in amplitude. All of this fits well with the arguments presented above.

3.7.2 Role of the magnetic field

In the presence of a magnetic field, the diffusive paths that return to their initial coordinates will enclose magnetic flux. To examine this behavior is only slightly more complicated than that of the above treatment, and we will take the magnetic field in the z-direction and in the Landau gauge $\mathbf{A} = (0, Bx, 0)$. We will also consider only a thin two-dimensional slab with no z variation at present, so that equation (3.18) can be written as

$$\left[\frac{\partial}{\partial t} - D\left(\nabla - i\frac{2e\mathbf{A}}{\hbar} \right)^2 + \frac{1}{\tau_\varphi} \right] W(t) = \delta(t)\delta(\mathbf{r}), \tag{3.23}$$

where we have incorporated the vector potential via the Peierls' substitution. In addition, the phase-breaking time has been incorporated. If we take this in the two-dimensional system of the normal heterostructure interface, there will be no z variation, and we can use the normal Landau gauge for the magnetic field. Finally, we assume that the diffusive propagator is a free plane wave in the y-direction. Then, the expanded form of the above equation becomes

$$\left[\frac{\partial}{\partial t} - D\frac{\partial^2}{\partial x^2} - D\left(ik - i\frac{2eBx}{\hbar} \right)^2 + \frac{1}{\tau_\varphi} \right] W(t) = \delta(t)\delta(\mathbf{r}). \tag{3.24}$$

If we now once again take the $x \to 0$ limit, then the result is a simpler one-dimensional diffusion equation in which the magnetic field and phase-breaking terms can be combined into the relationship

$$\frac{1}{\tau_\varphi} \to \frac{1}{\tau_\varphi} + Dk^2 = \frac{1}{\tau_\varphi} + (2n+1)\frac{eBD}{\hbar}, \tag{3.25}$$

where it has been assumed that the relevant momentum is the Fermi momentum which is defined by the cyclotron orbits that may result from the magnetic field. Of course, this treatment is oversimplified, and a more extensive exact treatment is required. Such a treatment is beyond the level that we are discussing here, but it has been shown that [57]

$$\Delta\sigma \sim \frac{e^2}{\pi\hbar} \left[\Psi\left(\frac{1}{2} + \frac{\hbar}{4eDB\tau_\varphi} \right) - \Psi\left(\frac{1}{2} + \frac{\hbar}{4eDB\tau} \right) + \frac{1}{2}ln\left(\frac{\tau_\varphi}{\tau} \right) \right], \tag{3.26}$$

where $\Psi(\cdot)$ is the digamma function. In figure 3.12, we plot magnetoresistance data, for a range of temperatures, taken from an AlGaAs/GaAs heterostructure sample [58]. The AlGaAs layer was 60 nm thick and uniformly Si doped, although a 10 nm undoped spacer layer was placed between the doped layer and the GaAs layer. A doped 20 nm GaAs cap layer was placed on the top surface to facilitate making ohmic contacts to the structure. This particular sample had a sheet density of 2.86×10^{11} cm^{-2}. The curves through the data are fits to the form of equation (3.26), illustrating the agreement that can be obtained between the theory and the experiment.

3.8 Graphene rings

Observing the AB effect in graphene rings has appeared in recent years. This is more complicated because of the degenerate pair of valleys in the conduction band, a zero energy gap, and the complications of the monolayer material. However, thanks to modern processing technology, this has now become possible. Rings fabricated by etching also have multiple problems due to scattering at the layer surface from the substrate and any overlayers as well as by edge roughness due to the etching. The

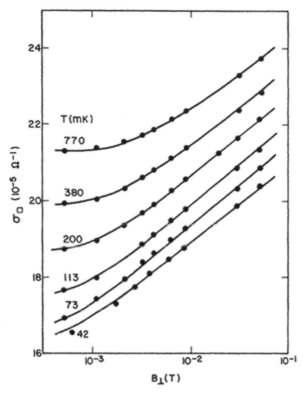

Figure 3.12. Conductance of a AlGaAs/GaAs heterostructure sample at various temperatures as a function of magnetic field. The dots are measured data and the lines are fits using equation (3.26). (Reprinted with permission from [58]. Copyright 1984 the American Physical Society).

former can be reduced by encapsulation with hexagonal BN layers on both top and bottom surfaces, and the rings have been formed by reactive ion etching of the BN/graphene/BN sandwich after using electron beam lithography to pattern the rings [59]. The dimensions of the ring were determined by scanning force microscopy to be an inner diameter of the ring of 405 nm and an outer diameter of 755 nm. A scanning force microscopy image of the graphene ring is shown in figure 3.13(a). Figure 3.13(b) shows the two-terminal (indicated as V in panel a) conductance G_{2W} as a function of the applied magnetic field. In this regime, the AB oscillations are fairly evident but ride upon a modulated background (the red curve in panel b). This red curve is obtained by filtering out the high frequency components and is thought to be conductance fluctuations (discussed in chapter 5). By subtracting this red curve (a different form is found for each gate voltage), the AB oscillations at different gate voltages are shown in panels c-d. In particular, the blue arrows in panel c suggest a periodicity of 5.2 mT for the AB oscillations, which is the expected periodicity for a ring with a mean diameter of 580 nm, which corresponds to the dimensions given above.

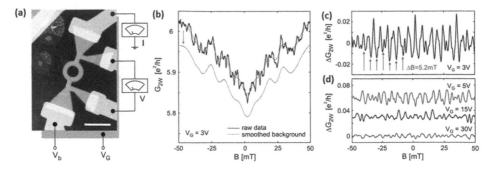

Figure 3.13. (a) Scanning force microscopy image of the graphene ring and contacts. The dimensions are given in the text. (b) The solid black line is the two-terminal conductance G_{2W} as a function of magnetic field. The red line shows the smoothed back-ground conductance which can be removed to high-light the AB oscillations. These are shown in panels (c) and (d). The blue arrows in (c) show the periodicity of the AB oscillations yielding a ring diameter of 580 nm, which agrees well with measurements of the ring. Reprinted with permission from [59], copyright 2017 by the American Physical Society.

Problems

1. Using the understanding in section 3.1, analyze the data in figure 3.5 to determine the size of the ring in figure 3.4 as well as the electrical width of the ring.

2. Consider a free electron in a magnetic field. Using the Peierls' substitution and the Landau gauge, show that the electron satisfies a Hamiltonian that has the form of a harmonic oscillator.

3. Consider a gated AB ring such as shown in figure 3.9. The ring is formed in an InAlAs/InAs heterostructure, in which the 2DEG is in the InAs. In the absence of the gate, the 2DEG has a density of 5×10^{11} cm^{-2}. When the gate voltage is applied, the density under the gate drops to 3×10^{11} cm^{-2}. What should the phase shift be in this situation?

4. In weak localization, the magnetic field dependence is often characterized by a correlation magnetic field B_c, which is the value of the magnetic field at which the weak localization correction has dropped to one-half of its peak value. Using the lowest temperature curve in figure 3.11, estimate the value of the peak correction to the resistance by using an estimated extension of the resistance curves at higher magnetic field down to zero field. (a) Estimate the correlation magnetic field. (b) Using the peak correction at zero magnetic field, estimate the phase-breaking time (the mobility of the device is estimated to be about 10^3 cm^2 V^{-1}s^{-1} at low temperatures).

Appendix C The gauge in field theory

Most students find their introduction to the concept of gauge in the study of electromagnetic fields, and so we shall start there as well. We begin with one of Maxwell's equations, which is written as

$$\nabla \times \mathbf{E} = -\frac{\partial \mathbf{B}}{\partial t}. \tag{C.1}$$

It is through this equation that we introduce the vector potential $\mathbf{A}$, which is related to the magnetic field via

$$\mathbf{B} = \nabla \times \mathbf{A}. \tag{C.2}$$

Using this formula in (C.1) leads to the result

$$\nabla \times \mathbf{E} = -\frac{\partial}{\partial t}(\nabla \times \mathbf{A}). \tag{C.3}$$

Many people jump to the conclusion that the electric field is just the partial derivative of the vector potential with respect to time, but this overlooks some important considerations. Strictly speaking, we have to integrate equation (C.3) to obtain the result

$$\mathbf{E} = -\frac{\partial \mathbf{A}}{\partial t} + \mathbf{C}, \ \nabla \times \mathbf{C} = 0, \tag{C.3}$$

where the last equation is required in order to satisfy equation (C.3) At this point, we assure ourselves of this last requirement by defining $\mathbf{C}$ as

$$\mathbf{C} \equiv -\nabla \varphi, \tag{C.5}$$

which introduces the scalar potential φ. These all now lead to one common form for the electric field

$$\mathbf{E} = -\frac{\partial \mathbf{A}}{\partial t} - \nabla \varphi. \tag{C.6}$$

We normally see this without the vector potential in our studies of semiconductor devices, but the vector potential term is important in, e.g., the AB effect. How we chose to represent the electric field, and the connections between the vector and scalar potentials, is referred to as a gauge condition.

To understand the conditions on the two potentials, we begin with the second of Maxwell's main equations and one of the constituent equations, as

$$\begin{aligned} \nabla \times \mathbf{B} &= \mu \varepsilon \frac{\partial \mathbf{E}}{\partial t} + \mu \mathbf{J} \\ \nabla \cdot \mathbf{E} &= \frac{\rho}{\varepsilon} \end{aligned}, \tag{C.7}$$

where μ and ε are the permeability and dielectric permittivity of the material in which the fields are present. Here, $\mathbf{J}$ is the current density flowing in the material and ρ is the charge density, which are the two quantities which can give rise to the fields. We transform these two equations by introducing the two potentials. Doing this with the first of equations (C.7) leads to

$$\nabla^2 \varphi + \frac{\partial}{\partial t}(\nabla \cdot \mathbf{A}) = -\frac{\rho}{\varepsilon}. \tag{C.8}$$

This form differs from that of the Poisson equation which is usually used to find the local potential and the electric field in self-consistent studies. We will recover the Poisson equation from this equation through a choice of gauge. We now turn to the second of equations (C.7), which gives

$$\nabla \times (\nabla \times \mathbf{A}) = -\mu\varepsilon\frac{\partial}{\partial t}\left(\frac{\partial \mathbf{A}}{\partial t} + \nabla\varphi\right) + \mu\mathbf{J}, \tag{C.9}$$

which can be rearranged to give

$$\nabla^2 \mathbf{A} - \mu\varepsilon\frac{\partial^2 \mathbf{A}}{\partial t^2} = -\mu\mathbf{J} + \nabla\left(\nabla \cdot \mathbf{A} - \mu\varepsilon\frac{\partial \varphi}{\partial t}\right). \tag{C.10}$$

If we could set the term in parentheses to zero, we would have the inhomogeneous wave equation for the vector potential with the current density as the driving term. At the same time, setting the relation to zero allows equation (C.8) to become the inhomogeneous wave equation for the scalar potential with the charge density as the driving term. Thus, this act uncouples the two potentials, and each of these will have its own driving term. This condition is known as the *Lorentz gauge*, or sometimes simply as the *gauge equation*

$$\nabla \cdot \mathbf{A} - \mu\varepsilon\frac{\partial \varphi}{\partial t} = 0. \tag{C.11}$$

We have to remember, however, that this is a choice of convenience and is not a required condition. The fact that it is usually made in the study of electromagnetic fields does not make it a firm fact or theorem. Nevertheless, when this gauge equation is imposed, it is a rigid constraint upon the solutions of the two wave equations.

There are further possibilities that can be imposed. For example, a further approximation is to invoke the Coulomb gauge, or the electrostatic gauge as it is sometimes called, in which we set

$$\nabla \cdot \mathbf{A} = 0, \tag{C.12}$$

for which equation (C.11) then leads to

$$\frac{\partial \varphi}{\partial t} = 0. \tag{C.13}$$

When this condition is assumed, we note that equation (C.8) now becomes the more familiar Poisson equation. Once more, we note that this familiar result arises from a choice of gauge; it is not automatically true and basic. So, when we solve the Poisson equation for a device, we are assuming that only low frequency effects are of interest, and that the potential and electric field instantaneously follow variations in charge (or that the propagation delays are much shorter than any time constant in the system). That is, in assuming (C.13), we are assuming that the charge density in the device is *static and time invariant*. Then, the scalar potential follows the charge density change instantaneously, in clear violation of relativity. So, if one wants to use the Coulomb gauge in device simulation, it must first be ascertained that any charge changes, due to the imposition of self-consistency, must be slow enough to validate use of this gauge.

Having discussed the various gauges that are commonly used above, we have come nowhere near exhausting the number of gauge choices that one can make. When we study the magnetic field effect on various mesoscopic devices, there are two more usual gauge choices that are made. These arise from the manner in which we can force the magnetic field and the vector potential to satisfy equation (C.2). We have already met one of these in chapter 3, where we introduced the *Landau gauge*

$$\mathbf{A} = Bx\mathbf{a}_y, \tag{C.14}$$

which could also have been written as

$$\mathbf{A} = -By\mathbf{a}_x. \tag{C.15}$$

The choice of which of these to use in a particular situation is one of convenience. But, in many applications, such as quantum dots to be seen in a later chapter, it is convenient to combine these two forms into the *symmetric gauge*

$$\mathbf{A} = \frac{1}{2}(-By\mathbf{a}_x + Bx\mathbf{a}_y). \tag{C.16}$$

In quantum mechanics, it is quite useful that we make the wave function gauge invariant, especially if we want to talk about the wave function as a field. Normally, we invoke gauge invariance through the condition that we create a new function Λ, which we use to change the vector potential through

$$\mathbf{A} \to \mathbf{A} + \nabla\Lambda. \tag{C.17}$$

In order to keep the gauge condition (C.11) satisfied, we then have to change the scalar potential as

$$\varphi \to \varphi - \frac{\partial\Lambda}{\partial t}. \tag{C.18}$$

In classical mechanics, as well as in quantum mechanics, the electromagnetic interactions are taken into account by a change in the momentum, in which we make the Peierls' substitution as

$$\mathbf{p} \to \mathbf{p} - e\mathbf{A}, \tag{C.19}$$

so that the introduction of a magnetic field to the system is accommodated by introducing the vector potential to the momentum in the Hamiltonian. Then, to keep the wave function gauge invariant, we require that a gauge shift such as equation (C.17) lead to a phase shift of the wave function according to

$$\psi(\mathbf{r}) \rightarrow e^{ie\Lambda/h}\psi(\mathbf{r}). \tag{C.20}$$

At this point, we have covered the normal discussion that is given to most graduate students on the use of gauge. However, it is important to understand that we have come nowhere close to covering the entire topic. Most discussions proceed with the use of the two potentials—the vector and scalar potentials. But, it has been known for a very long time that these two potentials can be determined from a single vector quantity that is both time and position varying [60], which is often called the Hertz potential. But, this can be supplemented by quantities such as the polarization P in discussions of the dielectric function. Many other potentials, some complex, have been introduced for particular applications, such as diffraction. Needless to say, there are many advanced forms of this, nearly all of which are beyond the level of discussion here. Another point worth discussing is the aforementioned dielectric function. It would be a major mistake to assume a constant for ε. In particular, we know that in condensed matter systems, there are many factors which contribute to the dielectric function [61], even if we evaluate this function within linear response. There are also nonlinear and inhomogeneous effects which cannot be handled easily and directly affect the manner in which electrons or wave propagate within the solid [62].

References

[1] Webb R A, Washburn S, Umbach C P and Laibowitz R B 1985 *Phys. Rev. Lett.* **54** 2696

[2] Ishibashi K, Takagaki Y, Gamo K, Namba S, Ishida S, Murase K, Aoyagi Y and Kawabe M 1987 *Sol. State Commun.* **64** 573

[3] Mankiewich P M, Behringer R E, Howard R E, Chang A M, Chang T Y, Chelluri B, Cunningham J and Timp G 1988 *J. Vac. Sci. Technol.* **B6** 131

[4] Aharonov Y and Bohm D 1959 *Phys. Rev.* **115** 485

[5] Imry Y 1986 *Directions in Condensed Matter Physics* ed G Grinsstein and E Mazenko (Singapore: World Scientific) pp 10363

[6] Chambers R G 1960 *Phys. Rev. Lett.* **5** 3

[7] Tonomura A, Matsuda T, Suzuki R, Fukuhara A, Osakabe N, Umezaki H, Endo J, Shinagawa K, Sugita Y and Fujiwara H 1982 *Phys. Rev. Lett.* **48** 1443

[8] Chadrasekhar V, Rooks M J, Wind S and Prober D E 1985 *Phys. Rev. Lett.* **55** 1610

[9] Datta S, Melloch M R, Bandyopadhay S, Noren R, Vaziri M, Miller M and Reifenberger R 1985 *Phys. Rev. Lett.* **55** 2344

[10] See, e.g.,; Sonin E B 2010 *J. Phys. A: Math. Theor.* **43** 354003

[11] Berry M V, Chambers R G, Large M D, Upstill C and Walmsley J C 1980 *Eur. J. Phys.* **1** 154

[12] Roux P, De Rosny J, Tanter M and Fink M 1997 *Phys. Rev. Lett.* **79** 3170

[13] Leonhardt U and Piwnicki P 1995 *Phys. Rev. Lett.* **51** 7679

[14] Ren S L, Heremans J J, Gaspe C K, Vijeyaragunathan S, Mishima T D and Santos M B 2013 *J. Phys.: Condens. Matter* **25** 435301

[15] Büttiker M 1990 *Phys. Rev. Lett.* **65** 2901

[16] Oliver W D, Kim J, Liu R C and Yamamoto Y 1999 *Science* **284** 299

[17] Oberholzer S, Henry M, Strunk C, Schönenberger C, Heinzel T, Ensslin K and Holland M 2000 *Physica* E **6** 314

[18] Hanbury Brown R and Twiss R Q 1954 *Phil. Mag., Ser. 7* **45** 663

[19] Splettstoesser J, Samuelsson P, Moskalets M and Büttiker M 2010 *J. Phys. A: Math. Theor.* **43** 1

[20] Takagaki Y and Ferry D K 1993 *Phys. Rev.* B **47** 9913

[21] Takagaki Y and Ferry D K 1993 *Phys. Rev.* B **48** 8152

[22] Bird J P, Ishibashi K, Stopa M, Aoyagi Y and Sugano T 1994 *Phys. Rev.* B **50** 14983

[23] Aharony A, Entin-Wohlman O, Tzarfati L H, Hevroni R, Karpovski M, Shelukhin V, Umansky V and Palevski A 2019 *Sol. State Electron.* **155** 117

[24] Kim H *et al* 2018 *Nano Lett.* **18** 6188

[25] Makk P, Handschin C, Tóvári E, Watanabe K, Taniguchi T, Richter K, Liu M-H and Schönenberger C 2018 *Phys. Rev.* B **98** 035413

[26] Zhu H, Richter C A, Yu S, Ye H, Zeng M and Li Q 2019 *Appl. Phys. Lett.* **115** 073107

[27] Krieg J, Giraud R, Funke H, Dufouleur J, Escoffier W, Trautann C and Toimil-Molares M E 2019 *J. Phys. Chem. Sol.* **128** 360

[28] Fowler A B 1985 *US Patent* 4550330

[29] Yacoby A, Heiblum M, Umansky V, Shtrikman H and Mahalu D 1994 *Phys. Rev. Lett.* **73** 3149

[30] Joe Y S and Ulloa S E 2003 *Phys. Rev.* B **47** 9948

[31] Park K, Lee S, Shin M, Lee E-II and Kwon H-C 1996 *Surf. Sci.* **361/362** 751

[32] Figielski T and Wosinski 1999 *J. Appl. Phys.* **85**

[33] Krafft B, Förster A, van der Hart A and Schäpers T 2001 *Physica* E **9** 635

[34] Lee T-H and Hu S-F 2014 *J. Appl. Phys.* **115** 123712

[35] Kozikov A A, Steinacher R, Rossler C, Ihn T, Ensslin K, Reichl C and Wegscheider W 2014 *New J. Phys.* **16** 053031

[36] Filusch A, Wurl C, Pieper A and Fehske H 2018 *J. Low Temp. Phys.* **191** 259

[37] Bandyopadhyay S, Datta S and Melloch M R 1986 *Superlatt. Microstruc.* **2** 539

[38] Batelaan A and Tonomura A 2009 *Phys. Today* **62** 38

[39] Walstad A 2010 *Int. J. Theor. Phys.* **49** 2929

[40] Walstad A 2017 *Int. J. Theor. Phys.* **56** 965

[41] Szafran B 2011 *Phys. Rev.* B **84** 075336

[42] Giulio P, Lu P-H, Tavabi A, Duchamp M and Dunin-Borkowski R E 2017 *Ultramicroscopy* **181** 191

[43] Kim Y-W and Kang K 2018 *New J. Phys.* **20** 103046

[44] Alt'shuler B L, Aronov A G and Spivak B Z 1981 *JETP Lett.* **33** 94

[45] Alt'shuler B L, Aronov A G, Spivak B Z, Sharvin D Y and Sharvin Y V 1982 *JETP Lett.* **35** 588

[46] Pannetier B, Chaussy J, Rammal R and Gandt P 1985 *JETP Lett.* **35** 588

[47] Verbruggen A H, Holweg P A M, Vloeberghs H, Van Haesendonck C, Romijn J, Radelaar S and Bruynseraede Y 1991 *Microelectron. Eng.* **13** 407

[48] Kawabata S and Nakamura K 1996 *J. Phys. Soc. Japan* **65** 3708

[49] Kawabata S and Nakamura K 1998 *Sol.-State Electron.* **42** 1131

[50] Heinzel T, Jäggi R, Ribeiro E, Waldkirch M v, Ensslin K, Ulloa S E, Medeiros-Ribeiro G and Petroff P M 2003 *Europhys. Lett.* **61** 674

[51] Lillianfeld R B, Kallaher R L, Herrmans J J, Chen H, Goel N, Chung J S, Santos M B, Van Roy W and Borghs G 2010 *Phys. Proc.* **3** 1231

[52] Han K, Tang N, Duan J-X, Lu F-C, Liu Y-C, Shen B, Zhou W-Z, Lin T, Sun L and Yu G-L 2011 *Chin. Phys. Lett.* **28** 087302

[53] Ferry D K 1991 *Semiconductors* (New York: Macmillan)

[54] Chakravarty S and Schmid A 1986 *Phys. Rep.* **140** 193

[55] Beenakker C W J and van Houten H 1988 *Phys. Rev.* B **38** 3232

[56] Bishop D J, Dynes R C and Tsui D C 1982 *Phys. Rev.* B **26** 773

[57] Altshuler B L, Khmelnitzkii D, Larkin A I and Lee P A 1980 *Phys. Rev.* B **22** 5142

[58] Lin B J F, Paalanen M A, Gossard A C and Tsui D C 1984 *Phys. Rev.* B **29** 927

[59] Dauber J, Oellers M, Venn F, Epping A, Watanabe K, Taniguchi T, Hassler F and Stampfer C 2017 *Phys. Rev.* B **96** 205407

[60] Stratton J A 1941 *Electromagnetic Theory* (New York: McGraw-Hill) section 1.11

[61] Ferry D K 2019 *Semiconductors: Bonds and Bands* 2nd edn (Bristol: IOP Publishing) ch 9

[62] Kline M and Kay I W 1965 *Electromagnetic Theory and Geometrical Optics* (New York: Wiley Interscience)

IOP Publishing

Transport in Semiconductor Mesoscopic Devices
(Second Edition)

David K Ferry

Chapter 4

Layered compounds

Since a single layer of graphene was first isolated and studied, it has been known to have some remarkable properties, primarily because of its Dirac-like linear bands with zero gap, which has led to new physics [1]. Graphene is a single layer of carbon atoms arranged in a hexagonal honeycomb structure, and is obtained by extracting such a single layer from bulk graphite, which is a layered compound. Earlier, there was interest in carbon nanotubes (CNTs), and this interest still remains, but the nanotube has been recognized as a rolled up layer of graphene. If we use a single layer of graphene, we obtain a single-walled nanotube; multi-layers of graphene give multi- walled nanotubes. Because of the promise of graphene, other layered materials, such as the transition-metal dichalcogenides (TMDC), have come back into the spotlight, and it has been recognized that these materials have considerable promise for electronic applications. In this chapter, we will examine graphene, CNTs, and some of the newer interest layered compounds.

4.1 Graphene

As we remarked above, graphene is a single layer of carbon atoms arranged in a hexagonal honeycomb structure. The proper unit cell contains two atoms per unit cell, which are denoted as atom A and atom B, as shown with two different colors in figure 4.1(a). This leads to an equivalent Brillouin zone with two minima at K and K', as shown in figure 4.1(b). These latter two points are known as the Dirac points for reasons that become clear below. The C–C nearest-neighbor distance is $a = 0.142$ nm, but the length of the primitive lattice vector shown in the figure is 0.246 nm (the two vectors are shown in the figure and have lengths $a(3, \pm\sqrt{3})/2$, while the reciprocal lattice vectors are $2\pi(1, \pm\sqrt{3})/3a$ in the x–y coordinates). Carbon atoms possess four valence electrons. Three of these form tight in-plane bonds, known as σ bonds, with three neighboring atoms in the graphene plane. The fourth bond is the

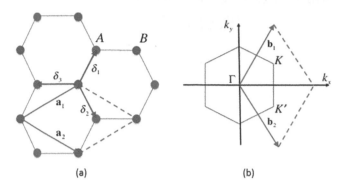

Figure 4.1. The lattice (a) and Brillouin zone (b) of graphene. The δ_i are nearest-neighbor vectors whose length is the distance $a = 0.142$ nm. The lattice vectors a_i and reciprocal lattice vectors b_i are also indicated.

Figure 4.2. The conduction (reddish) and valence (blue) bands of graphene. They have zero gap at the points K and K', identified in figure 4.1(b). There are six of these points around the hexagon.

p_z orbital, which is oriented normal to the lattice plane. The σ bonds form a deep valence band which will not play a role in the conductivity. The p_z orbitals form the π band and this constitutes the band of interest. The nature of graphene is such that the most common method of computing the band structure is simple tight-binding, in which the nearest-neighbor interaction is dominant, and is usually denoted by the general matrix element γ_0. Each atom has three nearest neighbors of the opposite type and six second neighbors of the same type, and the tight-binding formulation leads to [2]

$$E(k) = \pm \gamma_0 \sqrt{3 + 2\cos\left(\sqrt{3}\,k_y a\right) + 4\cos\left(\frac{\sqrt{3}}{2}k_y a\right)\cos\left(\frac{3k_x a}{2}\right)}. \tag{4.1}$$

This band structure is shown in figure 4.2. At the K and K' points the bands touch. Expanding around these points, for small wave vectors away from them, we find that the energies are approximately [1]

$$E = \pm \frac{3\gamma_0 a}{2}k. \tag{4.2}$$

The bands are linear bands; that is, they vary linearly with the wave vector k, as opposed to the quadratic behavior found in normal semiconductors. Moreover, these bands are chiral, in that the positive slope band has positive helicity and the negative slope band has negative helicity. The helicity arises from the pseudo-spin describing the two atomic contributions to the wave function, so that the wave function is a two component (spinor) entity. One component has a phase shift relative to the other which leads to the helicity; opposite helicities have opposite signs of this phase shift.

We can see a little more of the unusual nature of graphene when we note that these linear bands are Dirac-like in that we can write the energy

$$E = \pm\hbar v_F k, \tag{4.3}$$

as discussed in section 2.2.2 for the density of states. Here, the Fermi velocity, or effective 'speed of light', is given as

$$v_F = \frac{3\gamma_0 a}{2\hbar} \sim 8 \times 10^7 \text{ cm s}^{-1}. \tag{4.4}$$

This is obtained by fitting to angle-resolved photo-emission data [3]. Because of the zero gap, the carriers in graphene are often referred to as 'massless chiral Dirac fermions'. This connotation recognizes that there is no rest mass contribution to the Dirac-like bands, that the two atomic contributions to the wave functions impose chirality on these wave functions, and that they are indeed fermions. While the rest mass vanishes, this should not be construed as them being massless particles. Indeed, from equation (4.3), we can immediately determine the effective mass of, e.g., the electrons as [4]

$$\frac{1}{m^*} = \frac{1}{\hbar^2 k}\frac{\partial E}{\partial k} = \frac{v_F}{\hbar k}, \quad m^* = \frac{\hbar k}{v_F}. \tag{4.5}$$

This mass, and this band structure, give a different energy dependence for the normal density of states in two dimensions. The proper value was found in section 2.6, and for graphene, the number of carriers per unit area, per unit energy, is given as

$$\rho_2(E) = \frac{E}{\pi(\hbar v_F)^2}. \tag{4.6}$$

While the density of states and the effective mass both vanish as the energy moves to the Dirac point (where the two bands touch at the K and K' points), the density is not observed to vanish, although this is what one expects. Rather, it is found that a random potential exists in graphene which leads to electron–hole 'puddles' forming at energies near the Dirac point. Zhang et al [5] used STM to probe the local potential in graphene, and demonstrated the existence of these electron and hole 'puddles' near the Dirac point. These puddles were shown to be related to the impurities external to the graphene sheet. Gibertini et al [6] estimate, from their simulations, that the size of the puddles is a few nanometers. Deshpande et al [7] have used scanning tunneling spectroscopy, finding that the fluctuations of the surface topography show that puddle-like regions are of the order of 5–7 nm in extent. The simulations of Rossi and Das Sarma [8] suggest a similar size range for the puddles.

A typical conductivity curve for graphene is shown in figure 4.3, taken at room temperature. The graphene sheet was extracted from highly oriented pyrolytic graphite by mechanical exfoliation using the standard sticky tape approach. It was then deposited on an oxidized Si wafer with 290 nm of thermal oxide. Standard optical lithography was used to define the source and drain contacts, and the conductivity was then measured as the bias applied to a contact on the reverse side of the Si wafer was varied [9]. A constant 10 mV source–drain bias was applied. The variation of the back gate voltage is coupled to the density in the graphene sheet by the capacitance between the back gate electrode and the graphene. This variation can then sweep the Fermi energy throughout the graphene bands. There is a broad minimum around 45–50 V on the back gate, which is normally assumed to indicate the region where the Dirac point is located. The shift away from zero gate voltage is due to acceptor charges located either in the oxide or on the oxide surface under the graphene. The solid curve in the figure is a theoretical fit to the data which assumes that there are about 3.4×10^{12} cm^{-2} ionized acceptors, so that a positive gate bias is necessary to eliminate the associated holes in the graphene [10]. The fact that the conductivity does not go to zero at the Dirac point is direct evidence for the puddling discussed in the previous paragraph. Generally, the conductivity tends to rise linearly with gate voltage as one moves away from the Dirac point, which is mainly a result of the capacitive nature of the effect of the gate bias, where the density will increase linearly with the gate voltage. The linear rise of the conductivity then means that the mobility is relatively constant, at least within this density range.

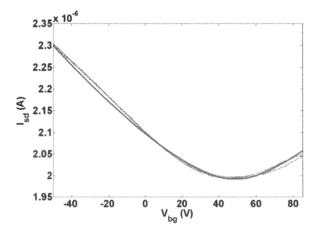

Figure 4.3. Conductivity in a sheet of graphene placed upon an oxidized Si wafer, with 10 mV bias applied. The minimum in the conductivity signals the location of the Dirac point. (Reprinted with permission from [9]. Copyright 2009 the American Vacuum Society.)

The conductivity and the mobility of the carriers in graphene are dominated by the various phonons in graphene [10], as well as by the Coulomb scattering from the impurities and the flexural modes of the rippled graphene sheet [11]. In addition, there can be scattering from point defects (short-range scatterers) [12, 13] as well as corrugations [14] and steps in the thin layer [15]. Very high values of mobility have been reported for free-standing graphene sheets, especially at low temperatures. However, when the graphene is placed on a substrate such as an oxidized Si wafer, or SiC or BN, the mobility is not found to be more than a few thousand $(cm^2(Vs)^{-1})$ at room temperature [16–20]. The variety of different types of non-intrinsic scattering mechanisms can make it difficult to understand individual mobility measurements.

Most experimental studies of graphene transport will utilize ribbons of the material, which are small slices of a nearly uniform width. When graphene is patterned into a desired width, there are two main terminations, which are referred to as the arm-chair and zig-zag edges. These two edges are illustrated in figure 4.4. Energetically, there is a difference in these two edges. Within the tight-binding formulation of the energy bands discussed above, the zig-zag edges will always retain their zero energy gap, and are termed metallic edges. The arm-chair edges will be different, however, and can have either metallic edges with zero gap or semiconductor edges in which a small energy gap opens. It has been shown that this gap will open except when the number of atoms in the ribbon width is given by $3p + 2$, where p is an integer [21]. The tight-binding formulation tends to keep the atoms in their perfect atomic alignment, but this is not how edges and surfaces of three-dimensional material behave. It is energetically favorable for a gap to open, as this lowers the energy of the electrons at the top of the valence band, and this gap usually is created by a relaxation or reconstruction of the atomic structure at the edge or bulk surface [22]. Relaxation occurs when the atoms move without changing the edge (or surface) unit cell, while reconstruction changes the unit cell. These changes can be calculated by introducing a set of molecular dynamics forces between the atoms, where these forces are computed from the band structure, usually by a technique developed by Feynman and Hellman [23, 24]. In general, this approach is more easily accomplished through the use of pseudopotentials to calculate the band

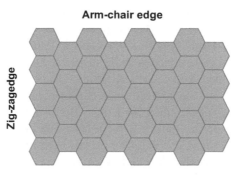

Figure 4.4. A small section of graphene illustrating the arm-chair (top and bottom) and zig-zag (left and right) edges for perfectly oriented material.

structure [4]. In any case, it is found that when this effect is accounted for, a gap always opens in the bands for a ribbon. It has been found, for example, that for armchair ribbons, the size of the gap varies with $E_{G,3p+1} > E_{G,3p} > E_{G,3p+2}$ [25]. Here, the smallest gap corresponds to the ribbon that would have preferred to remain metallic. In most materials, however, the formation of the ribbon does not produce perfect edges, and one does not have either of the two arising for the aligned lattice shown in the figure. Just as one can have voids and five- and seven-member rings within the lattice, these can occur at the edge as well [26].

As mentioned above, the zig-zag edges remain metallic, but this is not a simple conclusion. In fact, the metallic states are located at the edges of the ribbon, and in this region, the bands are almost flat so that the density of states is large [27]. The corresponding wave functions are thus localized at the ribbon edges. It is also found that the width (into the ribbon) increases as the ribbon is made wider. It is thought that the edge states in these zig-zag ribbons do not derive from either bulk graphite nor from the dangling bonds. Rather, they seem to be a general property of π electron networks with a zig-zag edge [27].

Creating constrictions in graphene with which to observe conductance quantization is a nebulous process. The reason is the lack of a band gap in the bulk graphene. Normally, the Schottky gate depletes the electron gas by pushing that region into the band gap of the material. However, in graphene, the gate merely replaces electrons by holes, or vice versa. This leads to tunneling right through the voltage-induced barrier by Klein tunneling [28] (appendix D), a process which has apparently been experimentally observed in graphene [29, 30]. Moreover, the mobility in graphene laid on most substrates is quite low, even at low temperatures, which tends to preclude the existence of the quasi-ballistic transport needed to see the coherent conductance quantization. Nevertheless, attempts have been made to make gate defined constrictions [31], but the evidence of conductance quantization is limited. Another approach is to use bilayer graphene, as an electric field applied vertically between the two layers can open a gap in the energy spectrum of both layers. This allows one to try to create actual depletion layers more effectively, and gate controlled constrictions have been created effectively with this approach [32–34]. Confinement and quantum dot behavior have been observed, but the mobility issue has not been overcome. The use of suspended graphene is more difficult, but can overcome the mobility problem. In one such approach [35], a polymer was placed between the oxidized Si wafer and the graphene sheet. In this case, the oxide was about 500 nm thick, and the polymer was about 1 μm thick. The polymer was removed in areas where the graphene is to be suspended and the sheet is patterned by lithographic methods. This forms a constriction of the order of 250–280 nm wide, as estimated electrically. Then, the conductance can be measured as the back gate bias is varied to change the density in the graphene ribbon. Interestingly, the conductance seemed to be quantized at the values $2Ne^2/h$ (N is an integer), whereas twice this value would be expected for graphene. Quantization in multiples of $2e^2/h$ were seen in chapter 2 for GaAs, and other III–V materials, where there is only spin degeneracy to worry about. However, graphene has the valley degeneracy between the K and K' points, and this should produce an extra factor of two, although it is known that a magnetic field can lift both degeneracies [36, 37]. It has been conjectured that the states in a nanoribbon will hybridize the two valley wave functions [38], this may be the process at work here to lead to the reduced value of the conductance steps.

The AB effect has been observed in graphene rings [39]. Here, we illustrate more recent work [40], where the graphene flake was deposited on an oxidized Si wafer in which the oxide was 295 nm thick. Then, electron-beam lithography and reactive-ion etching were used to pattern the graphene ring, shown in figure 4.5(a). Raman scattering was used to provide the evidence that the graphene was a single layer, with the characteristic G and $2D$ lines (shown in figure 4.5(b)) providing the evidence. The resistance of the flake is shown in figure 4.5(c) as a function of the back gate (Si layer) voltage. The ring had an inner radius of about 200 nm and an outer radius of about 350 nm, as can be observed in the image of the ring. The leads to the ring are graphene ribbons of approximately 150 nm width. In figure 4.6(a), the four-terminal resistance across the graphene ring is shown as a function of the magnetic field. This resistance consists of several parts, including the two leads and the ring itself. This resistance was obtained at a back gate voltage of -5.8 V, which corresponds to a hole density of 1.2×10^{12} cm^{-2}. The AB oscillation signal is obtained by subtracting the background resistance, and this is shown in figure 4.6(b). This signal has a period of about 17.9 mT, as indicated by the vertical lines. The oscillatory signal was Fourier transformed, with the spectrum shown in figure 4.6(c). There is a dominant single peak around 60 mT^{-1} which corresponds reasonably well to the periodicity seen in the oscillatory signal. It is interesting that the width of this peak is narrower than might be expected from the size of the ring, shown as the gray shaded region in this last figure. This suggests that the electrical width of the ring is less than the physical width, which is not unusual.

Weak localization is also seen in graphene [41] (see figure 4.7). Here, we discuss measurements made on high quality epitaxial graphene grown on the silicon face of

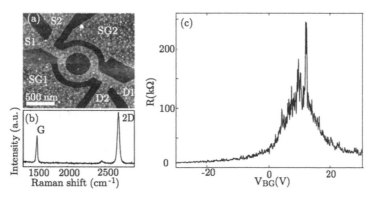

Figure 4.5. (a) A scanning-force micrograph of the patterned graphene layer, showing the ring and its leads and side gates. (b) The Raman spectrum observed from the same flake prior to patterning, establishing the single-layer nature of the material. (c) Four-terminal resistance of the ring structure with back gate bias at 0.5 K. (Reprinted with permission from [40]. Copyright 2010 IOP Publishing and Deutsche Physikalische Gesellschaft.)

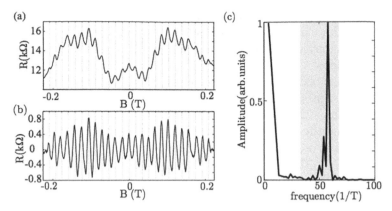

Figure 4.6. (a) Four-terminal resistance measurement across the ring as a function of magnetic field. (b) Oscillatory signal after the background has been removed. The lines indicate a rough period of 17.9 mT. (c) The Fourier transform of the oscillatory signal. The main peak near 60 mT−1 is much narrower than that expected from the size of the ring (gray shaded area). (Reprinted with permission from [40]. Copyright 2010 IOP Publishing and Deutsche Physikalische Gesellschaft.)

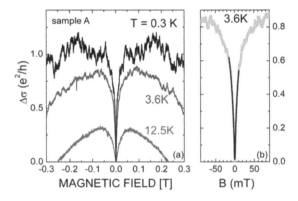

Figure 4.7. (a) Magnetoconductance at three temperatures. (b) A fit (black curve) to the weak localization signal. (Reprinted with permission from [42]. Copyright 2011 IOP Publishing and Deutsche Physikalische Gesellschaft.)

6H–SiC substrates [42]. Conventional photolithography was used to pattern the Hall-bar structures used in the experiments. The sample had an electron density of 6×10^{11} cm^{-2} and a mobility of about 10 000 cm^2 V^{-1}s^{-1} at low temperature. In figure 4.7, the relative conductivity $\Delta\sigma(B) = \sigma_{xx}(B) - \sigma_{xx}(0)$ is shown at three selected temperatures [42]. The conductivity is deduced from the actual resistivity measured for a constant current through the sample. Weak localization is observed at all three temperatures. In order to observe the effect, a very slow sweep of the magnetic field had to be employed, basically of the order of 100 min per tesla. Any faster sweep rate in magnetic field was found to reduce the amplitude of the weak localization peak. In panel (b) of the figure, the peak around zero field has been fit to an analytic form described by McCann *et al* [43], which is a modification of that

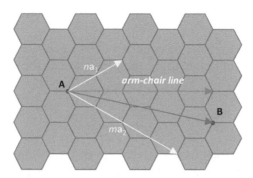

Figure 4.8. The rolling vector is $\mathbf{R} = \mathbf{r}_A - \mathbf{r}_B$, which in turn is defined as multiples of the unit vectors (see figure 4.1) according to (4.7). The wrapping angle is the angle between this vector and the arm-chair line.

given in the last chapter which is felt to be more appropriate to graphene. This allowed the authors to extract a phase-breaking time which was temperature-independent below about 1–1.5 K, and then decreased as $T^{-1/2}$ at higher temperatures. This contributed to a phase-breaking length of just over 1 μm at low temperature.

4.2 Carbon nanotubes

Another interesting material that has been around longer than graphene is the CNT, having been first discovered in 1991 [44]. But its properties derive from those of graphene. If we take the sheet of graphene that is depicted in figure 4.4, and roll it into a cylinder, then we obtain a CNT. If we use a single layer of graphene, then we obtain a single-walled CNT. But, if we use multiple layers of graphene, then we obtain a multi-walled CNT. Like a sheet of paper, however, there are many ways in which to roll up the sheet of graphene, so we need a method of characterizing just how the nanotube is rolled. We start to define this method by picking an arbitrary point on the graphene lattice, and we denote this as point A, as shown in figure 4.8. In figure 4.1, the lattice vectors for the graphene lattice were defined as extending, for example, from one A type atom to the nearest other A type atoms on the hexagon (the nearest neighbors are B type atoms). Hence, we can define multiples of these lattice vectors using the integers n and m as shown in figure 4.8. Suppose we roll the graphene sheet so that point A goes to point B (admittedly, this is a very tightly rolled tube, but it is the example that is important here). We can define the rolling vector, or chiral vector as

$$\mathbf{R} = \mathbf{r}_A - \mathbf{r}_B = n\mathbf{a}_1 + m\mathbf{a}_2 \to 2\mathbf{a}_1 + 4\mathbf{a}_2. \tag{4.7}$$

where the last form is for the particular case shown in figure 4.8. The axis of the CNT is perpendicular to the rolling vector, so it would be a vector normal to $\mathbf{R}$. Since it is usually impossible to count the number of unit cells in a large sheet of graphene, or a large CNT, another useful quantity is the wrapping angle. We define this wrapping angle as the angle between the arm-chair line (that defined by either lattice vector) and the chiral vector $\mathbf{R}$, as [45]

$$\varphi = \angle_{\text{armchair}}^{\mathbf{R}}. \tag{4.8}$$

The arm-chair vector is the case of equation (4.7) when $n = m$. Hence, we can find the cosine of the wrapping angle using the properties of the two vectors, as defined by n and m, to show that

$$\cos(\varphi) = \frac{2n + m}{2\sqrt{n^2 + m^2 + nm}}. \tag{4.9}$$

There is, of course, an ambiguity about the sign of the wrapping angle, but a proper choice follows from the symmetry of the tube. If you assign one sign of the angle, just interchanging the two ends of the CNT changes the sign. Since the physics of the CNT does not depend upon the orientation of it, the sign is thus not particularly important until we apply something like a magnetic field which will break the symmetry.

Just as the width of a graphene ribbon introduces some changes to the band structure of the flake, the length of the rolling vector also introduces some changes. In computing the density of states in chapter 2, the periodicity of the lattice was important in setting the values of the momentum wave vectors. The rolling vector has the same effect in a CNT, as this defines the periodicity of the structure (as we move around the circumference of the tube). The crucial factor here is the difference between n and m. If $n - m = 3p$, where p is any integer (or zero), then the CNT will retain the graphene band structure and have a zero energy gap between the conduction and valence bands. These tubes will then be metallic. If $n - m \neq 3p$, then the CNT will have a gap between the two bands and will be a semiconducting tube. Nevertheless, the density of states for the CNT will have the peculiar behavior of a Q1D conductor given by equation (2.29). However, one has to distinguish here whether or not we are dealing with a metallic tube or a semiconducting tube, as we have to be careful with the conversion from wave vector to energy. We note that the derivation of equation (2.29) involves the fact that the density of states in one dimension is given as

$$\rho_1 = \frac{1}{\pi}\frac{dk}{dE}, \tag{4.10}$$

which includes a factor of two for spin, but no extra term for valley degeneracy has been added yet. So, if we have a metallic tube which retains the graphene band structure, we have the valley degeneracy and the Dirac bands, which lead to

$$\rho_{1,\text{metallic}} = \frac{2}{\pi \hbar v_F} = \frac{2k}{\pi E}. \tag{4.11}$$

On the other hand, if the gap is opened, then we have semiconducting behavior with an easily determined effective mass, and we obtain equation (2.29)

$$\rho_{1,\text{semicon.}} = \frac{1}{\pi \hbar}\sqrt{\frac{2m^*}{E}}, \tag{4.12}$$

although a factor of two has been added for valley degeneracy. In both cases, there is a divergence of the density of states as one approaches the band edge. In the semiconducting tubes, the gap depends upon the diameter of the CNT, with the latter given by

$$d = \frac{|\mathbf{a_1}|}{\pi}\sqrt{n^2 + m^2 + nm}, \tag{4.13}$$

with the length of $\mathbf{a_1}$ given as 0.246 nanometers. If either n or m are zero, then the CNT is called a zig-zag nanotube. As noted above, if $n = m$, then the CNT is called an arm-chair tube.

In a metallic tube, the quantization of the wave number discussed above will lead to the condition that [45]

$$\Delta k R = 2\pi s, \tag{4.14}$$

where R is the magnitude of the chiral vector equation (4.7) and s is an integer, which defines the particular band, as it derives from the graphene band structure. Each value of s defines a line (in one dimension) of allowed k vectors that contributes to one occupied π-band. In semiconducting tubes, we have a similar behavior except that there is a gap opening. We can redefine the above condition on n and m to be

$$n - m = 3p + \nu. \tag{4.15}$$

If $\nu = 0$, then we have a metallic tube. But, if ν is ± 1, then we have a semiconducting tube. The integer p is then related to the band index s, and we can generalize equation (4.14) to handle the semiconducting tubes as well by using

$$k_s R = 2\pi\left(s + \frac{\nu}{3}\right) \rightarrow \frac{2}{d}\left(s + \frac{\nu}{3}\right), \tag{4.16}$$

where, as before, d is the tube diameter. Now, the wave number k is the momentum vector for motion around the tube. We recall that the presence of a magnetic field changes the momentum, described by the Peierl's substitution in earlier chapters. Indeed, the Hamiltonian is modified by the vector potential according to this substitution, and this leads to a gauge variation that applies a phase shift to the wave function. Because we have applied additional quantization to the wave vector in creating the tube, this magnetic field will cause a variation in this quantization. Hence, a magnetic field, applied along the tube axis will lead to a modification of equation (4.16) to account for this AB phase, as [46]

$$k_\perp \rightarrow k_\perp(\varphi) = \frac{2}{d}\left(s + \frac{\nu}{3} + \frac{\Phi}{\Phi_0}\right), \tag{4.17}$$

where $\Phi = B_\parallel \pi d^2/4$ is the flux enclosed by the tube and $\Phi_0 = h/e$ is the quantum flux unit as before. Thus, we see that the magnetic field modulates the energy bands of the CNT [47]. The magnetic field can change a metallic tube to a semiconducting tube, and vice versa. Varying the magnetic field gives this transition as a function of the field. To see this effect, however, requires enormous magnetic fields due to the extremely small cross-section of a CNT. For example, a 10 nm diameter CNT would require more than 50 T to complete one cycle of the modulation, but this is a doable field as we will see.

The first experimental search for AB oscillations in a CNT was apparently by Bachtold *et al* [48], and these authors did find magnetic modulation of the conductivity when the magnetic field was aligned with the tube axis. But, these authors found a magnetic period of $h/2e$, more related to the disorder-induced AAS behavior of rings discussed in the last chapter [49]. The outer current-carrying ring (of a multi-walled CNT) was measured to have a radius of approximately 8 nm by atomic force microscopy (AFM), and the $h/2e$ behavior gave an inferred radius of approximately 8.6 nm. Several more recent studies continued to find these AAS oscillations [50, 51]. The first clear evidence of the AB effect was apparently the work of Coskun *et al* [52], who observed it in CNT quantum dots. In this latter case, the authors used a multi-walled CNT with an outer radius of about 15 nm, but placed the CNT on top of the metallic contacts; this tends to produce tunneling barriers so that the interior of the CNT was effectively an isolated quantum dot. A back-contact gate could be used to modulate the CNT conductance, and the authors observed a modulation of the quantum dot conductance diamonds (we discuss these in a later chapter) with the magnetic field, which had the proper h/e periodicity. At about the same time, another group used a single-walled CNT of about 1 nm radius, and saw some signatures of the AB effect, but could not see oscillations in this small diameter tube [53].

As indicated above, it is useful to have a very high magnetic field available for these studies. Lassagne *et al* [54, 55] were able to bring a 55 T magnet to the study of a multi-walled CNT 10 nm in diameter. The tubes were placed on an oxidized Si wafer so that the voltage applied to the Si would serve as a back gate. Pd metallic electrodes, spaced by 200 nm, were deposited on top of the CNT, which usually produces ohmic contacts. Nevertheless, some evidence of Schottky barrier behavior was observed. Experiments suggested that their tubes were in the ballistic regime so that good phase coherence could be expected. In figure 4.9, we display the conductance as a function of the back gate voltage and the magnetic field. The data suggest an oscillatory behavior with a period of 48 T, which is strong evidence of the h/e modulation, as the 10 nm diameter tube would be expected to show a period of 50 T, as discussed above. Subsequent experiments [55] used an 18 nm diameter CNT with a contact spacing of 150 nm, so that more periods of the oscillation could be observed in the available 55 T. The measured conductance is shown in figure 4.10 for several values of the back gate voltage. Here, it is clear that good h/e oscillations are seen in the data, even at the elevated temperature of 100 K. These oscillations persist over basically all the values of the gate voltage, even though the peak positions do shift with the bias, as expected for modulation of the energy bands by the magnetic field. In all of these experiments, it is evident that the dominant conduction is in the outer shell of a multi-wall CNT.

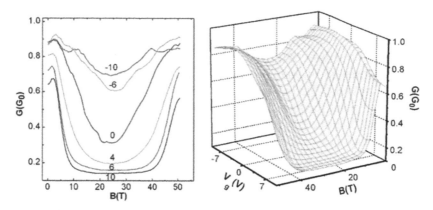

Figure 4.9. Left panel: the magnetoconductance at 100 K for several values of gate voltage as a function of the magnetic field. At 10 V, the tube is basically biased to a point with the Fermi level in the gap, and the magnetic field produces the necessary change to the bands to give the h/e modulation. As discussed in the text, 50 T is just about enough to produce one complete oscillation for this 10 nm tube. The right panel is a 3D representation of the conductance. (Reprinted with permission from [54]. Copyright 2007 the American Physical Society.)

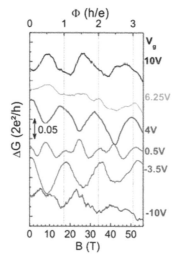

Figure 4.10. Magnetoconductance observed in an 18 nm diameter multi-walled CNT at 100 K, for various back gate voltages. The curves have been offset for clarity. (Reproduced with permission from [55]. Copyright 2009 Elsevier Masson SAS.)

Weak localization is also seen in the CNT, especially when the magnetic field is oriented normal to the tube axis [50, 56]. Because of the complicated periodic band structure of the CNT, however, the peak in resistance does not always occur at $B = 0$, as it can depend upon gate bias and other properties of the CNT.

4.3 Topological insulators

Topological insulators are materials in which a surface or interface provides a localized energy structure that has the Dirac-like bands of graphene [57], but generated with some additional properties. One prototypical material system is a heterostructure between HgTe and CdTe. In most semiconductors with a direct band gap at the Γ point (center of the Brillouin zone), the bottom of the conduction band is composed of atomic S orbitals from the cations, and often denoted as the Γ_6 or Γ_1 (the former is the so-called double-group notation used when the spin–orbit interaction is included) band. On the other hand, the top of the valence band is usually composed of the anion P orbitals, and denoted as the Γ_8 or Γ_{15} band. In HgTe, these two roles are reversed, so that the Γ_6 band lies below the Γ_8. Hence, HgTe is often referred to as having a negative band gap. In the interface between these two materials, these bands must cross as they reverse their roles from HgTe to CdTe. Now, this property has been known for quite some time as the HgTe/CdTe superlattice band structure was studied at least as early as 1979 [58, 59]. But, for a topological insulator, one wants more to assure that the zero gap and its properties are topologically protected from disorder. In figure 4.11, we draw schematically how the interface bands extend from the bulk bands to provide the Dirac-like bands. One would like the bands to be such that, for example, one had spin up and the other spin down. In a topological insulator, the spin of one branch is locked at a right angle to their momentum (termed spin-momentum locking), so that carriers in the other branch have different spin and back-scattering is then forbidden. Time-reversal

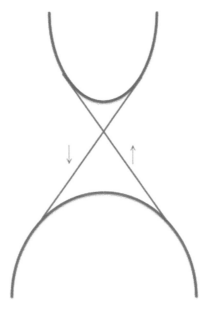

Figure 4.11. The bulk bands are shown in blue and characterize a normal semiconductor. The surface bands in red lead to the topological insulator on the surface. The nature of the spin and time-reversal symmetry leads to the effect.

symmetry was predicted to lead to these type of edge states in quantum wells of HgTe placed between layers of CdTe in 1987 [60], and it was experimentally observed in 2007 [61]. More recently, this type of surface band structure was predicted to occur in three-dimensional materials such as some Bi compounds [62], although the general basis for the topological insulator seems to have been put forward more generally in 2005 [63, 64]. The first experimental evidence for a three-dimensional topological insulator was in bismuth antimonide, in 2008 [65].

The interest here, of course, is in mesoscopic effects as they may occur in these materials. And, it is clear that the AB effect can be observed in topological insulators [66], and has been seen in Bi_2Se_3 nanoribbons [67]. Bi_2Se_3 is a layered compound with a rhombohedral phase, but with covalent bonding in the layer and weak van der Waals bonding between the layers, much like a nanoribbon, as it is on these surfaces that the protected state exists (the bulk is an insulator) [67]. The authors plotted the magnetic field position of each resistance minimum as a function of its index (a counting of the position away from $B = 0$). The linearity of this curve, along with the index as a multiple of h/e supports the interpretation in terms of the AB effect. The Fourier transform of the resistance trace illustrated a single dominant AB peak. An interesting observation was the presence of weak anti-localization at $B = 0$, as the resistance shows a drop instead of the peak expected for weak localization [67]. We recall from the previous chapter that weak localization occurs when the electrons can move around two time-reversed paths and constructively interfere with back-scattering. However, when the spin–orbit interaction is important, as it is in the topological insulators, the spin is coupled to the momentum, as mentioned above. Hence the spins of the carriers in the two time-reversed paths are opposite to each other. As a result, the two paths interfere in such a manner that it results in a reduction in the resistance, and weak anti-localization [68]. The magnetic field still leads to breaking the time-reversal symmetry and rapid decay of the signal.

More recently, a field-effect transistor has been fabricated with Bi_2Se_3 for which the AB effect has been observed at low temperature [69]. In this latter case, the AB effect was somewhat anomalous in that the AB period corresponded to a minimum in the magnetoresistance, rather than a maximum, presumably due to the spin–orbit interaction. We see the effect in figure 4.12. Panels (a) and (b) show the transistor action at various magnetic fields. There is an obvious turn-on of the transistor at a particular gate characteristic as well as no real magnetic field modulation of the output current. Panel (c) shows the presence of the AB oscillations as the magnetic field is swept from −9 to 9 T, as well as a strong weak localization peak (note the increase in resistance at zero magnetic field. Finally, panel (d) shows the Fourier transform illustrating a clear h/e period.

A variation of the topological insulator is the topological crystalline insulator (TCI) which refers to materials in which the crystalline symmetry leads to topologically protected surface states with a chiral spin texture (see figure 4.11). Materials such as SnTe and $Pb_{1-x}Sn_xSe$ or $Pb_{1-x}Sn_xTe$ grown along the (001) direction develop non-trivial surface states [70, 71]. These protected states can even form at step edges of the grown layer [72]. Indeed, enhanced current has been

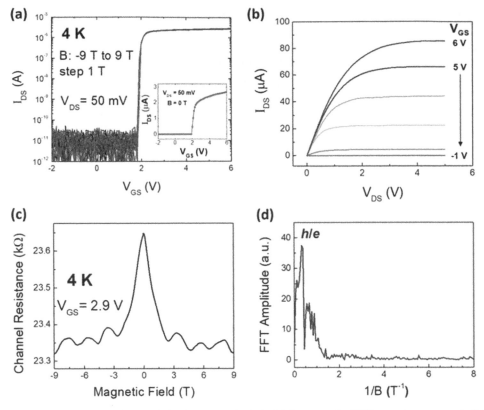

Figure 4.12. (a) Transfer characteristics ($I_{DS} - V_{GS}$) of the Bi_2Se_3 nanowire FET at various magnetic field intensities ranging from −9 T to 9 T. V_{DS} was maintained at 50 mV. Inset: The linear-scale transfer characteristics. (b) Output characteristics ($I_{DS} - V_{DS}$) of the Bi_2Se_3 nanowire FET under different V_{GS}. (c) Oscillatory features of the magnetoresistance obtained by sweeping the magnetic field between ±9 T. (d) FFT analysis showing the amplitude as a function of 1/B. The location of h/e was labeled. Reprinted with permission from [69], copyright 2019 by AIP Publishing.

observed at step edges in epitaxially grown Bi_2Te_3 [73]. Since this effect is not observed at step edges of graphite, the authors suggest that there may be a possible interaction between the spin–orbit coupling and the topological nature of the edge states.

4.4 The metal chalcogenides

Following the rapid increase of interest in graphene, people began to wonder if there were layered compounds in which a single layer could be exfoliated and which had a real band gap. The need for a band gap, of course, arises from the desire to incorporate these materials into active semiconductor devices, and that meant a need to be able to turn the device off, something that cannot be easily done with graphene. It turns out that there are some materials which have these properties, and these are the transition metal dichalcogenides (TMDC). The chalcogenides have a large number of stoichiometries and phases, but the best known for transport purposes are the dichalcogenides of the form MX_2, where M is a transition metal, typically Mo or Ta, and X is usually S, Se, or Te. The compounds with Mo or W are semiconductors with a band gap usually near or larger than that of Si. However, the Nb and Ta based compounds are usually metals. These materials form in a layered compound, where each layer is typically a layer of the transition metal with a layer of the chalcogenide both above and below the metal layer [74]. The layer to layer bonding is weak, as it is in graphene, so that individual layers can be removed by exfoliation, either mechanically or via a liquid-based procedure, and then placed on a convenient substrate.

These materials actually have a rather longer history than graphene, as they were pursued heavily in the mid-1970s and later in a search for charge density waves. The understanding of this phenomenon begins with the Peierls distortion, in which a one-dimensional chain of atoms is actually unstable. Peierls showed that a commensurate, periodic distortion of the chain that coincided with the Fermi surface would lower the overall energy of the electron gas [75]. A metallic one-dimensional material in which there is one electron per atom fills the band up to $\pi/2a$, which leads to a natural factor of two distortion. Here, two neighboring atoms move slightly closer together while having a slightly larger distance from their neighboring pairs. It was later shown that even Q2D materials could undergo this distortion. If a distortion momentum vector could be found that spanned the Fermi surface, and which was also some integer multiple of the lattice vectors, then a charge density wave could be formed. Here, the two opposite wave vectors lead to a standing density wave composed of two counter propagating electron waves [76]. This couples to the Peierls distortion of the lattice. Most interestingly, it is the TMDCs that were suggested as being the proper type of layered compound in which the charge density wave could be observed [77]. One of the earliest observations of the existence of the charge density wave was in TaS_2 [78]. The electron density of the charge density wave in TaS_2 and $TaSe_2$ has been directly imaged by the use of an STM [79].

As an example, let us look at MoS_2, which is a layered compound in where each layer is composed of Mo atoms at the center of the layer and S atoms displaced above and below this center. In bulk form, it has an indirect band gap, which becomes direct only in the monolayer limit [80, 81]. The nature of the three atoms per unit cell leads to the lack of inversion symmetry in the monolayer. When a second monolayer is added to the first, it is reversed in direction of the unit cell so that inversion symmetry is restored. Hence, the bulk is actually stacks of bi-layers. The band extrema of the monolayers are located at the K and K' points of the Brillouin zone, so that there are two minima for the conduction and valence band. Usually, by the time the second monolayer is placed on the first, the minimum of the conduction band moves to the T (or Q) valleys, which lie midway between Γ and K, and the material is indirect. In the monolayer subsidiary valleys of the conduction arise from the residual valleys of what was the indirect gap. These valleys, referred to as the T valleys [82]. The conduction band mass is approximately $0.45m_0$ in the K valleys and $0.57m_0$ in the subsidiary valley [81]. These bands are non-parabolic, and this has to be taken into account for transport. With certain limitations, the material parameters, phonon energies, and coupling constants are all given in the work of [80].

One natural question is about the polar LO modes which one would expect in this material, due to the dissimilar atoms. These have been claimed to exist [80], but there are some anomalies with them. First, because of the 3 atoms per unit cell, there are 9 phonon modes, so there are two LO-TO pairs. The polar interaction normally would arise from the LO–TO splitting at the Γ point, but this splitting vanishes at this point for both the LO_1–TO_1 and the LO_2–TO_2 modes [80, 83]. As the electromagnetic waves couple only at the Γ point, this lack of splitting means that these phonons do not couple to the electromagnetic wave, and thus give rise to no polarization at this point. Hence, the dielectric constant is continuous at this energy, as required by the Lyddane–Sachs–Teller relation, and the polar mode is not generated. On the other hand, the two modes are split at the K point, but this phonon momentum would be an intervalley interaction, and this is dominated by the deformation potential coupled LO mode [80]. In particular, MoS_2 and WS_2 have been studied for their electrical properties, but as of this writing, no mesoscopic studies have surfaced.

It is also possible to form trichalcogenides such as $NbSe_3$ or MoS_3. Typically, these materials form what are often called 'triangle poles', which are six chalcogenide atoms surrounding a central metal atom. The pole is a triangular prism of chalcogenide atoms, one above the metal atom and one below the metal atom [84]. These structures tend to form chains of such poles, where the poles are built up parallel to one another, although they typically have several different phases [85]. These chains tend to make the material look like whiskers. Importantly, these materials can be superconducting, and can show Q1D charge density wave behavior [86]. But, it has also been observed that these materials can form layered compounds which can be effectively exfoliated [87]. Importantly, these Q1D wires, or ribbons, can be joined to form what are called *topological crystals* (see the previous section). That is, the end of the ribbon can be attached to the beginning of the ribbon to form a loop, which can be relatively wide [88]. Interestingly, the ends can be joined with a twist of $\pm\pi$ to create a real world Möbius strip, and therefore have only a single surface, hence the name topological crystal.

As in the dichalcogenides, the trichalcogenides support charge density waves, so that the mesoscopic effects can be somewhat different. Tsubota *et al* [89] have grown TaS_3 in a ring geometry without any twist. In fact, the material grows into what may better be described as tubes, but these were sliced using a focused ion beam to produce a ring about 27 μm in diameter and with a 1×0.1 μm cross-section. Au electrodes were deposited and current passed through the ring, with a magnetic field normal to the plane of the ring, so that AB oscillations could be observed. The results of the measurements are shown in figure 4.13. AB oscillations with a period of $h/2e$ were observed in the oscillatory current (constant bias voltage is applied). The rings are large enough that the AAS disorder oscillations can be ruled out [89]. In fact, the AB oscillations are observed up to temperatures as high as 79 K. It is believed that the current induces a sliding charge density wave of soliton nature which carries an effective charge of approximately $2e$, with the measurements giving an actual effective charge of $\sim 1.9e$. The model of this behavior is that the charge density wave is a correlated electron system that can transport the electrons through

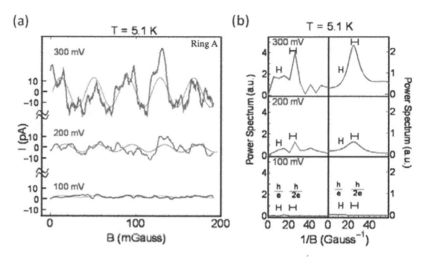

Figure 4.13. (a) Variation of the current as the magnetic field is changed for three different bias voltages, for a ring of TaSe3. (b) The power spectra of the Fourier transform by two methods: the DTF (left) and maximum entropy method (right). Adapted with permission from Tsubota *et al* [89], copyright 2012 IOP Publishing.

a Q1D material in groups, and the time correlated soliton model that follows from this is based upon coherent, Josephson-like tunneling of microscopic solitons of charge $2e$ [90].

Problems

1. Use a small k expansion around the Dirac point in graphene to show that the bands are linear in this region.
2. The conductivity of graphene is observed to increase linearly as the back gate voltage is varied (for either electrons or holes). As the back gate voltage has a capacitance effect on the graphene, the density of electrons or holes must also vary linearly with the voltage. Using the density of states for graphene, determine what the energy variation of the mobility must be to give linear conductivity changes.
3. Graphene can have phonons whose vibration lies either in the plane or normal to the plane. Explain why the phonons whose vibration is normal to the plane cannot scatter electrons in the first-order perturbation theory that is usually used.
4. For the weak localization signals of figure 4.7, use the theory of the last chapter to estimate the phase coherence time.
5. In multi-walled CNTs, it is usually found that the current is flowing mainly in the outer layer of the tube. What physical effect could cause this to be the case?
6. Explain the difference between a charge density wave and a spin density wave.

References

[1] Castro Neto A H, Guinea F, Peres N M R, Novoselov K S and Geim A K 2009 *Rev. Mod. Phys.* **81** 109
[2] Wallace P R 1947 *Phys. Rev.* **71** 622
[3] Bostwick A, Ohta T, Seyller T, Horn K and Rotenberg E 2006 *Nat. Phys.* **3** 37
[4] Ferry D K 2020 *Semiconductors: Bonds and Bands* 2nd edn (Bristol: IOP Publishing) section 2.6
[5] Zhang Y, Brar V W, Girit C, Zettl A and Crommie M F 2009 *Nat. Phys.* **5** 722
[6] Gilbertini M, Tomadin A, Guinea F, Katsnelson M I and Polini M 2012 *Phys. Rev.* B **85** 201405
[7] Deshpande A, Bao W, Miao F, Lau C N and LeRoy B J 2009 *Phys. Rev.* B **79** 205411
[8] Rossi E and das Sarma S 2011 *Phys. Rev. Lett.* **107** 155502
[9] Shishir R S, Chen F, Xia J, Tao N J and Ferry D K 2009 *J. Vac. Soc. Technol.* B **27** 2003
[10] Shishir R S, Chen F, Xia J, Tao N J and Ferry D K 2009 *J. Comp. Electron.* **8** 43
[11] Castro E V, Ochoa H and Katsnelson M I *et al* 2007 *Phys. Rev. Lett.* **105** 266601
[12] Rutter G M, Crain J N, Guisinger N P, Li T, First P N and Stroscio J A 2007 *Science* **317** 219
[13] Ni Z H *et al* 2010 *Nano Lett.* **10** 3868
[14] Katsnelson M I and Geim A K 2008 *Philos. Trans. R. Soc. London* A **366** 195
[15] Low T, Perebeinos V, Tersoff J and Avouris P 2012 *Phys. Rev. Lett.* **108** 096601
[16] Dorgan V E, Bae M-H and Pop E 2010 *Appl. Phys. Lett.* **97** 082112
[17] Lafkioti M, Drauss B, Lohmann T, Zschieschang U, Klauk H, Klitzing K v and Smet J H 2010 *Nano Lett.* **10** 1149
[18] Lin Y-M, Farmer D B, Jenkins K A, Wu Y, Tedesco J L, Myers-Ward R L, Myers-Ward C R, Gaskill D K, Dimitrakopoulos C and Avouris P 2011 *IEEE Electron Dev. Lett.* **32** 1343
[19] Robinson J A, Trumbull K A, LaBella M III, Cavalero R, Hollander M J, Zhu M, Wetherington M T, Fanton M and Snyder D W 2011 *Appl. Phys. Lett.* **98** 222109
[20] Kim E, Jain N, Jacobs-Gedrim R, Xu Y and Yu B 2012 *Nanotechnology* **23** 125706
[21] Son Y-W, Cohen M L and Louie S G 2006 *Phys. Rev. Lett.* **97** 216803
[22] Ferry D K 1991 *Semiconductors* (Reading, MA: Macmillan) section 5.10
[23] Hellmann H 1937 *Einführung in die Quantenchemie* (Leipzig: Franz Deuticke) p 285
[24] Feynman R P 1939 *Phys. Rev.* **56** 340
[25] Fischetti M V, Kim J, Narayanan S, Ong Z-Y, Sachs C, Ferry D K and Aboud S J 2013 *J. Phys. Condens. Matter* **25** 473202
[26] Terrones H, Lv R, Terrones M and Dresselhaus M S 2012 *Rep. Prog. Phys.* **75** 062501
[27] Nakad K, Fujita M, Dresselhaus G and Dresselhaus M S 1996 *Phys. Rev.* B **54** 17954
[28] Katsnelson M I, Novoselov K S and Geim A K 2006 *Nat. Phys.* **2** 620
[29] Gorbachev R V, Mayorov A S, Savcheko A K, Horsell D W and Guinea F 2008 *Nano Lett.* **8** 1995
[30] Standard N, Huard B and Goldhaber-Gordon D 2009 *Phys. Rev. Lett.* **102** 026807
[31] Nakaharai S, Williams J R and Marcus C M 2011 *Phys. Rev. Lett.* **107** 036602
[32] Goossens A M, Driessen S C M, Baart T A, Watanabe K, Taniguchi T and Vandersypen L M K 2012 *Nano Lett.* **12** 4656
[33] Allen M T, Martin J and Yacoby A 2012 *Nat. Commun.* **3** 394
[34] Dröscher S, Barraud C, Watanabe K, Tanaguchi T, Ihn T and Ensslin K 2012 *New J. Phys.* **14** 103007

[35] Peres N M R, Castro Neto A H and Guinea F 2006 *Phys. Rev.* B **73** 195411
[36] Luk'yanchuk I A and Bratkowsky A M 2008 *Phys. Rev. Lett.* **100** 176404
[37] Yang Z and Han J H 2010 *Phys. Rev.* B **81** 115405
[38] Brey L and Fertig H A 2006 *Phys. Rev.* B **73** 235411
[39] Russo S, Oostinga J B, Wehenkel D, Heersche H B, Sobhani S S, Vandersypen L M K and Morpurgo A F 2008 *Phys. Rev.* B **77** 085413
[40] Huefner M, Molitor F, Jacobsen A, Pioda A, Stampfer C, Ensslin K and Ihn T 2010 *New J. Phys.* **12** 043054
[41] Morozov S V, Novoselov F, Katsnelson M I, Schedin F, Ponomarenko L A, Jiang D and Geim A K 2006 *Phys. Rev. Lett.* **97** 0168801
[42] Pan W, Ross A J III, Howell S W, Ohta T, Friedmann T A and Liang C-T 2011 *New J. Phys.* **13** 113005
[43] McCann E, Kechedzhi K, Fal'ko V I, Suzuura H, Ando T and Altshuler B L 2006 *Phys. Rev. Lett.* **97** 146805
[44] Iijima S 1991 *Nature* **354** 56
[45] Charlier J-C, Blasé X and Roche S 2007 *Rev. Mod. Phys.* **79** 677
[46] Ajiki H and Ando T 1993 *J. Phys. Soc. Japan* **62** 1255
[47] Roche S, Dresselhaus D, Dresselhaus M S and Saito R 2000 *Phys. Rev.* B **62** 16092
[48] Bachtold A, Strunk C, Salvetat J-P, Bonard J-M, Forró L, Nussbaumer T and Schönenberger C 1999 *Nature* **397** 673
[49] Alt'shuler B L, Aronov A G and Spivak B Z 1981 *JETP Lett.* **33** 94
[50] Strunk C, Stojetz B and Roche S 2006 *Semicond. Sci. Technol.* **21** 538
[51] Stojetz B, Roche S, Miko C, Triozon F, Forró L and Strunk C 2007 *New J. Phys.* **9** 56
[52] Coskun U C, Wei T-C, Vishveshwara S, Goldbart P M and Bezryadin A 2004 *Science* **304** 1132
[53] Zaric S, Ostojic G M, Kono J, Shaver J, Moore V C, Strano M S, Hauge R H, Smalley R E and Wei X 2004 *Science* **304** 1129
[54] Lassagne B, Cleuziou J-P, Nanot S, Escoffier W, Avriller R, Roche S, Forró L, Racquet B and Broto J-M 2007 *Phys. Rev. Lett.* **98** 176802
[55] Nanot S, Escoffier W, Lassagne B, Broto J-M and Raquet B 2009 *C. R. Phys.* **10** 268
[56] Langer L, Bayot V, Grivei E, Issi J-P, Heremans J P, Olk C H, Stockman L, Van Haesendonck C and Bruyneseraede Y 1996 *Phys. Rev. Lett.* **76** 479
[57] Moore J 2011 *IEEE Spectr.* **48** 38
[58] Schulman J N and McGill T C 1979 *J. Vac. Sci. Technol.* **16** 1513
[59] Schulman J N and McGill T C 1979 *J. Vac. Sci. Technol.* **17** 1118
[60] Pankratov O A, Pakhomov S V and Volkov B A 1987 *Sol. State Commun.* **61** 93 66
[61] Konig M, Wiedmann S, Brüne C, Roth A, Buhmann H, Molenkamp L W, Qi X-L and Zhang S-C 2007 *Science* **318** 7
[62] Fu L and Kane C L 2007 *Phys. Rev.* B **76** 045302
[63] Hatsugai Y 2005 *J. Phys. Soc. Japan* **74** 1374
[64] Hastings M B 2005 *Europhys. Lett.* **70** 824
[65] Hsieh D, Qian D, Wray L, Xia Y, Hor Y S, Cava R J and Hasan M Z 2008 *Nature* **452** 970
[66] Bardarson J H and Moore J E 2013 *Rep. Prog. Phys.* **76** 056501
[67] Peng H, Lai K, Kong D, Meister S, Chen Y, Qi X-L, Zhang S-C, Shen Z-X and Cui Y 2010 *Nat. Mater.* **9** 225
[68] Hikami S, Larkin A I and Nagaoka Y 1980 *Prog. Theor. Phys.* **63** 707

[69] Zhu H, Richter C A, Yu S, Ye H, Zeng M and Li Q 2019 *Appl. Phys. Lett.* **115** 073107

[70] Dziawa P *et al* 2012 *Nat. Mater.* **11** 1023

[71] Liu J, Hsieh T H, Wei P, Duan W, Moodera J and Fu L 2014 *Nat. Mater.* **13** 178

[72] Sessi P *et al* 2016 *Science* **354** 1269

[73] Macedo R J, Harrison S E, Dorofeeva T S, Harris J S and Kiehl R A 2015 *Nano Lett.* **15** 4241

[74] Wang Q H, Kalantar-Zadeh K, Kis A, Coleman J N and Strano M S 2012 *Nat. Nanotech.* **7** 699

[75] Peierls R E 1955 *Quantum Theory of Solids* (Oxford: Clarendon) section 5.3

[76] Grüner G 1988 *Rev. Mod. Phys.* **60** 1129

[77] McMillan W L 1975 *Phys. Rev.* B **12** 1187

[78] Wilson J A, DiSalvo F J and Mahajan S 1975 *Adv. Phys.* **24** 117

[79] Slough C G, McNairy W W, Coleman R V, Drake B and Hansma P K 1986 *Phys. Rev.* B **34** 994

[80] Kaasbjerg K, thygesen K S and Jacobsen K W 2016 *Phys. Rev.* B **85** 115317

[81] Li S-L, Tsukagoshi K, Orgiu E and Smori P 2016 *Chem. Soc. Rev.* **45** 118

[82] Slater J C 1965 *Quantum Theory of Molecules and Solids* vol 2 (New York: McGraw-Hill) p 347ff

[83] Liu W, Sarkar D, Kang J, Cao W and Banerjee K 2015 *ACS Nano* **9** 7904

[84] Renard M 1980 *The Physics and Chemistry of Low Dimensional Solids* ed L Alcácer (Berlin: Springer) pp 293–303

[85] Lévy F and Berger H 1983 *J. Cryst. Growth* **61** 61

[86] See, e.g., ed Seeger K, Mayr W and Philipp A 1985 *Advances in Solid State Physics* vol 25 ed P Grosse (Berlin: Springer) pp 175–9

[87] Fedorov V E, Artemkina S B and Grayfer E D *et al* 2013 *Proc. Int. Conf. on Nanomaterials: Applications and Properties* vol 2 01NTF34

[88] Toshima T, Inagaki K and Tanda S 2005 *Topology in Ordered Phases* ed S Tanda *et al* (Singapore: World Scientific) pp 1114–18

[89] Tsubota M, Inagaki K, Matsumura T and Tanda S 2012 *Europhys. Lett.* **97** 57011

[90] Miller J H Jr, Wijesinghe A I, Tang Z and Guloy A M 2013 *Phys. Rev.* B **87** 115127

IOP Publishing

Transport in Semiconductor Mesoscopic Devices
(Second Edition)

David K Ferry

Chapter 5

Localization and fluctuations

In the previous chapters, we have discussed the role of disorder primarily as creating scattering centers and affecting phase coherence. But, disorder can also lead to localization of the carriers and to an effect known as conductance fluctuations. These are not unrelated effects, although many have thought that they were. In earlier days, it was assumed that localization, which is often called strong localization in distinction from the weak localization discussed in previous chapters, was caused by strong disorder. At the same time, it was believed that conductance fluctuations were caused by phase interference in the potential landscape created by the weak disorder arising from a large impurity or defect density. In fact, the two are closely related, which should have been obvious from the fact that the same theoretical model is used to simulate both effects! This model is the Anderson model [1], which we will discuss in the next section.

The fundamental problem with disorder is to determine how it affects the allowed energy levels in the material. Normally, we think of band semiconductors as having a region of allowed energy states, in which the electron waves are free to propagate, and a band gap, in which the electron waves become localized and non-propagating. With disorder in the band semiconductor, the edge between the band and the gap becomes a dispersed gray area. Usually the width of the band increases allowing states to broaden into the gap area, which is called band tailing. But many of the band states in this tail, as well as somewhat into the normal band region, can become localized. Here, the long-range order of the crystal can be compromised, with the ultimate limit producing amorphous material. Because of the extending of the band region and its interaction with localized states, a new parameter arises which plays the role of the normal band edge, and this is the mobility edge. The mobility edge now separates the localized states from the conducting, or long-range ordered, states.

doi:10.1088/978-0-7503-3139-5ch5

Because of the disorder in the potential, which is assumed to be random, many small AB-like loops of various size can exist in the normally conducting states. In addition, many back-scattering loops, like those in weak localization, can also exist. These various phase coherent loops will randomly appear and disappear as one varies the Fermi energy or a magnetic field. This then leads to a fluctuation in the conductance as the interferences within a phase coherent area vary with these external variations. For some time, it was assumed that these conductance fluctuations were basically different from the localization phenomena, and that the former were universal, in which they had an amplitude that did not vary with the disorder strength, or did not vary with choice of perturbation—Fermi energy variation or magnetic field variation. Today, we know that this is not the case. Rather, the amplitude of the fluctuations vary with the size of the disorder, just as the localization behavior does, up to a critical value where a single Landauer channel is being switched on or off. At that point the amplitude saturates, which would lead one to believe in a universality if the strength of the disorder is sufficiently large. Hence, both effects derive from the same source—disorder in the crystal. But, very high quality material, such as the super high mobility GaAs/AlGaAs heterostructure generally does not display any conductance fluctuations.

So, in this chapter, we will begin with the Anderson theory of how disorder affects the electronic states and leads to localization. Then, we see how this leads to fluctuations, and how these vary with the strength of the disorder. This leads us to a discussion of the phase coherent area and the phase-breaking time.

5.1 Localization of electronic states

The concept of a rapid transition, at some critical energy, from a set of strongly localized states with only short-range order to a set of extended states with long-range coherence of their wave function, is remarkable. Normally, in band semiconductors, this is just the band edge and we give it no further thought. Perhaps the reason lies in the fact that most people are not taught about the complex band structure with the continuum of localized states that exist throughout the band gap [2]. But, this idea of such a transition warrants further investigation. Here, we will follow the approach of Anderson [1], using one version of the several he discusses. The model adopts a set of atomic sites in a crystal, in which the atomic energies are randomly, but uniformly, distributed over a fixed energy range, which is commonly denoted as W. This means that the energy is a random function with a probability distribution function given by $1/W$ for energies within a given range, usually denoted as $-W/2 \leqslant E \leqslant W/2$. This will have the effect of shifting the zero of energy to the center of the band, whereas we normally associate it with one of the band edges. The reason for this will become clear later. We will find that there is a critical value of the width, such that if W is greater than this critical value, all the states in the band will be localized. The value of W at this critical value is normally associated with the Anderson transition.

5.1.1 The Anderson model

In this approach, each atomic site (which may not lie on a lattice site in the disorder model) has a 'site' energy, which corresponds to the atomic energy level of that particular atom, and an 'overlap' energy describing the interaction of the wave function of that atom with the wave function of neighboring atoms. For example, in the previous chapter, our model of graphene assumed the site energy was zero and the overlap energy was the parameter γ_0. There are two general approaches to disorder that have developed with other approaches to this topic. One is to consider primarily site disorder, as described by the energy distribution discussed above, or some other equivalent form. The second is bond disorder which primarily treats a random variation in the overlap energy. The Anderson model used here is primarily a site disorder model, but it is mapped into a system in which the lattice is regular with the atoms located at the lattice sites.

In the presence of the other atoms, one can use normal lowest order perturbation theory to write the Schrödinger equation in terms of the wave function and the perturbation of the neighboring sites as

$$i\hbar\frac{\partial \psi_s}{\partial t} + H\psi_s = E\psi_s = E_{0,s}\psi_s + \sum_{s' \neq s} V_{ss'}\psi_{s'}, \tag{5.1}$$

where the subscript s refers to the site and the on-site energy $E_{0,s}$ is a random variable as described above. The quantity $V_{ss'}$ is the overlap energy between neighboring sites, and the sum runs over only those nearest-neighbor sites. This latter energy is usually relatively constant. The last term on the right-hand side of equation (5.1) is just the first-order perturbation term, and higher-order terms lead to a series expansion for either the wave function or the energy. In such a series, the random site energy is just the zero-order term, while the actual final energy can be expressed as

$$E = E_{0,s} + \sum_{s' \neq s}\frac{V_{ss'}V_{s's}}{E_{0,s} - E_{0,s'}} + \sum_{s', s'' \neq s}\frac{V_{ss'}V_{s''s}V_{s's''}}{(E_{0,s} - E_{0,s'})(E_{0,s} - E_{0,s''})} + \dots. \tag{5.2}$$

Unless the energy E_s is real, the wave function decays with time since the wave function is normally connected to the energy via a term $\exp(iE_st/\hbar)$. Thus, the nature of the states will be investigated by examining the convergence properties of the infinite series in equation (5.2). If this series converges, then the energy is real and the state is localized at that site. On the other hand, if the series does not converge, it must be assumed that the energy lies within an extended band of energies which correspond to wave functions that are extended over the entire crystal.

The series in equation (5.2) is a stochastic series since the zero-order on-site energies are a random variable uniformly distributed between $-W/2$ and $W/2$. Consequently, we can examine the convergence of the series in a statistical sense. Each term in the series contains V_{L+1}, where L is an integer giving the order of the term in the series. In a general lattice, each atom has Z nearest neighbors, so that there are ZL contributions to the Lth term, a point we shall use later. The general Lth term has the form

$$T_{s'} = \frac{V}{E_{0,s} - E_{0,s'}}. \tag{5.3}$$

If the values of the site energies on the primed subscripted sites are statistically uncorrelated, the magnitude of the contribution of a product of such terms can be estimated by taking the average of its logarithm, as

$$\langle ln | T_s T_{s'} \ldots T_{s''} | \rangle \sim L \langle \ln(|T|) \rangle. \tag{5.4}$$

This form of the terms tells us that the series is to be interpreted as a geometric series, which is a special form of a power series that will converge provided that the ratio of subsequent terms approaches a limit which is less than unity. Thus, we require that if the coefficients of the series are A_L, the series will converge if

$$\lim_{L \to \infty} \left| \frac{A_{L+1}}{A_L} \right| x < 1, \tag{5.5}$$

where x is the argument of the series, which here is our energy. We now apply this convergence criterion to the perturbation series for the energy (5.2). This leads us to the conclusion that the series will converge if

$$Z \exp(\langle \ln(|T|) \rangle < 1). \tag{5.6}$$

We now introduce the probability distribution function discussed above in order to evaluate the expectation value. This leads to

$$\langle \ln(|T|) \rangle = \frac{1}{W} \int_{-W/2}^{W/2} \ln \left| \frac{V}{E_{0,s} - E_{0,s'}} \right| dE_{0,s'}$$

$$= 1 - \frac{1}{2} \left[\ln \left| \frac{4E_{0,s}^2 - W^2}{4V^2} \right| + 2\frac{E_{0,s}}{W} \ln \left| \frac{2E_{0,s} + W}{2E_{0,s} - W} \right| \right]. \tag{5.7}$$

It is obvious from the above result that the value of W required for localization depends upon the energy in which one is interested. The energy in equation (5.7) is now a smooth variable. If we take the center of the band, where the energy is 0, then only the first term in the larger parentheses survives, and

$$Z \exp \left(1 - \ln \left(\frac{W}{2V} \right) \right) < 1 \tag{5.8}$$

or

$$\frac{W}{2V} > Z \exp(1) \sim 10.87 \tag{5.9}$$

is required to completely localize the band (we have taken $Z = 4$ for the tetrahedrally coordinated semiconductors). In general, we can say that $W > 2.72\Delta E$, where $\Delta E = 2VZ$ will totally localize the band, so that no extended states exist. At the other extreme, we can take $W = 0$, and we arrive at the equivalent inequality that says states will be localized if

$$E_{0,s} > \Delta E/2. \tag{5.10}$$

But, this is just the normal band requirement, as the bandwidth is basically VZ.

It is of interest to also determine the point at which an energy at the edge of the distribution, $E_{0,s} = W/2$ is localized. We can then combine the two logarithm terms, to find that when $W > \Delta E \exp(1)/4$, we start to develop localized states at the edge of the band. Hence, there is a critical size of the disorder that must be exceeded before the band states near the edge begin to be localized. These different conditions are sketched in figure 5.1.

We can illustrate the localizing effect of disorder and the random potential in another way, and that is to simulate a standard mesoscopic material such as the AlGaAs/GaAs heterostructure. The conducting layer is a Q2D electron (or hole) gas located at the interface between the two materials. The simulation is a study of the conductance of this electron layer, following the approaches of sections 2.7 and 2.8. In this approach, we discretize the two-dimensional layer with a grid size of 5 nm, and consider a range of widths, in the range of 0.2–0.6 μm, and lengths, in the range of 0.3–0.8 μm. We use the Anderson model to impose a random potential at each grid point according to the uniform distribution discussed above. This gives a random potential whose peak-to-peak amplitude is W. While this random potential induces states below the lowest energy of the band (and above the highest energy in the band), it also localizes a fraction of the normally transmitting states. A critical energy separating the localized modes and the propagating modes defines what has been called the *mobility edge* [2]. Since we are going to vary the amplitude of the random potential, it is useful to see how this localizes the conductance. We consider a density of 4×10^{11} cm^{-2}, which gives a Fermi energy about 15 meV in GaAs. Since we have a finite width of sample, the system is quantized in the transverse direction

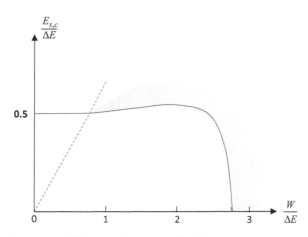

Figure 5.1. The critical energy at which states become localized (red curve). The localized state are shown as green shading, while the low disorder range where there is no formation of localized states is indicated by the dashed line.

into a set of modes, and the conductance is determined by the number of these modes whose transverse eigen-energies lie below the Fermi energy. Then, we determine the fraction of these modes that actually propagate in the presence of the random potential. This is plotted in figure 5.2, and it may be seen that the behavior suggested by figure 5.1 is certainly followed. The error bars represent an average over many sizes of sample, all of which have different implementations of the random potential. While there may be 30–60 modes propagating normally, once the disorder becomes sufficiently large, only a very few remain propagating. This is a significant point, as we will see later in discussions of fluctuations. Nevertheless, the lowest data point in the figure lies at a peak-to-peak disorder potential of $0.1E_t$, where E_t is the hopping energy (2.54). Even though this potential reaches peaks more than twice the Fermi energy, almost all of the modes remain propagating. So, it takes a significant amount of disorder to really affect the band states.

5.1.2 Deep levels

The levels of interest above were the shallow levels, as they are sometimes called. In these levels, the impurity ionization energies were only a few, or a few tens, of meV. These levels are usually treated within the effective mass approximation as a kind of *hydrogenic impurity*, and this continues in the localized states. In some defects, however, the impurity creates a significant lattice distortion about the core of the

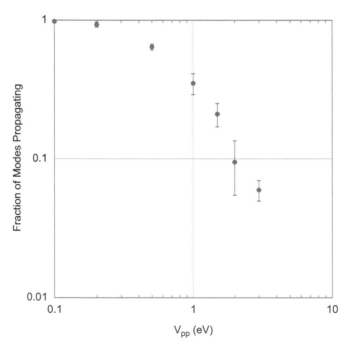

Figure 5.2. Fraction of modes that remain propagating as a function of the peak-to-peak amplitude (W) of the random potential. The values are averaged over a number of samples of various sizes. The amplitude of the random potential is normalized to the hopping energy.

impurity site. The difference between the case above and these so-called *deep levels* lies in the central cell potential—the core potential that interacts on a short range (whereas the Coulombic potential is relatively slowly varying over the unit cell size). When this core potential is sufficiently strong to affect the electrons, this affect is much stronger. An electron bound to the deep levels is largely localized by the strength of the potential to the range of the local unit cell. These levels are often near the middle of the band gap. Two effects create a sizable contribution to the understanding of these deep levels and their energy levels. These are the Jahn–Teller distortion and the Franck–Condon effect.

The Jahn–Teller theorem states that any electronic system with energy levels that are multiply degenerate may be split when the confining potential is affected by a different symmetry. Thus, if we put a deep level atom into a crystal such that the electronic ground state is multiply degenerate, there can be distortion of the lattice around this atom that raises this degeneracy. For example if an As atom sits on a Ga site (a so-called anti-site defect), there are two extra electrons bound to the As atom in the neutral state, and these two atoms have degenerate energy levels. This will lead to a local distortion around the As atom that raises this degeneracy, an effect that has been seen experimentally [3].

There is another distortion, however, that can also arise. The strength of the central cell potential can significantly affect the actual band structure in the neighborhood of the defect atom. The role that this interaction plays is considerably different when the defect is neutral than when it is ionized and becomes charged. Thus, we expect the local band structure and the defect energy levels to change as the defect is ionized by removing an electron. This means that the optical transition energy is different for excitation and for relaxation, and this is the Franck–Condon effect. According to this latter effect, the energy of the localized state will change as the charge level of the defect is modified, and this arises from a local polarization of the lattice. In other words, the atoms in the vicinity of the local defect will relax to a new set of positions, in much the same way in which a surface relaxes. This relaxation is quite local in the lattice and therefore occupies a large part of the Brillouin zone; hence the relaxation is often accompanied by the excitation of a large number of phonons. For this reason, the optical transitions for a deep level are often accompanied by a number of 'phonon sidebands'—transitions that differ from one another by a phonon energy.

The distortion that accompanies the Franck–Condon effect can lead to another interesting phenomenon. When an electron is excited from a localized deep level, the lattice relaxation can lead to a reduction in the conduction band energy that gives a configuration of *lower* energy. The electron can no longer recombine with the charged deep level, since it no longer has sufficient energy to make the transition back; in essence there is an energy barrier preventing the electron from recombining, as it must now not only have the energy to recombine but must also drive the lattice relaxation back to its former state. If this kinetic barrier is sufficiently large, we can have the effect of persistent photoconductivity, which is observed in GaAlAs at low temperatures [4]. In this latter case, it has been hypothesized that this effect is related to a donor-vacancy complex, referred to as the *D-X* center. Optical absorption excites electrons from the deep levels, which produces a higher conductivity. When the light is removed, the semiconductor remains conducting until the temperature is raised to the order of 200 K or so, where the excited carriers have sufficient thermal energy to surmount the barrier and recombine.

In both physics and chemistry, the details of the Franck–Condon effect and deep impurities is based upon what are called reaction coordinates. One of the earliest papers in this area actually dealt with electron transfer from a donor or acceptor in which there was a lattice relaxation described by a reaction coordinate [5]. As discussed above, the local state of the lattice around the impurity changes with the charge on the impurity, and this is described by the reaction coordinate. This also changes the energy level, so that optical transitions can change depending upon the charge state. In this approach, the interactions can be studied by a spin-boson model that provides a pragmatic, yet realistic formulation for the role of dissipation in the electron transfer [5]. The role of a driving electric field was also considered in the study. Others have also studied electron transfer when coupled to collective boson degree of freedom, including the study of the fluctuations in the system [6]. Again, these latter studies also employed the reaction coordinate for the impurity system.

Let us now describe the approach with the reaction coordinate model. The idea is about the same whether we deal with a molecule or with condensed matter. For an impurity, say a donor, the upper state corresponds in the condensed matter system with the conduction band, while the lower state is the actual impurity level (in the molecule, this might actually be an extended state with dispersion). When the electron is excited from the donor impurity, the local potential changes primarily due to the short range local potential around the impurity. The potential change induces a lattice relaxation around the impurity, and this in turn changes the level of the impurity relative to the conduction band [7], but this is usually treated as a change in the conduction band. There are two contributions in this interaction. First, the Jahn–Teller theorem tells us that any electronic system with multiply degenerate ground states is unstable against a distortion that removes the degeneracy. This leads to the lattice relaxation, which is of course much more prevalent in deep donors as opposed to shallow impurity levels. As a result, the atom is actually displaced in position, and this is the reaction coordinate. But, there is a second distortion, and that is the change in the band structure between the neutral and the ionized states, as mention above, and this is the Franck–Condon effect, and this results in a difference between the optical excitation energy and the thermal excitation energy of the impurity [3, 8]. The overall situation, as mentioned in the previous paragraph, is best described by a reaction coordinate diagram, as shown in figure 5.3. The two potential curves cross at a point x^*, which is the point at which electron transfer occurs. Once the electron is into the second potential V_2, it must gain an energy greater than ε_0 in order to recombine with the donor. Actually, if we consider tunneling, the electron only needs to be excited to positive energy, where it can tunnel to V_1. In semiconductors, this leads to persistent photoconductivity at low temperatures. Here, the Hamiltonian can be written as [5, 6]

$$H = H_{EL} + H_{RC} + H_B, \tag{5.11}$$

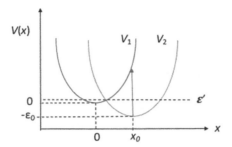

Figure 5.3. A diagram of the reaction coordinate x and the various potentials. V_1 represents the neutral impurity, whose energy level is below those shown here. The potential V_2 represents the shift of the conduction band when the neutral impurity level is taken as the reference point for the energy. The energy $(\varepsilon_0 + \varepsilon')$ represents the energy the electron must gain in order to recombine with the charged impurity.

where the bare electronic system is written in terms of a two-state pseudo–spin system as

$$H_{EL} = -\frac{1}{2}[V_1(x) + V_2(x)]\hat{\sigma}_z + \frac{\hbar\Delta}{2}\hat{\sigma}_x. \tag{5.12}$$

The two pseudo-spin operators are given by

$$\begin{aligned} \hat{\sigma}_z &= |1\rangle\langle 1| - |2\rangle\langle 2| \\ \hat{\sigma}_x &= |1\rangle\langle 2| + |2\rangle\langle 1|. \end{aligned} \tag{5.13}$$

The two states are those of the impurity and band combinations. The coupling term Δ is assumed to be independent of the reaction coordinates. The reaction coordinate term is given by

$$H_{RC} = \left[\frac{p^2}{2m} + V_1(x) + V_2(x)\right]\begin{bmatrix} 1 & 0 \\ 0 & 1 \end{bmatrix}. \tag{5.14}$$

The two harmonic potentials, described in figure 5.3, are given as

$$\begin{aligned} V_1(x) &= \frac{m\omega_0^2}{2}x^2 \\ V_2(x) &= \frac{m\omega_0^2}{2}(x - x_0)^2 - \varepsilon_0, \end{aligned} \tag{5.15}$$

and the bare term H_B is the energy of the non-interacting particles. Here, the so-called reorganization energy is $V_1(x_0)$, and the excitation energy indicated by the red arrow in figure 5.3 is the sum of this energy and the offset energy ε_0. One notes that the reaction coordinate x is the primary variable in the Hamiltonian and the resulting problem being studied. The important point is that when the electron is transferred from the 'donor' molecule (state $|1\rangle$) to the 'acceptor' molecule (state $|2\rangle$), the relaxation process makes this a 'one way' reaction, as the barrier $(\varepsilon_0 + \varepsilon')$ to the back reaction is created by the molecular relaxation characterized by the reaction coordinate. This is what leads to the persistent photoconductivity.

5.1.3 Transition metal dichalcogenides

We have already shown the role of localization in a GaAs/AlGaAs system in figure 5.2. A more complicated form can be found in the transition metal dichalcogenides (TMDCs), discussed in section 4.4. The TMDCs form interesting two dimensional semiconductor layers that have been shown to be useful for electronic applications. However, they currently suffer from the presence of defects that act as both deep levels and random potentials leading to localization. In particular, it has been shown that charge impurities such as S vacancies in MoS$_2$ and impurities from the dielectric environment not only have significant impact on the mobility of the MoS$_2$ devices [9], but also can lead to the formation of an impurity band tail within the band gap region as well as localized states above the conduction band edge [10]. The TNDCs tend to have a significant number of chalcogenide vacancies (e.g., missing S in this case). These vacancies lead to both effects. First, the vacancy itself tends to produce a deep level that lies near mid-gap in the material. Secondly, the missing atomic potential produces a modulation of the crystal potential formed from the pseudo-potentials of the atoms themselves. This modulation creates the random potential that leads to the Anderson model and the resulting localization.

To gain a quantitative understanding of the effects of these impurity states, we examine the room temperature device characteristics using a model proposed by Zhu *et al* [10]. From studying extensive capacitance voltage measurements between a back gate on the TMDC and the layer itself (using a SiO$_2$ insulator grown on the Si back gate), these authors find that there is a mid-gap state that has a population of approximately 10^{12} cm^{-2}. This is the deep level, although it is not known if this level has any lattice relaxation associated with it. They also find an extremely broad distribution of localized states, as in the Anderson model, and an extremely high mobility edge for the conduction band. They then adapt a simulation model to fit the data. In this model, the impurity band tail is incorporated by the following density of states distribution:

$$
D_n(E) = \begin{cases} \alpha D_0 \exp\left[\dfrac{E}{\varphi}\right], & -\dfrac{E_G}{2} < E < 0, \\[2ex] D_0 - (1-\alpha)\exp\left[-\dfrac{E}{\varphi'}\right], & E > 0. \end{cases}
\tag{5.16}
$$

As before, the energies here are referenced to the conduction band edge, *in the absence of the disorder*. Here, D_0 is the normal two-dimensional density of states given by (2.21), using an effective mass of 0.4 m_0. The $(1 - \alpha)$ term for positive energy accounts for the states that are removed from the conduction band to form the band tail states of the first line. The parameter φ is the characteristic width of the localized states that form the band tail discussed above and is found to be $\sim$100 meV, and φ' is chosen so that the two piece-wise functions have a continuous gradient at $E = 0$. The parameter α is found to be $\sim$0.33. From this model and the

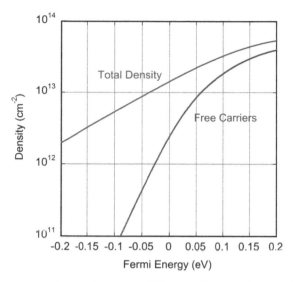

Figure 5.4. Localized states trap electrons into non-mobile states. Here are the total electron density and the free carrier density for a MoS$_2$ monolayer at 300 K. The parameters are discussed in the text.

measured conductance, the mobility edge is found to lie 10 meV above the conduction band edge.

To calculate the free carrier density, we consider only the extended states above the mobility edge energy E_M with respect to the conduction band edge:

$$n_{\text{free}} = \int_{E_M}^{\infty} D_n(E) P_{FD}(E) dE. \tag{5.17}$$

In different experiments, reported by Xiao *et al* [11], a slightly different value of the band tailing, $\varphi \sim 110$ meV, and a positively charged oxide impurity density of 3.5×10^{12} cm^{-2} were found. As previously, these were found for a layer of MoS$_2$ placed upon an oxidized Si wafer, so that the Si could be used as a back gate to vary the electron concentration. Using this data and parameters, we plot the total induced electron concentration as well as the free carrier (those above the mobility edge) concentration at room temperature in figure 5.4. The difference in these two densities yields the number of carriers which are trapped in the localized states.

5.2 Conductivity

As indicated in the previous section, the role of disorder is to create band tails as well as to localize various states, and we saw this in the numerical example for GaAs. We indicate this in figure 5.5. Here, the normal conduction and valence bands are shown in blue. Since the total number of states in the band is constant, the presence of disorder must move some states from the band to the tail in order to have band tailing. This produces the new band edges shown in red. As is clear from figure 5.1, there is a threshold in the disorder strength required to produce the band tailing effects. Normal impurities in the semiconductor do not produce enough disorder to

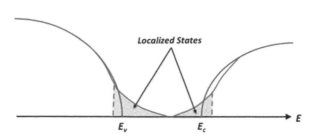

Figure 5.5. The normal bands are shown in blue, along with the localized states that arise in the presence of significant disorder. The mobility edges are shown in green.

cause the band tailing, unless the density is approaching the degeneracy level. In figure 5.5, we show the localized states that arise from the disorder in brown. They extend throughout the tails and into the band proper, as was evident in the example of figure 5.2. The boundary between the localized states and the extended states is termed the mobility edge, shown as the dashed green vertical lines in figure 5.5. The localized states now have wave functions that are spatially localized and are not sufficiently dense to cause overlap of these wave functions. The degree of tailing and position of the mobility edge both depend upon the degree of disorder in the system. In many amorphous semiconductors, the tailing from the conduction and valence bands actually overlaps so that there is no real band gap. However, excitation studies, such as those for photoconduction, actually measure the mobility edges, as the localized states do not contribute to normal conduction. Hence, the optical gaps can be larger than the normal band gaps in crystalline semiconductors.

In crystalline (ordered) semiconductors, the activation energy for the conductivity arises from the presence of the band gap; that is, the intrinsic electron and hole population (in non-degenerate material) is given by

$$n_i = \sqrt{N_C N_V} \exp\left(-\frac{E_{\text{gap}}}{2k_B T}\right),$$ (5.18)

where N_C and N_V are the effective density of states for the conduction and valence bands, respectively. The band gap variation arises because the density of states in the ordered material has a sharp cutoff at the band edges. However, in disordered material, the existence of the tails of localized states below E_C and above E_V would appear to eliminate the possibility of a sharp activation energy. This is not really the case, because of the basic difference in the nature of the extended and localized states. In the extended states, the wave functions have some coherence (at least in amplitude) over many 'unit cells' of the material, so that one might expect some of the concepts of effective mass and mobility to be applicable. In this sense, the mobility exists and may be not too much smaller than in the ordered state. In the localized states, on the other hand, the wave functions do not have any long-range coherence and do not overlap the neighboring atoms, or 'unit cells', to any great degree. Therefore, the concept of nearly free motion is just not applicable, and one must think of the electron moving from one cell to the next by a tunneling process, or 'hopping' process, in which the carrier is thermally activated over the potential barrier between the atoms, only to be retrapped at the next atom. There is thus a large barrier preventing motion among the atoms in the localized states, and the mobility is essentially zero (very small) on the scale of the extended states.

Thus, at the transition energy between the localized and the extended states, it is to be expected that the mobility will fall by several orders of magnitude. In the disordered semiconductors, the boundaries between these two types of states are therefore the boundaries between states for which the carriers are free to move and those in which the movement is largely forbidden. The disordered semiconductor therefore has regions of allowed and forbidden mobility, as opposed to allowed and forbidden states, and the activation energy is that required to take the electron from an extended state in the valence band to an extended state in the conduction band. As we mentioned above, the activation energy will usually be larger than the band gap, as part of the otherwise allowed states in the conduction and valence bands will in fact have forbidden mobilities due to the disorder-induced localization.

When the energy is in the range of the localized states, conductivity proceeds mainly by a mechanism of hopping, whereby an electron jumps from one site to a neighboring site. Repeated hopping leads to the possibility of the carrier transiting through the entire sample, but the conductivity is reduced by the need to be excited over a barrier which has to be surmounted at each individual hop, and this barrier can also be a random function. Since the wave functions are localized on a single site, the probability of the jump, either by a phonon-assisted transition over the barrier or tunneling through the barrier, is proportional to the overlap of the wave function on the two neighboring sites, which falls off as $\exp(-\alpha R)$, where R is the hopping distance and α is a decay constant (which for tunneling can be calculated if the details of the potential barrier are known). Such hopping is known as *nearest-neighbor hopping* [12] and is often found in the case of impurity conduction in highly doped semiconductors. The conductivity is proportional to the density of states at the Fermi level and the width of the Fermi–Dirac distribution, the difference in the probabilities for forward and backward hopping when a field is present, and an effective velocity that is approximately the distance times a 'hop frequency'. The latter is related to the phonon frequency in phonon-assisted hopping. Thus, the current density is given by [13]

$$J \sim \sum_{\pm} ek_BT\rho(E_F)R\nu_{ph}\exp\left(-2\alpha R - \frac{\Delta E \pm eFR}{k_BT}\right)$$
$$= 2ek_BT\rho(E_F)R\nu_{ph}\exp\left(-2\alpha R - \frac{\Delta E}{k_BT}\right)\sinh\left(\frac{eFR}{k_BT}\right),$$

(5.19)

where ΔE is the effective barrier height, ν_{ph} is the phonon-induced attempt frequency, and F is the electric field. For weak fields, the hyperbolic sine function can be expanded, and this leads to

$$\sigma = 2e^2\rho(E_F)R^2\nu_{ph}\exp\left(-2\alpha R - \frac{\Delta E}{k_BT}\right).$$

(5.20)

Nearest-neighbor hopping is expected mainly when $\alpha R \gg 1$. In this case, the conductance is greatly reduced by the exponential factor. In this limit, the energy levels that are allowed are expected to be rather widely spaced, and an estimate of the barrier may be obtained from the density of states $\rho(E_F)$, which is the number of states per unit volume (in d-dimensional space). Then, the number of states per unit energy is just $R^d \rho(E_F)$, and the average energy spacing per state is $\Delta E \sim [R^d \rho(E_F)]^{-1}$. But, if this spacing is that observed at a single site, it is comparable to the separation of the atomic levels at that site and hence is of the order of the bandwidth. This leads us to the conclusion that such nearest-neighbor hopping is expected only in the case for which all levels in the conduction band (or the impurity band) are localized, so that any mobility edge lies in a higher-lying band [12].

On the other hand, it is often the case that $\alpha R \leqslant 1$, so that the hopping range may extend well beyond that of the nearest neighbor. This is the regime of variable-range hopping. Here, the distance over which the hop is expected to occur increases with increasing temperature, and the exponential argument is a very low power of the inverse temperature [13, 14]. At a given temperature, the electron will 'hop' to a site that lies somewhere inside a radius of order R, which gives $\gamma_d (R/R_0)^d$ available sites for the hop (R_0 is an average distance for the hops), where γ_d is a numerical factor that depends upon the dimensionality (this factor is $4\pi/3$ in three dimensions, π in two dimensions, and 2 in one dimension). The hop generally occurs to a site for which the activation energy is the lowest. The latter is given by the same argument of the previous paragraph, and is $\Delta E \sim [\gamma_d R_0^d \rho(E_F)]^{-1}$. Thus, the effective barrier depends upon the distance over which the hop can occur. The probability of a hop is proportional to

$$\exp\left(-2\alpha R - \frac{\Delta E}{k_B T}\right) \tag{5.21}$$

as before. Because of the definition of the average energy barrier, we can find a maximum value in the exponential argument when

$$\alpha = \frac{\Delta E}{2R k_B T} = \frac{1}{2\gamma_d R_0^{d+1} \rho(E_F) k_B T}. \tag{5.22}$$

The prefactor for this exponential will involve the average hop distance, which involves the factor R as well, giving

$$\langle R \rangle \sim \frac{d}{d+1} R, \tag{5.23}$$

and, as before, we can write the conductivity as

$$\sigma = 2e^2 \rho(E_F) \langle R \rangle^2 \nu_{ph} \exp\left(-2\alpha R - \frac{B}{T^{1/(d+1)}}\right), \tag{5.24}$$

where

$$B = \frac{4\alpha^{d/(d+1)}}{[2\gamma_d \rho(E_F) k_B]^{1/(d+1)}}. \qquad (5.25)$$

The result of equation (5.24) provides the famous $T^{-1/4}$ behavior of the conductivity that is often found in disordered material (in three dimensions). In two dimensions, the expected behavior is $T^{-1/3}$, and this has been found in activated transport conductivity in the inversion layer of MOSFETs at low temperatures [15, 16]. In these experiments, Na was intentionally introduced into the oxide of the MOSFET. Then the Na atoms could be induced to drift through the oxide by a large gate voltage, and this allowed one to control the scattering effect of the Na atoms, and thereby the degree of disorder introduced into the channel of the transistor. That is, the charge in the oxide provides the random potential seen by the electrons in the channel and leads to the disorder. The strength of the disorder is controllable by the distance the charge resides from the oxide–semiconductor interface. The Na doped transistor provides a unique opportunity for the evaluation of various models for hopping conduction because it allows one to vary a number of parameters in the transport model by the use of substrate bias, gate bias, and the position of the oxide charge [17]. In figure 5.6, we show the results on the temperature dependence of the conductivity in such a MOSFET with a gate oxide thickness of 100 nm and a channel length of 10 μm [17]. Here, the Na concentration at the interface was 7.5×10^{11} cm^{-2} and the carrier density was 2.8×10^{11} cm^{-2}. The various curves are

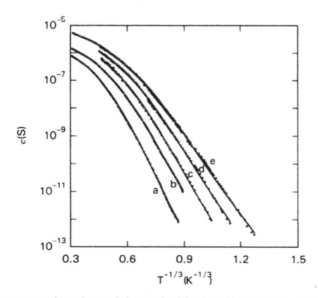

Figure 5.6. The temperature dependence of the conductivity in a Si MOSFET with Na in the oxide, as described in the text. The various curves have different values of the disorder bandwidth ΔE and the value of α^{-1} found from fitting to the theory (solid lines). These values are, respectively, a: 7.38 meV and 4.11 nm; b: 5.49 meV and 4.49 nm; c: 4.36 meV and 4.15 nm; d: 3.74 meV and 4.56 nm; e: 3.21 meV and 4.73 nm. The substrate biases for these five curves were (in order) 8.0, −3.0, −1.0, 0.0, and 0.5 V, respectively. (Reprinted with permission from [17]. Copyright 1986 the American Physical Society.)

for different values of the substrate bias. It is clear that the straight line portions of the various curves fit to the temperature dependence expected for variable range hopping in this Q2D system.

5.3 Conductance fluctuations

In our discussion of weak localization in an earlier chapter, it was clear that the existence of time-reversed paths can lead to quantum interference that causes a correction to the conductivity. This was also apparent in the AB effect. These paths are modified by the change in the Fermi energy (changes in the wave phase velocity) and in the magnetic field (similar changes in the momentum occur through the vector potential). In fact, the magnetic field destroyed the weak localization while it made possible the oscillatory terms in the AB effect. So, the effect of the magnetic field can be more complicated than that of the variation in the Fermi energy. If a sample is composed of a great many such loops (and not necessarily time-reversed loops), with each contributing a phase-dependent correction to the conductivity, then the summation over these loops may or may not ensemble average to zero, depending upon the number of such loops contained within the sample (and, hence, upon the size of the sample). On the other hand, in mesoscopic systems, where the number of such loops is relatively small, ensemble averaging can fail, and these loops lead to the presence of conductance fluctuations [18, 19]. Since the impurity distribution is reasonably fixed for a given sample configuration, the particular oscillatory pattern is often thought of as a fingerprint of an individual sample. That is, the impurity distribution is different from one sample to another, and therefore the detailed nature of the interference pattern seen in any one sample is a characterization of its unique impurity distribution. In figures 5.7 and 5.8, we show experimental data for fluctuations in AlGaAs/GaAs heterostructures [20, 21].

One view in estimating the amplitude of these fluctuations arises from the Landauer formula, in which the conductance is quantized for each possible channel through the sample. Here, the conductivity is given by [22]

$$G = \frac{2e^2}{h} \sum_{i,j=1}^{N} |t_{ij}|^2, \tag{5.26}$$

which differs from equation (2.36) in the possibility that the disorder induces scattering from one channel (mode) to another channel. This is characterized by the transmission t_{ij} which describes the transmission of the wave entering in channel j and leaving in channel i. In the case of no such inter-channel scattering, the sum reverts to the value N found in equation (2.36).

It has been suggested that the amplitude of these fluctuations is universal in nature and arises directly from the universality of the Landauer formula. This suggests that we could write the amplitude of the fluctuations (determined from the variance of the conductance) as [23, 24]

$$\delta G = \sqrt{\text{var}(G)} \sim \frac{g}{2} \frac{Ce^2}{\sqrt{\beta}h}, \tag{5.27}$$

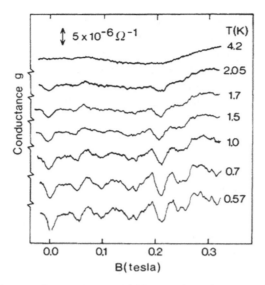

Figure 5.7. Magnetoconductance for a narrow sample for a variety of temperatures and a gate voltage of −3.01 V. At zero magnetic field, the base conductance was 10^{-4} s and varied little over these temperatures. (Reprinted with permission from [20]. Copyright 1987 the American Physical Society.)

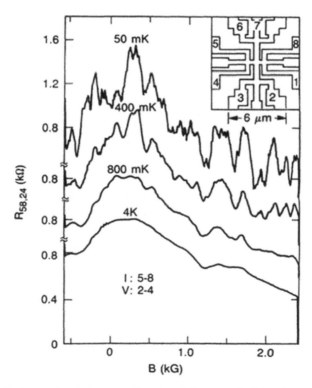

Figure 5.8. Longitudinal magnetoresistance as a function of the magnetic field at several temperatures. The sample is depicted in the inset. (Reprinted with permission from [21]. Copyright 1988 the American Physical Society.)

where g is a factor accounting for spin and valley degeneracy (we use only spin degeneracy in equation (2.36)), C is a constant thought to be about 0.73 for a narrow channel with $L \gg W$, and β is unity in zero magnetic field but takes a value of 2 when the magnetic field lifts the time-reversal symmetry of the system. It would be nice to adopt this view, but this concept of universality is contrary to our intuition. As we learned above, the role of disorder comes on gradually as the size of the disorder is increased. Certainly, in the absence of disorder one would not expect any local-ization or fluctuations, and there are no reports of conductance fluctuations in the very high mobility AlGaAs/GaAs heterostructures discussed in chapter 1. Indeed, in our own numerical studies of conductance fluctuations, it was found that the amplitude of the fluctuations increased gradually as the size of the random potential was increased [25]. Certainly, the amplitude tended to saturate once a critical value of disorder was reached, and this does have a relation to the Landauer formula. If we consider that the phase interferences have a maximum value when a single channel is being randomly switched on or off, then the peak-to-peak value of the fluctuation should be of the order of $2e^2/h$ for a spin degenerate channel. Naturally, the peak amplitude is one-half of this value and the rms value is reduced further. This leads to a value for the effective amplitude of

$$\delta G = \sqrt{\text{var}(G)} \sim \frac{g}{2} \frac{\sqrt{2}\,e^2}{2h} \sim 0.7\frac{2e^2}{h}, \tag{5.28}$$

for a spin degenerate channel. And, in a high magnetic field, this would be reduced further by a factor of two as the spin degeneracy is removed, although this is not seen in some early experiments [26]. One problem, of course, lies in the fact that most studies on high mobility heterostructures are not able to perform Fermi energy sweeps, as they are ungated samples. This was not the case in early studies in Si MOSFETs where there is a natural gate voltage that can be swept. We will see equation (5.28) again below in connection with studies of the fluctuations in graphene.

While the above gives a reasonable approach for the general semiconductor, it is possible to use advanced simulation techniques to study the fluctuations. Let us demonstrate this with a study of the conductance fluctuations seen in a graphene nanoribbon [27, 28]. In the theoretical approach, the disorder arose from impurities located in the SiO_2 sheet upon which the graphene rested. The local potential arising from these impurities was modeled self-consistently for each different siting of the impurities, with an impurity density of 3×10^{12} cm^{-2}. This local potential was then imposed onto an atomic-basis simulation of transport in the graphene nanoribbon [29], following the approach of section 2.5. The graphene nanoribbon sample used in our calculation has a width of 199 atoms and a length of 200 columns (100 slices). So the area of the sample will be around 24.3 nm (width) $\times$ 42.4 nm (length) = 1030.32 nm^2 (we refer to this as the 'normal' size). This means that the number of charged impurities in the area of the ribbon is 31. Then, these impurity charges are randomly distributed throughout the area of the graphene nanoribbon. In addition, the distance between the graphene layer and the impurity charge layer is set to $d = a_0$

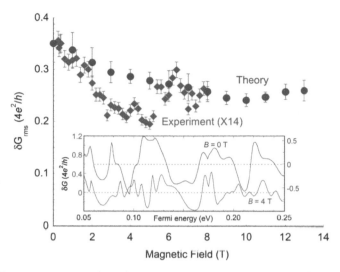

Figure 5.9. Amplitude of the conductance fluctuations obtained with a sweep of the magnetic field for our calculations (blue circles) and from the experiments of [27] (red diamonds). The latter has been multiplied by a factor of 14 to account for averaging over phase coherent regions in the experiment. The inset shows actual fluctuations versus Fermi energy for two different magnetic fields. Reprinted with permission from [28]. Copyright 2016, Institute of Physics Publishing.

and the screening coefficient is $\xi = 1/(10 \times a_0)$, corresponding to a screening length of $10 \times a_0$. Here, $a_0 = 0.142$ nm is distance between two adjacent carbon atoms in the nanoribbon. This random potential provides the landscape in which the transport is computed by the above prescriptions. Because the conductance fluctuations are random, they will change with each implementation of the impurity potential. Hence, we use several samples, each of which has the same impurity density.

In the main panel of figure 5.9, we plot the RMS amplitude (δG_{rms}) of the CF induced by the Fermi energy sweeps, undertaken in the presence of various static magnetic fields oriented normal to the graphene plane. In the inset to this figure, we also plot the CF as a function of the Fermi energy for two different values of the magnetic field. It can clearly be seen that the amplitude of the fluctuation is significantly smaller with the presence of the magnetic field. Also plotted in the same figure are the experimental data from [27], which have been multiplied by a factor of 14. In the experiments, studies of the correlation function of the fluctuations observed in magnetic field sweeps were used to estimate the phase coherence length l_φ, which was found to be about 200 nm [27]. This distance is much smaller than the size of the sample over which the measurements were made, and smaller than the estimated thermal diffusion length. Hence, it is expected that the fluctuations observed experimentally should be considerably smaller than the theory due to statistical averaging of many phase coherent regions. The averaging length L in the measurements is about 2 μm. For the conditions discussed here, the reduction in amplitude is expected to be about [24] $6(l_\varphi/L)^{3/2}$, which suggests that the experimental values should be smaller than the zero-temperature calculation by a factor of 12.9. Here, we focus upon a qualitative comparison with the experiment. Hence, we have arbitrarily used a factor of 14 in figure 5.9 to match the experimental and theoretical data at zero magnetic field, as this gives agreement for the values of theory and experiment at this point.

Referring to the field-induced reduction in the magnitude of the computed fluctuations in figure 5.9, we may conclude that this is not due to any symmetry breaking process (such as that due to the breaking of time reversal symmetry or the lifting of spin or valley degeneracy), since such mechanisms are not included in our calculations. Indeed, spin degeneracy should be lifted also when the magnetic field is applied in the plane of the graphene, but no such evidence of such symmetry breaking was found in the study of [27]. So, there must be another explanation for the magnetic-field-induced reduction in the amplitude of the conductance fluctuations. In this case, it is likely related to the fact that, in a large magnetic field, there can be a suppression of both elastic and inelastic (phase breaking) back scattering [30]. As edge states begin to form in the magnetic field, most of the perceived resistance is found at the current contacts, and the forward and backward edge states are unequally occupied. Reflection from impurities cannot induce back scattering of the topologically-protected edge states, and the magnetic field also leads to a reduction in the number of channels that actually propagate through the sample. The result of these two effects should be to suppress the amplitude of the fluctuations in the conductance, and we speculate that it is this behavior that is seen in figure 5.9. However, we caution that the edge states are nowhere near fully formed in these samples. The fact that the experimental data falls more rapidly than our calculations suggest that the reduction of back scattering may well be more effective in the presence of additional scattering mechanisms, and when there are several phase coherent regions that lead to the averaging seen in the experimental data.

In figure 5.10, we plot the amplitude of the conductance fluctuations observed in experimental and simulated magnetic field sweeps as a function of the electron density. The calculations are actually done as a function of the Fermi energy, while the experiments are performed by varying the back-gate voltage. We have converted both variations to one in terms of carrier density, so as to have a common parameter for comparison. Again, the experimental data has been multiplied by a factor of 14 to account for the multiple phase coherent area averaging [27]. A similar reduction in the amplitude of the fluctuations with density has been seen for both gate voltage sweeps and magnetic field sweeps [31]. It may be seen from this figure that the amplitude of the fluctuations is smaller than that in figure 5.9 for the Fermi energy sweeps. The difference is of the order of a factor of 3, which has also been seen in simulations for conductance fluctuations in GaAs [32]. This is a clear indication that the CF do not possess any ergodic properties, as such a property would lead to equal amplitude fluctuations for different perturbations, such as sweeps in energy versus sweeps in magnetic field, or for different samples. In addition, it may be observed that the amplitude of the computed fluctuations decreases somewhat as the carrier density is raised, behavior that is captured in experiment also. This might be explained in the experiments by a decreasing phase breaking time at higher densities [33, 34], which would lead to a reduction of the phase coherence length. But, this cannot be the case in the theory, as the phase coherence length is assumed to be set

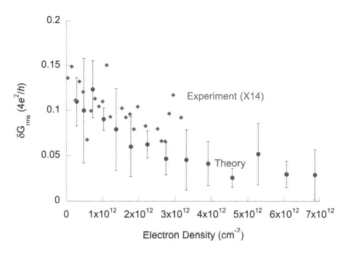

Figure 5.10. The average value of the conductance fluctuations that arise in a Fermi energy (or gate voltage in the experiment) sweep as a function of the electron density. The experimental data (red diamonds) has been multiplied by a factor of 14 (discussed in the text). Reprinted with permission from [28]. Copyright 2016 by the Institute of Physics Publishing.

by the simulation size. The observation, in both experiment and theory, of the density-dependent variations in the fluctuation amplitude points, once again, to the non-ergodic character of these fluctuations.

When the fluctuations are fully developed, their amplitude is essentially just that expected by the turning on or off of a single transverse mode in the sample. Thus, we expect that this will lead to a maximum rms value of the conductance fluctuation as

$$\delta G_{rms} = \frac{\eta_{valley}\eta_{spin}}{2\sqrt{2}} \frac{e^2}{h},$$
(5.29)

where the η are the degeneracy factors for valley and spin, and this equation is just a generalization of equation (5.28). This value actually agrees well with the earlier perturbation theory estimates, but is based more on empirical observations than on analytical methods. It should be remarked that the peak rms amplitude seen in graphene can be less than this, with this peak value of $0.35(4e^2/h)$ being seen only at zero magnetic field in figure 5.9.

5.4 Correlation functions

In our discussion of weak localization in chapter 3, the magnetic field led to a decay in the effect. In the AB effect, the magnetic field led to oscillatory conductance. We have pointed out that the conductance fluctuations arise from similar considerations, and that the interactions of many such phase coherent loops lead to the fluctuations. Thus, we might expect that variation of the magnetic field or the Fermi energy will affect the amplitude of the interaction and thus lead to the fluctuations. The randomness of the fluctuations can be examined by the use of a correlation function of the fluctuations. Variation in the position of the Fermi energy in the density of states, whether caused by a gate bias or by a magnetic field, should also cause the correlation among the phase coherent loops to decay (or at least to change). This can be examined with the correlation function, which is often defined by

$$C(\Delta E, \Delta B) = \langle G(E + \Delta E, B + \Delta B) G(E, B) \rangle - \langle G(E, B) \rangle^2, \tag{5.30}$$

although this must be normalized properly (the value of the correlation function when ΔE and ΔB are zero must be 1). Normally, the variables are not continuous, but are a discrete set of data when the data are gathered by a computer. Hence, a more useful definition is to work with the form

$$C(\Delta E, \Delta B) = \sum_i G(E_i + \Delta E, B_i + \Delta B) G(E_i, B_i) - \left[\sum_i G(E_i, B_i) \right]^2. \tag{5.31}$$

Again, this must be properly normalized. It is obvious that the magnetic field variation causes a change in the correlation function. The energy variation arises from the fact that the impurities cause an inhomogeneous energy surface and that different current paths will see different local variations in the Fermi energy. This means that small changes in the overall Fermi energy, due to an applied gate voltage, for example, will cause larger variations in the local potential.

Another useful connection is the variance of the conductance. The variance is the traditional rms value defined as

$$\text{var}(G) = \langle G^2(E, B) \rangle - \langle G(E, B) \rangle^2. \tag{5.32}$$

This becomes important as it is thought that the fluctuation is given as [24]

$$\delta G_{rms} = \sqrt{\text{var}(G)}. \tag{5.33}$$

In figures 5.9 and 5.10, the fluctuation itself is plotted as a function of magnetic field and gate voltage (density), respectively. The correlation function is slightly different from the variance, but the decay of the fluctuation in these figures is related to the correlation energy ΔB_c and the correlation energy (or gate voltage, reduced by the effective lever arm between a voltage change and the Fermi energy change) ΔE_c. These values are determined by the energy or magnetic field for which the correlation function has fallen to a value of 0.5 of its initial value (which should be unity).

The value of the correlation field ΔB_c is generally related to the increment of flux enclosed within a phase coherent area. Some argue that since we are not using two time-reversed paths, but only a single loop, the magnetic field coupled to the loop is reduced by a factor of two from that found for weak localization reduction. But, from the ideas of the AB experiments, this would still lead to the flux being coupled through a single phase coherent area. Thus, we expect that the correlation magnetic field would be related to this through

$$\Delta B_c \sim \frac{h}{e W l_\varphi}, \qquad W \ll l_\varphi \ll L, \tag{5.34}$$

where W and L are the width and length of the sample, respectively, and l_φ is the phase coherence length. More complicated arguments and equations have appeared in the literature in place of equation (5.34) [23, 24, 35]. Hence, some versions simply put a factor of β in front of the right-hand side of equation (5.34), with β having a numerical value of 0.55 commonly being observed, although others have suggested a value closer to 3 [36]. However, in most cases, determining the value of the phase coherence length, or the related phase-breaking time, is quite difficult from the experimental structures, so that a wide range of results can be found in the literature. For example, a value of ~0.25 was found for β in a study of a variety of long wires fabricated in high mobility material [37].

In computational experiments, such as discussed above, the lack of a phase-breaking process within the 'sample' area means that the phase coherent area is that of the entire sample. That is, in the zero temperature limit where phase breaking occurs only at the sample contacts, it is expected that the phase coherence length in equation (5.34) is replaced by the sample length itself. In these experiments, it becomes possible to estimate the validity of equation (5.34) directly. In figure 5.11, we plot the values for ΔB_c as a function of the amplitude of the random potential [32]. Here, it may be seen that relation (5.34) holds fairly well except at the lowest values of the random potential. There is no statistical difference in the energy at which the magnetoconductance sweep is made. Again, there is a distribution of values at each choice of the random potential and energy, but when a sufficiently

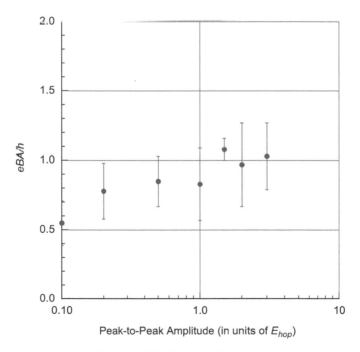

Figure 5.11. Values of the effective flux enclosed in the sample determined from the correlation magnetic field in fluctuation correlation functions.

large ensemble of systems (with varying size and energy) is chosen, the average lies very close to the value suggested in equation (5.34).

The correlation energy ΔE_c is usually discussed in terms of a phase coherence time, or a diffusion time across the sample. Thus, there should be some correlation between its value and the sample size. It has been suggested that one can define the correlation energy as the energy difference for the phase difference achieved after diffusing a length $l_1 = \sqrt{D\tau_1}$ is unity. Here, the correlation time has been introduced, which is the time over which the acquired phase difference between two paths of energy difference δE, is of order unity. With this complicated set of quantities, the correlation energy may be expressed as [24]

$$\Delta E_C = \frac{\hbar D}{l_1^2} \sim \delta E. \tag{5.35}$$

This all seems to be very complicated, if not confusing. The end result seems to be that the correlation energy is given approximately by the mean energy separation of the various energy levels in the device, a quantity that has often been accused of various sins, but seldom has proved to be of much value in estimating the properties of mesoscopic devices. In the simulations discussed here [32], it was generally found that the value for the correlation energy lay in the range of 0.15–$0.25\ \mu eV$, where the lower value occurred at smaller values of the random potential. Typically, this value is a small fraction of the mean energy separation of the modes in the simulation. Moreover, the observed values seemed to have little correlation with the sample size, which is quite different from the behavior observed for the correlation magnetic field.

On the other hand, Thouless [38] suggested that the critical time was that time required for a particle to diffuse across the entire system in the current direction. If we take this view, then we can arrive at a correlation energy as [23, 39]

$$\Delta E_C = \frac{\hbar \pi^2 D}{L_z^2}. \tag{5.36}$$

This suggests a strong correlation between the observed correlation energy and the length of the system ($L_z = L$ used earlier). Yet, as remarked above, this does not seem to be found in our simulations, and the dearth of reports of experimental values for the correlation energy seem significant. The problem lies with the fact that the correlation energy increases weakly with the amplitude of the random potential. Such an increase requires a corresponding increase in the diffusivity according to either of the above two equations. In turn, this requires an increase in the scattering time as the amplitude of the random potential is increased, which is counter-intuitive. In fact, both of these suggested equations rely upon the Thouless argument, and neither depends upon the amplitude of the random potential. Hence, it is unlikely that any of these extrinsic arguments apply here. Thus, it appears that the correlation energy arises from an intrinsic property of the material, probably related to the scattering and decoherence processes introduced by the random potential itself, and suggests that the correlation energy is related to the

Fermi energy change needed to turn on or off a single Landauer mode in the sample, as suggested by (5.28).

5.5 Phase-braking time

In mesoscopic physics, the size of various effects often depends upon the phase breaking, or phase coherence, time τ_φ, or less directly upon the phase coherence length $l_\varphi = \sqrt{D\tau_\varphi}$. In chapter 3, these quantities were crucial to the description of weak localization, while just above they appeared in the discussion of the magnetic correlation ΔB_c. But, there were many descriptions of how to fit the weak localization peak, depending upon the dimensionality and the relative sizes of the coherence length, the mean free path, the thermal length and the various dimensions of the device. Similarly, there are various formula for ΔB_c depending upon these same variables in length. Nevertheless, one can estimate the values for the coherence length and the phase-breaking time from these measurements, especially as the various forms of the equations do not vary by orders of magnitude.

There are a variety of contributors to the decay of the coherence length itself with temperature. In many cases, however, these various mechanisms are not separated in an experimental measurement. Rather, only the change in the coherence length is measured by measuring, for example, the change in the behavior of the weak localization or the change in the amplitude of the fluctuations. In this section, we review a few of the measurements of the coherence length and, more importantly, the phase-breaking time. It is important to remember, however, that this length is not itself a key factor, but that it summarizes a variety of temperature variations due to the diffusion 'constant' and to the phase coherence time. In figure 5.12, we show the inelastic, or phase-breaking, time found in an AlGaAs/GaAs heterostructure from measurements of the weak localization [40]. The data are taken for three different carrier densities (three different samples) in these modulation doped

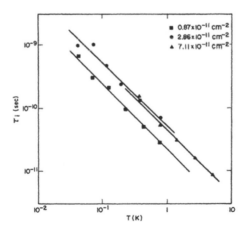

Figure 5.12. The temperature dependence of the inelastic, or phase-breaking, time in an AlGaAs/GaAs heterostructure for three different densities (three different samples). (Reprinted with permission from [40]. Copyright 1984 the American Physical Society.)

heterostructures. The values are extracted from a least-square fit to the digamma function fit to the weak localization lineshape. The solid lines are fits to the data and have the slope of $1/T$, which is the usual behavior for Q2D systems except at very low temperatures.

In general, one sees a saturation of the phase coherence length and the phase-breaking time at low temperatures, both in metals [41] and in semiconductors [42]. Usually, the transition from the temperature-independent behavior and the temperature-dependent behavior occurs in the vicinity of 0.1–1.0 K. Where the transition occurs seems to depend upon the quality of the material and the carrier density. There is just a hint that this might be 'occurring for the top curve in figure 5.12. This can be seen somewhat better in figure 5.13, in which the temperature dependence of the phase-breaking scattering rate ($1/\tau_\varphi$) is plotted for a pair of narrow GaAs wires formed in an AlGaAs/GaAs heterostructure [43]. As usual, the heterostructure is grown by molecular-beam epitaxy, and the wires are created by electron-beam lithography. Data for two wires, with nominal widths of 90 nm and 260 nm, are shown in the figure. The data are consistent with the phase-breaking rate depending linearly on the temperature, but the line intercepts the $T = 0$ edge at a value well above zero phase-breaking rate. This infers that there is a saturation in this quantity at the lowest temperatures. Rather than use the standard formulas, these latter authors Fourier transformed the combination of the weak localization and conductance fluctuation signals, with the magnetic field as the principal variable. By connecting the amplitude an of each Fourier component f_n with the effective area S_n that arises from the Fourier component through [43]

$$f_n(B) = \frac{2e}{h}\mathbf{B} \cdot \mathbf{S}_n, \tag{5.37}$$

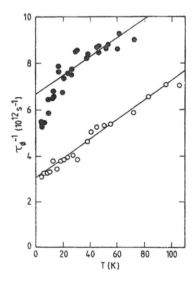

Figure 5.13. The temperature dependence of the phase-breaking rate for two narrow GaAs ribbon devices. (Reprinted with permission from [43]. Copyright 1988 Elsevier.)

they can determine the phase-breaking length from

$$a_n(T) = \beta \exp\left(-\frac{2a\sqrt{S_n}}{l_\varphi T}\right), \tag{5.38}$$

where $\alpha \sim 4.5$ is a parameter determined from a statistical analysis for the size of the loops most likely to contribute to conductance fluctuations.

Recognizing that the correlation function is not always well behaved, it has been suggested that one use the inflection point of the curve rather than the value at half-height [36]. These latter authors argue that the half-width is not always the best choice of metric, and that the inflection point, where the second derivative of the correlation point is zero, is the field separation at which the correlations change the fastest [36]. In numerical studies, they have found that the value of the phase-breaking time found from the inflection point is about a factor of four smaller than that determined from the half-width. In figure 5.14, we illustrate this with a correlation function determined from studies of fluctuations in graphene [44]. As usual, a monolayer of graphene was exfoliated onto an oxidized Si wafer, and the transport studied at low temperatures. From the conductance fluctuations, the correlation function could be computed. In figure 5.14, both the correlation function and the derivative (dotted curve) are shown as a function of the separation magnetic field. From the derivative, it is easy to find the inflection point of the correlation function, and this is indicated by a vertical dashed line. The same approach has also been applied to studies of the weak localization and consistent results are obtained there as well. In figure 5.15, the phase-breaking rates determined from this inflection point analysis are shown for the graphene sample. Two curves are shown here, and the difference is that an in-plane magnetic field of 6 T has been applied for the lower curve. It is thought that the difference lies in the role of magnetic defects in, or on, the graphene. Magnetic defects have been studied for some time [45, 46]. By applying the in-plane magnetic field, it is assumed that these magnetic defects are polarized and their random potential is thereby removed [44], leading to the lower value of dephasing rate. It is also clear that there is a saturation in the dephasing rate at the lowest temperatures (below about 0.1 K in this case), which is supported by

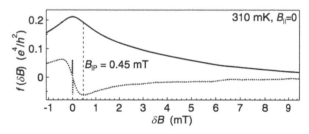

Figure 5.14. A typical autocorrelation function for the magnetic-field-induced conductance fluctuations (solid curve). The derivative is shown as the dotted curve, and the inflection point of the correlation function is indicated by the vertical dashed curve. (Reprinted with permission from [44]. Copyright 2013 the American Physical Society.)

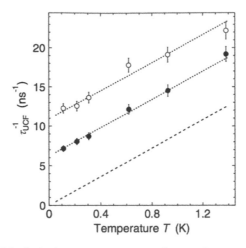

Figure 5.15. Dependence of the dephasing rate on temperature for a graphene sample. The open circles are for the absence of an in-plane magnetic field, while the filled circles have an in-plane magnetic field of 6 T. The dotted line shows the expected results if there were no saturation in the dephasing rate at low temperatures. (Reprinted with permission from [44]. Copyright 2013 the American Physical Society.)

some theories of the carrier–carrier interactions [47]. But, as we remarked earlier, the saturation at low temperatures occurs in nearly all materials that have been studied to date.

Generally, the conductance fluctuations in a magnetic field are studied at lower values of the magnetic field. As one increases the magnetic field, a breakdown in the scaling relationships has been reported to occur [48]. This generally is seen by an increase in the correlation magnetic field [49], as shown in figure 5.16. In this study, narrow samples of GaAs wires, formed in an AlGaAs/GaAs/AlGaAs double heterostructure via electron-beam lithography, were used to study the fluctuations in a magnetic field. Results for several different temperatures are shown in the figure, but they seem to be relatively similar. Just above $\omega_c \tau = 1$, the correlation magnetic field begins to increase. Here, $\omega_c = eB/m^*$ is the cyclotron frequency and τ is the scattering time inferred from the mobility (about 8600 cm^2 V^{-1} s^{-1}) of the material. In this case, the GaAs quantum well was 10 nm thick and the wires were about 700 nm wide. In the standard analysis, this increase in the correlation magnetic field would lead one to conclude that the phase-breaking time was increasing, but this is not correct due to the breakdown of the scaling at this magnetic field. However, this allows a different approach to determine the phase-breaking time.

Generally, in diffusive samples with relatively low mobility, in the higher magnetic fields, the correlation magnetic field increases with increasing magnetic field without any increase in the amplitude of the fluctuations. In high mobility material, however, the increase in the correlation magnetic field is accompanied by an increase in the amplitude of the fluctuations [50, 51]. Normally, one thinks that the diffusion constant, part of the phase coherence length, decreases as the magnetic field increases, but this would lead to a reduction in the coherence length and an

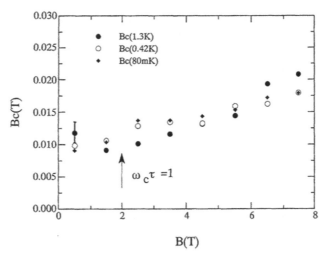

Figure 5.16. Variation in the correlation magnetic field at high values of the magnetic field, with data shown for three different temperatures. (Reprinted with permission from [49]. Copyright 1993 IOP Publishing. Reproduced with permission. All rights reserved.)

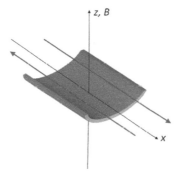

Figure 5.17. Schematic for a parabolic potential in the y-direction and a magnetic field normal to this.

increase in the correlation magnetic field, it should also lead to a reduction in the amplitude of the correlations, which is not observed. There is enough rigor in the theoretical arguments [39] if one reinterprets the meaning of the phase coherent areas, which are represented in equation (5.34) by the product Wl_φ. The Nottingham group has argued that there are (at least) two kinds of phase coherent areas to be considered [52]. One of these is the bulk regions of the entire sample, which is important at lower magnetic fields. However, once one reaches $\omega_c\tau = 1$, the current-carrying trajectories start to be pushed into edge states, and the regions important for the fluctuations get squeezed into small areas at the edges of the sample. Let us first examine the edge-state formation.

Consider the diagram shown in figure 5.17. Here, we have a narrow strip of semiconductor with confinement in the y-direction, indicated by the blue quadratic potential. The axis of the semiconductor lies along the x-direction, and we will consider a magnetic field in the z-direction, as indicated. We have chosen the quadratic confinement potential for ease of calculation, as we are familiar with the harmonic oscillator of appendix B. We want to include the magnetic field in this formulation, so we introduce the field via the Landau gauge, with a vector potential

$$\mathbf{A} = -By\mathbf{a}_x. \tag{5.39}$$

As we can see in (C.19) of appendix C, this goes into the momentum via Peierls' substitution to give the resulting Schrödinger equation

$$\frac{1}{2m}(-i\hbar\nabla - e\mathbf{A})^2\psi(x, y) + \frac{1}{2}m\omega^2 y^2\psi(x, y) = E\psi(x, y), \tag{5.40}$$

in the two dimensions of an assumed heterostructure interface electron gas. Inserting equation (5.39) and expanding the first quadratic term leads to

$$\left[-\frac{\hbar^2}{2m}\left(\frac{\partial^2}{\partial x^2} + \frac{\partial^2}{\partial y^2}\right) - \frac{ie\hbar B}{m}y\frac{\partial}{\partial x} - \frac{e^2B^2y^2}{2m}\right]\psi(x, y)$$
$$+ \frac{1}{2}m\omega^2 y^2\psi(x, y) = E\psi(x, y). \tag{5.41}$$

Let us now introduce the hybrid frequency

$$\Omega^2 = \omega^2 + \omega_c^2, \tag{5.42}$$

so that we can reduce (5.41) to

$$\left[-\frac{\hbar^2}{2m}\left(\frac{\partial^2}{\partial x^2} + \frac{\partial^2}{\partial y^2}\right) - \frac{ie\hbar B}{m}y\frac{\partial}{\partial x}\right]\psi(x, y) + \frac{1}{2}m\Omega^2 y^2\psi(x, y) = E\psi(x, y). \tag{5.43}$$

At this point, we postulate that the electrons are free to move in the x-direction, and can be represented as plane waves in this direction. Hence, we assert that the wave function can be written as

$$\psi(x, y) = e^{ikx}\psi(y), \tag{5.44}$$

Now, equation (5.43) can be rewritten as

$$\left[-\frac{\hbar^2}{2m}\frac{\partial^2}{\partial x^2} + \hbar\omega_c ky + \frac{1}{2}m\Omega^2 y^2\right]\psi(y) = \left[E - \frac{\hbar^2 k^2}{2m}\right]\psi(y). \tag{5.45}$$

By completing the square for the potential terms, we can write this as

$$-\frac{\hbar^2}{2m}\frac{\partial^2\psi(y)}{\partial x^2} + \frac{1}{2}m\Omega^2(y + y_0)^2\psi(y) = \left[E - \frac{\hbar^2 k^2}{2m}\left(1 - \frac{\omega_c^2}{\Omega^2}\right)\right]\psi(y), \tag{5.46}$$

where

$$y_0 = \frac{\hbar k \omega_c}{m\Omega^2} \xrightarrow[\omega_c \gg \omega]{} \frac{\hbar k}{eB}. \tag{5.47}$$

In this last high magnetic field limit, the correction term to the energy in the square brackets of equation (5.46) vanishes. This term y_0 represents a shift of the electron trajectory in the magnetic field. Remember that the sign on the electronic charge is negative so that for electrons moving in the positive x-direction ($k > 0$), the carriers are shifted in the negative y-direction, and for electrons moving in the negative x-direction, the shift is in the positive y-direction. If one thinks about the drift of the electrons (we have not specified any longitudinal electric field or potential) arising from the electromagnetic $\mathbf{E} \times \mathbf{B}$ forces, the electric fields are given by the gradients of the confinement potential, and point toward the center of the potential well. That is, for $y < 0$, the electric fields point in the positive y-direction, and conversely for the other side of the potential well. Hence, the $\mathbf{E} \times \mathbf{B}$ forces produce the velocities in the directions shown, when one remembers that the charge is negative. Another way of thinking about this is that the electrons circulate around the magnetic field in a cyclotron motion. Those shifted in the positive y-direction hit the wall and are specularly reflected which results in a bouncing orbit moving in the direction indicated. The net result is that the left-hand side of equation (5.46) gives rise to the harmonic oscillator energy levels, but the center of the wave functions are shifted to one side of the potential or the other, depending upon their motion in the x-direction.

With multiple quantized levels full, the multiple trajectories lead to interference and conductance fluctuations just as in the case of normal quantum interferences. As mentioned above, the interferences in the high magnetic field case lead to a different type of phase coherent area, as shown in figure 5.18, where the multiple trajectories are pushed to the side of the sample. The Nottingham group has carried out simulations which suggest that the length of the phase coherent area is changed very little, but the width of this area is now determined by the cyclotron diameter [52]. As a result, the correlation magnetic field, which depends upon this phase coherent area, will increase as the magnetic field increases, and in a linear fashion.

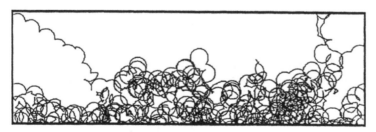

Figure 5.18. Numerical simulation of classical electron trajectories showing their confinement along the edge of the sample, which leads to enhanced diffusion for a large magnetic field. (Reprinted with permission from [52]. Copyright 1993 the American Physical Society.)

Using the ideas from the Nottingham group, we can construct the effective phase coherent area in a slightly different manner. We note that the phase coherence length is given simply as before as

$$l_\varphi = \sqrt{D\tau_\varphi}, \tag{5.48}$$

and the inelastic mean free path is

$$l_{\text{inel}} = v_F\tau = l_\varphi\sqrt{\frac{\tau_\varphi}{\tau}}. \tag{5.49}$$

In the systems of interest, we have $d = 2$ for most purposes; that is, we are primarily considering Q2D systems. For relatively high mobility material, the length of the average phase retaining trajectory is of the order of $N\pi r_c^2/2$, where N is the number of bounces along the edge of the sample in the high magnetic field case [53]. Here, the length of this path is $N\pi r_c \sim l_{\text{inel}}$. This leads to the replacement of the phase coherent area as

$$Wl_\varphi \rightarrow \frac{r_c l_{\text{inel}}}{2} = \frac{\hbar^2 k_F^2 \tau_\varphi}{2meB}, \tag{5.50}$$

where we have used $r_c = k_F l_B^2 = \hbar k_F/eB$. Then, using equation (5.34), the correlation magnetic field in these high magnetic field regions is approximately

$$\Delta B_c = \frac{2mheB}{e\hbar^2 k_F^2 \tau_\varphi} = \frac{hB}{E_F\tau_\varphi}. \tag{5.51}$$

A slightly different form is found by the Nottingham group [52], although both forms retain the linear increase with magnetic field. Hence, by determining the correlation magnetic field as a function of the magnetic field itself, one can determine directly the value for the phase-breaking time. The density for the sample in figure 5.16 was about 3.7×10^{11} cm^{-2}, which corresponds to a Fermi energy of about 13.3 meV, and this gives a phase-breaking time of about 1.2×10^{-10} s.

Problems

1. An alternative to the uniform distribution of energies assumed in section 5.1 is to take a Lorentzian distribution as

$$P(E_i) = \frac{1}{\pi}\left(\frac{\Gamma}{E_i^2 + \Gamma^2}\right).$$

Repeat the arguments of that section and show that the result is a smooth band with a width of ΔE when $\Gamma = 0$, and a band with localized edges leading to a mobility edge given by

$$E_{\text{mob}} = \pm\sqrt{\left(\frac{\Delta E}{2}\right)^2 - \Gamma^2}$$

for $\Delta E > \Gamma > 0$.

2. From the data in figure 5.6, estimate the hopping range, as a function of temperature, for each of the curves shown. Work only in the range below 10^{-8} s, where the curves are linear in the plot.
3. In figure 5.8, the magnetoresistance curves seem to be somewhat asymmetric around B = 0. Explain why this may be the case.

References

[1] Anderson P W 1958 *Phys. Rev.* **109** 1492
[2] Ferry D K 2019 *Semiconductors: Bonds and bands* (Bristol: IOP Publishing)
[3] Jaros M 1982 *Deep Levels in Semiconductors* (Bristol: Adam Hilger)
[4] Lang D V and Logan R A 1977 *Phys. Rev. Lett.* **39** 635
[5] Goychuk I, Hartmann L and Hänggi P 2001 *Chem. Phys.* **268** 151
[6] Ankerhold J and Lehle H 2004 *J. Chem. Phys.* **120** 1436
[7] Kittel C 1966 *Introduction to Solid State Physics* 7th edn (New York: John Wiley)
[8] Baranowski J M, Grynberg M and Porowski S 1982 *Handbook on Semiconductors* vol 1 ed W Paul (Amsterdam: North-Holland) ch 6
[9] Ma N and Jena D 2014 *Phys. Rev.* X **4** 011043
[10] Zhu W J *et al* 2014 *Nat. Commun.* **5** 3087
[11] Xiao Z, Song J, Ferry D K, Ducharme S and Hong X 2017 *Phys. Rev. Lett.* **118** 236801
[12] Mott N F and Davis E A 1979 *Electronic Processes in Non-Crystalline Materials* (Oxford: Clarendon)
[13] Mott N F 1968 *J. Non-Cryst. Solids* **1** 1
[14] Mott N F 1969 *Phil. Mag.* **19** 835
[15] Hartstein A and Fowler A B 1976 *J. Phys. C: Solid State Phys.* **8** L249
[16] Fowler A B and Hartstein A 1980 *Phil. Mag.* B **42** 949
[17] Timp G, Fowler A B, Hartstein A and Butcher P N 1986 *Phys. Rev.* B **34** 8771
[18] Beenakker C W J and van Houten H 1988 *Phys. Rev.* B **37** 6544
[19] Chandrasekhar V, Santhanam P and Prober D E 1990 *Phys. Rev.* B **42** 6823
[20] Thornton T J, Pepper M, Ahmed H, Davies G J and Andrews D 1987 *Phys. Rev.* B **36** 4514
[21] Chang A M, Timp G, Cunningham J E, Mankiewich P M, Behringer R E, Howard R E and Baranger H U 1988 *Phys. Rev.* B **37** 2745
[22] Landauer R 1957 *IBM J. Res. Develop.* **1** 223
[23] Lee A and Stone A D 1985 *Phys. Rev. Lett.* **55** 1622
[24] Beenakker C W J and van Houten H 1991 *Solid State Phys.* vol 44 ed H Ehrenreich and D Turnbull (New York: Academic) pp 1–228
[25] Grincwajg A, Edwards G and Ferry D K 1996 *Physica* B **218** 92
[26] Kaplan S B and Hartstein A 1986 *Phys. Rev. Lett.* **56** 2403
[27] Bohra G, Somphonsane R, Aoki N, Ochiai Y, Akis R, Ferry D K and Bird J P 2012 *Phys. Rev.* B **86** 161405
[28] Liu B, Akis R, Ferry D K, Bohra G, Somphonsane R, Ramamoorthy H and Bird J P 2016 *J. Phys. Condens. Matter* **28** 135302
[29] Liu B, Akis R and Ferry D K 2014 *J. Comput. Electron.* **13** 950
[30] Büttiker M 1988 *Phys. Rev.* B **38** 9375
[31] Ojeda-Aristizabal C, Monteverde M, Weil R, Ferrier M, Guéron S and Bouchiat H 2010 *Phys. Rev. Lett.* **104** 186802
[32] Liu B, Akis R and Ferry D K 2014 *J. Phys. Cond. Matter* **25** 395802

[33] González J, Guinea F and Vozmediano M A H 1996 *Phys. Rev. Lett.* **77** 3589

[34] Hwang E H, Hu B Y-K and Das Sarma S 2007 *Phys. Rev. Lett.* **99** 226801

[35] Chakravarty S and Schmid A 1986 *Phys. Rep.* **140** 193

[36] Lundeberg M B, Renard J and Folk J A 2012 *Phys. Rev.* B **86** 205413

[37] Haucke H, Washburn S, Benoit A D, Umbach C P and Webb R A 1990 *Phys. Rev.* B **41** 12454

[38] Thouless D J 1977 *Phys. Rev. Lett.* **39** 1167

[39] Lee P A, Stone A D and Fukuyama H 1987 *Phys. Rev.* B **35** 1039

[40] Lin B J F, Paalanen M A, Gossard A C and Tsui D C 1984 *Phys. Rev.* B **29** 927

[41] Washburn S *et al* 1988 *Physics and Technology of Submicron Semiconductor Structures* ed H Heinrich, G Bauer and F Kuchar (Berlin: Springer) pp 98–107

[42] de Graaf C, Caro J and Radelaar S 1992 *Phys. Rev.* B **46** 12814

[43] Taylor R P, Leadbeater M L, Whittington G P, Main P C, Eaves L, Beaumont S P, McIntyre I, Thoms S and Wilkinson C D W 1988 *Surf. Sci.* **196** 52

[44] Lundeberg M B, Yang R, Renard J and Folk J A 2013 *Phys. Rev. Lett.* **110** 156601

[45] Bergmann G 1986 *J. Magn. Magn. Mater.* **54–57** 1433

[46] Amarai V S 1990 *J. Phys.: Condens. Matter* **2** 8201

[47] Ferry D K, Goodnick S M and Bird J P 2009 *Transport in Nanostructures* 2nd edn (Cambridge: Cambridge University Press) section 8.3.9

[48] Geim A K, Main P C, Beton P H, Eaves L, Beaumont S P and Wilkinson C D W 1992 *Phys. Rev. Lett.* **69** 1248

[49] Ochiai Y, Onishi T, Kawabe M, Ishibashi K, Bird J P, Aoyagi Y and Sugano T 1993 *Japan. J. Appl. Phys.* **32** 528

[50] Timp G, Chang A M, Mankiewich P, Behringer R, Cunningham J E, Chang T Y and Howard R E 1987 *Phys. Rev. Lett.* **59** 792

[51] Bird J P, Ishibashi K, Ochiai Y, Lakrimi M, Grassie A D C, Hutchings K M, Aoyagi Y and Sugano T 1995 *Phys. Rev.* B **52** 1793

[52] Brown C V, Geim A K, Foster T J, Langerak G M and Main P C 1993 *Phys. Rev.* B **47** 10935

[53] Ferry D K, Edwards G, Yamamoto K, Ochiai Y, Bird J, Ishibashi K, Aoyagi Y and Sugano T 1995 *Japan. J. Appl. Phys.* **34** 4338

IOP Publishing

Transport in Semiconductor Mesoscopic Devices
(Second Edition)

David K Ferry

Chapter 6

The quantum Hall effect

In mesoscopic systems, a static magnetic field may have a profound effect on the electronic and transport properties. The application of high magnetic fields to these mesoscopic devices is an invaluable degree of freedom available to the experimentalist in probing the system, and also to the theoretician. Magnetic fields give rise to new fundamental behaviors not observed in bulk-like systems, such as the quantum Hall effect [1]. The fundamental quantity characterizing a magnetic field is the magnetic flux density, B, which in MKS units is measured in Teslas. In the study of semiconductor mesoscopic devices, low fields usually correspond to fields less than 1 T, which is the regime in which low-field magnetotransport experiments such as Hall effect measurements are usually performed. Magnetic intensities of 10–20 T are obtainable in the usual university and commercial research environments, using superconducting alloy coils with high critical magnetic fields (superconductivity can be destroyed by a high magnetic field) immersed in a liquid He Dewar. Higher magnetic fields are obtainable only at a few large-scale facilities scattered around the world.

6.1 The Shubnikov–de Haas effect

When we consider mesoscopic systems such as quantum wells, wires, and dots, the effect of a magnetic field may be roughly separated into two cases. In the first, the magnetic field is parallel to one of the directions of free-electron propagation; in the other, it is perpendicular to the free-electron motion of the system. Qualitatively, when free particles are subject to a magnetic field, they experience a Lorentz force

$$\mathbf{F} = e(\mathbf{E} + \mathbf{v} \times \mathbf{B}), \tag{6.1}$$

where $\mathbf{v}$ is the velocity of the carrier and $\mathbf{E}$ is the electric field. Since the force is always perpendicular to the direction of travel of the particle, its motion in the absence of other forces is circular, with angular frequency given by the cyclotron frequency, which we have seen in earlier chapters, given as

$$\omega_c = \frac{eB}{m^*}.$$

(6.2)

Quantum mechanically, the circular orbits associated with the Lorentz force must be quantized in analogy to the orbital quantization occurring about a central potential, for example, about the nucleus of an atom. Since the particles execute time-harmonic motion similar to the motion in a harmonic oscillator potential, the energy associated with the motion in the plane perpendicular to the magnetic field is expected to be quantized. If we now consider the magnetic field applied perpendicular to the plane of a 2DEG, then the entire free-electron-like motion in the plane parallel to the interface is quantized, and the energy spectrum becomes completely discrete. Indeed, we saw this in the last chapter, where we coupled the magnetic motion to that of a harmonic oscillator. If we take equation (5.46) and remove the harmonic oscillator by letting $\omega \to 0$, $\Omega \to \omega_c$, we then have

$$-\frac{\hbar^2}{2m}\frac{\partial^2\psi(y)}{\partial x^2} + \frac{1}{2}m\omega_c^2(y+y_0)^2\psi(y) = E\psi(y)$$

(6.3)

where

$$y_0 = \frac{\hbar k}{eB}$$

(6.4)

is the position offset. By comparison with equation (B.1) in appendix B, we see that the magnetic field introduces its own harmonic oscillator. The energy is then given, in comparison with equation (B.11), to be

$$E_n = \left(n + \frac{1}{2}\right)\hbar\omega_c, \quad n = 0, 1, 2....$$

(6.5)

The various energy levels of this harmonic oscillator are termed the Landau levels. As the electron is traveling in a closed circular orbit, we can refer to the radius of this orbit as the cyclotron radius

$$r_c = \sqrt{\left(n + \frac{1}{2}\right)\frac{2\hbar}{eB}} = k_F\frac{\hbar}{eB} \equiv k_F l_B^2.$$

(6.6)

This latter form introduces the magnetic length as well.

An important point about the formation of the Landau levels is the effect they have on the density of states. Let us consider simply a two-dimensional system in which the magnetic field is oriented normal to the plane of this system. Normally, in such a two-dimensional system, the density of states is uniform at $m^*/\pi\hbar^2$, and the energy is continuous. The formation of the Landau levels in energy means that the density of states becomes non-uniform, with these states migrating to the momentum space orbits of the electrons. This leads to a localized density of states which is generally of the form [2]

$$\rho(E, E_n) \cong \frac{1}{2\pi l_B^2} \sum_n \frac{1}{\sqrt{1 + \left(\frac{E - E_n}{\Gamma}\right)^2}},$$ (6.7)

where E_n is given by equation (6.5) and the broadening is generally given by

$$\Gamma^2 = \frac{2\hbar^2 \omega_c}{\pi\tau}.$$ (6.8)

Hence, we see that, in general, the broadening of the levels is proportional to the square root of the magnetic field. But, as we increase the magnetic field, the levels move further apart and the density of states peak in each level increases linearly with the magnetic field.

As the magnetic field is raised, each of the Landau levels rises to a higher energy. However, the Fermi energy remains fixed, so at a critical magnetic field, the highest filled Landau level will cross the Fermi energy. In general, at a given magnetic field and Fermi energy, the Landau levels are filled up to some $n_{\max}$ determined by the density as

$$n_s r_c^2 = 2 \sum_{n=0}^{n_{\max}} \left(n + \frac{1}{2}\right) \frac{\hbar}{m^* \omega_c} \;\rightarrow\; r_c^2 = \sum_{n=0}^{n_{\max}} (2n + 1) \frac{\hbar}{eBn_s}.$$ (6.9)

As the magnetic field increases in value, the number of available states in each Landau level increases correspondingly. Thus, when the highest occupied Landau level pushes up through the Fermi energy, the remaining carriers drop into the lower Landau levels, and this reduces the number of terms in the sum in equation (6.9). This can only occur due to the increase of the density of states in each Landau level, so a given Landau level can hold more electrons as the magnetic field is raised. As a consequence, the average radius (obtained from the squared average of the radius) is modulated by the magnetic field, going through a maximum each time a Landau level crosses the Fermi level and is emptied. From equation (6.9), it appears that the radius is periodic in the inverse of the magnetic field. At least in two dimensions, this periodicity is proportional to the areal density of the free carriers and can be used to measure this density. The effect, commonly called the Shubnikov–de Haas effect, is normally applied by measuring the conductivity in the plane normal to the field. In figure 6.1, we show schematically how this emptying of the Landau levels occurs. As shown in the figure, the Fermi energy appears to oscillate as the magnetic field increases.

The separation of the Landau levels is $\hbar\omega_c$, and all the states in the range $E_n \pm \hbar\omega_c/2$ are coalesced into the nth Landau level. The density of states in two dimensions is $m^*/\pi\hbar^2$, so the number of carriers in each Landau level is the product of this density of states and $\hbar\omega_c$, or $m^*\omega_c/\pi\hbar = 1/\pi l_B^2$. Thus, the number of filled Landau levels is $\pi l_B^2 n_s$ and this leads to a periodicity of

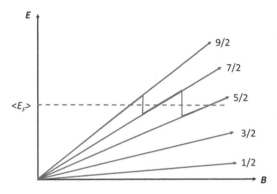

Figure 6.1. The Landau levels 'fan out' as the magnetic field is increased, and the electrons drop from the topmost filled level into lower levels as the states become available.

$$\Delta\left(\frac{1}{B}\right) = \frac{e}{\pi\hbar n_s}. \tag{6.10}$$

To understand the conductivity oscillations, it is necessary to reintroduce the scattering process. Without this process, the electrons remain in the closed Landau orbits. However, the scattering process can cause the electrons to 'hop' slowly from one orbit to the next in real space by randomizing the momentum. This leads to a slow drift of the carriers in the direction of the applied field (we will see below that the edge states are mainly responsible for this). The drift is slower than in the absence of the magnetic field because the tendency is to have the carriers remain in the orbits. Here the scattering induces the motion instead of retarding the motion as in the field-free case (one can compare figure 5.18, for example). Thus the conductivity is expected to be less in the presence of the magnetic field. When the Fermi level lies in a Landau level, away from the transition regions, there are many states available for the electron to gain small amounts of energy from the applied field and therefore contribute to the conduction process. On the other hand, when the Fermi level is in the transition phase, the upper Landau levels are empty and the lower Landau levels are full. Thus there are no available states for the electron to be accelerated into, and the conductivity drops to zero in two dimensions. In three dimensions it can be scattered into the direction parallel to the field (the z-direction), and this conductivity provides a positive background on which the oscillations ride. A sample of these oscillations in two dimensions are shown in figure 6.2, which is a typical measurement of the longitudinal resistance. In this case, the sample is a quantum well of AlGaAs grown on GaAs. The density of the 2DEG is approximately 3×10^{11} cm^{-2} and the mobility was 3×10^5 cm^2 V^{-1} s^{-1}, and the measurement was made at 1.5 K. The Landau index n is shown for several of the minima. The small dip between the $n = 2$ and $n = 3$ minima is the onset of a spin split level [3], which will be discussed further in the next section.

It may be noted that the resistance goes to zero at certain values of the magnetic field, but this does not indicate superconductivity. Rather, it is a result of the conductance going to zero at these magnetic fields. In the Q2D system, the

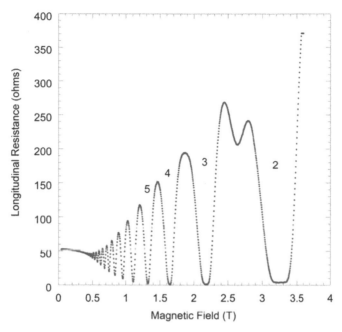

Figure 6.2. Typical Shubnikov–de Haas measurements in an AlGaAs/GaAs heterostructure. The various oscillations arise from the emptying of the Landau levels for two different subbands as the magnetic field is increased. The integers refer to the particular Landau level.

conductivity is a matrix due to the transverse electric fields that arise from the Hall effect. This matrix may be written as

$$[\sigma] = \begin{bmatrix} \sigma_{xx} & \sigma_{xy} \\ -\sigma_{xy} & \sigma_{xx} \end{bmatrix},$$ (6.11)

and when this is inverted, the longitudinal resistivity becomes

$$\rho_{xx} = \frac{\sigma_{xx}}{\sigma_{xx}^2 + \sigma_{xy}^2}$$ (6.12)

and the non-zero nature of the transverse (off diagonal) terms leads to zero resistivity when the longitudinal conductivity vanishes. This just means that the electric field is perpendicular to the current flow, which leads to vanishing dissipation, although the material is not a superconductor.

Equation (6.10) has an important point to tell us, and that is that the Shubnikov–de Haas resistance, the longitudinal resistance of the sample, is periodic in the inverse magnetic field. To illustrate this, we re-plot the data in figure 6.2 as a function of the inverse magnetic field in figure 6.3. One can almost pick out the periodicity of this curve. If there are more than a single subband, it becomes more complicated. First, the oscillations arising from the second subband have a different periodicity in an inverse magnetic field, and this oscillation causes interference with that of the lowest subband. As a result, one must Fourier transform data like

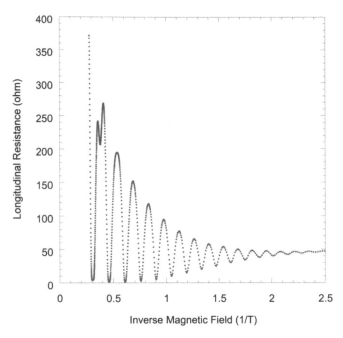

Figure 6.3. The data from figure 6.2 are re-plotted here as a function of the inverse magnetic field, corresponding to (6.10). Now, the periodicity is clearly evident.

figure 6.3 and examine the various frequencies that are present in the signal. With luck, only two frequencies will be present. Hence, the Fourier transform is one of the most useful tools to use in data analysis.

The second problem that arises is that we have assumed that the density of states is spin degenerate. At the high magnetic fields that are used in this experiment, this is not always the case. The Zeeman energy splits the Landau levels into two different levels, one with spin up and the other with spin down. This splitting is given by

$$\delta E = \pm \frac{1}{2} g \mu_B B, \tag{6.13}$$

where μ_B is the Bohr magneton ($e\hbar/2m_0 = 58\ \mu\text{eV T}^{-1}$) and g is the Landé factor, but is usually a 'fudge' factor that differs for electrons in various semiconductors [4]. In fact, it is usually found that the g factor dramatically increases with magnetic field [2, 5]. As the density in each Landau level increases, the electron–electron interaction gets stronger and this is thought to lead to the enhanced g factor. In practice, however, as we will see in the next section, the enhanced g factor leads to the spin-split Landau levels being spaced between the Landau ladder of levels, but the totality of the levels are not equally spaced. Because of this doubling of the number of levels at high magnetic fields, a different symbol is usually used, and this is ν. Thus, $\nu = 1$ and 2 refer to the two spin-split levels that emanate from the $n = 0$ landau level. $\nu = 3$ and 4 refer to the two spin-split levels that emanate from the $n = 1$ landau level, and so on. We will attach more significance to this parameter in the next section.'

6.2 The quantum Hall effect

In the previous chapter, we treated a conducting channel which had a harmonic oscillator potential in the transverse direction and a magnetic field normal to the channel. In this situation, we discovered that the motion of the electrons was offset to one side of the channel or the other depending upon the sign of the momentum. This led to what we called edge states. In that situation, the energy of a particular level was given by

$$E_n = \left(n + \frac{1}{2}\right)\hbar\omega_c + \frac{\hbar^2 k^2}{2m^*}\left(\frac{\omega^2}{\Omega^2}\right), \tag{6.14}$$

where $\Omega^2 = \omega^2 + \omega_c^2$. When the magnetic field is large, the levels are primarily Landau levels, but they continue to rise in energy as they approach the sides of the confining potential. The confinement energy raises the energy of the Landau levels. We find essentially the same behavior if the confinement is due to the physical confinement of a finite width sample of semiconductor material. The potentials at the edges of the ribbon sample are basically the work function of the material, and decay rapidly into the sample from the edge. Nevertheless, as we near the edge of the sample, the Landau levels rise above their values in the interior of the sample. This is shown schematically in figure 6.4, in which we have identified by the various energy levels by the index ν which means that these are spin slit levels. When the Fermi energy lies in one of the bulk levels, then the conductivity is high and one sees the peaks that appear in figures 6.2 and 6.3. But, when the Fermi level lies as shown in

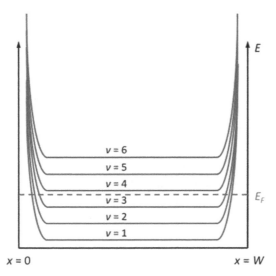

Figure 6.4. Schematic view of spin-split Landau level behavior near the edge of a finite width sample. While these levels are shown equally spaced, this is not the real case, but this will not impact the edge state picture.

figure 6.4, the bulk levels are completely full, and electrons exist up to where the levels cross the Fermi energy at the edges of the sample. We see that the index ν not only indicates the upper most filled bulk level, but also indicates the number of edge states that cross the Fermi energy.

The three edge states that are indicated in figure 6.4 are essentially one-dimensional channels that flow along the edges of the sample. From the discussion in chapter 2, and the Landauer formula, we would expect each of these spin-split channels would produce a conductance of e^2/h, but as we noted in the previous section, the net conductivity is zero. We could think about this as arising from the fact that the currents are oppositely directed on the two sides of the sample and therefore cancel, but this would be misleading. In fact, the conductance vanishes as there is no longitudinal electric field when we are in the edge-state regime. The voltage that may be exist between the ends of the sample must appear entirely as the Hall voltage.

We can follow the same logic that led to equation (6.10) by noting that the spacing of the spin-split levels is approximately $\hbar\omega_c/2$ so that the density of carriers in each level is given by $m^*\omega_c/h = eB/h = 1/2\, l_B^2$. The Hall resistance is given by the product of the Hall constant and the magnetic field (basically E_y/J_x in the normal configuration in which the current flows in the x-direction, and the Hall voltage is measured in the y-direction). If we put this density into the Hall constant, then we find the Hall resistance as

$$R_H = E_y/J_x = \frac{h}{\nu e^2}, \tag{6.15}$$

where ν is our index of the highest filled level and the number of edge states. In figure 6.5, we plot the Hall resistance for the data used in figures 6.2 and 6.3. There are clear plateaus in the curve, which are the levels given by equation (6.14). The highest plateau shown here is for $\nu = 4$ (in the lowest subband). Of course, this result occurs only when the Fermi energy is not in one of the bulk levels, but the result is phenomenal in that the Hall resistance is the ratio of two fundamental constants and a pure number. This is the quantum Hall effect, discovered by Klaus von Klitzing in 1980 [1], and for which he won the Nobel prize. Since 1990, the quantum Hall effect has been adopted as the standard value for the SI ohm [6] by nearly all countries in the world. The quantity $h/e^2 = 25\,812.807\ \Omega$ (this has now been standardized by the redefinition of the SI units in 2019) is now referred to as the von Klitzing constant (R_K) and experiments have demonstrated that this value is known to 3.5 parts in 10^{10}. That is, it is known to better than nine significant digits. It is remarkable as well that this is independent of the material being studied.

One might ask why the Hall resistance is so stable. This is because the set up that leads to the quantum Hall effect is topologically stable. It is independent of the size of the sample under test. It is not affected by the surface conditions or the manner of isolating the device. We have already remarked that it is not dependent upon the material being studied. Because of the topology of the experiment, we are led to the result that the Hall resistance is a result of the gauge invariance and depends only

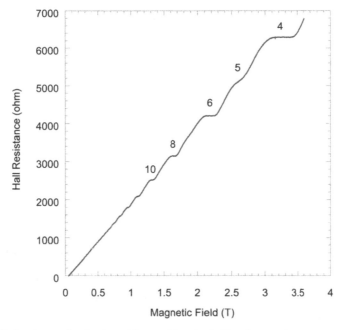

Figure 6.5. The Hall resistance for the data of figures 6.2 and 6.3. The plateaus that appear are a manifestation of the quantum Hall effect [1]. The integers ν have been inserted by several of the plateaus. Only one spin-split level is observable.

upon the ratio of these fundamental constants [7]. The stability of the quantum Hall effect is remarkable, and because of this it has become the international standard for resistance. One important aspect of equation (6.17) is that the value depends only upon fundamental constants, and efforts have been taken to assure that its value is universal among various materials [8], as mentioned above.

6.3 The Büttiker–Landauer approach

In section 2.3, we developed the Landauer equation for the transport of carriers through a Q1D system in which the transport was dominated by channels or modes. The Shubnikov–de Haas effect and the quantum Hall effect both raise the need for a multi-contact, or multi-probe version of this approach. This extension of the channel idea was primarily as derived by Büttiker [9, 10]. As a generalization of the two contact case, current probes are considered phase-randomizing agents that are connected through ideal leads to reservoirs (characterized by a chemical potential μ_i, for the ith probe) which emit and absorb electrons incoherently. In general, the 'probes' need not be physical objects. They can be any phase-randomizing entity, such as inelastic scattering centers distributed throughout the sample [10]. Here, we will assume that voltage probes are phase randomizing, although there is some disagreement over the validity of this assumption of phase randomization.

To simplify the initial discussion, we first assume that the leads contain only a single channel with two states at the Fermi energy corresponding to positive and negative velocity. The various leads are labeled $i = 1, 2,\ldots$, each with corresponding chemical potential μ_i. A scattering matrix may be defined which connects the states in lead i with those in lead j. As discussed in chapter 2, this relationship is $T_{ij} = |t_{ij}|^2$ and is the transmission coefficient into lead i of a particle incident on the sample in lead j, while $R_{ii} = 1 - |t_{ii}|^2$ is the probability of a carrier incident on the sample from lead i to be reflected back into that lead. In the general case, there may be a magnetic flux Φ present which penetrates the sample. The elements of the scattering matrix must satisfy the following reciprocity relations due to time-reversal symmetry

$$t_{ij}(\Phi) = t_{ji}(-\Phi). \tag{6.16}$$

Because of this symmetry in the transmission itself, both the mode reflection and mode transmission possess this same symmetry due to the circuit reciprocity theorem.

In order to simplify the discussion, an additional chemical potential is introduced, μ_0, which is less than or equal to the value of all the other chemical potentials so that all states below μ_0 can be considered filled (at $T_L = 0$ K). With reference then to μ_0, the total current injected from lead i is given as $(2e/h)\Delta\mu_i$, where $\Delta\mu_i = \mu_i - \mu_0$ (again assuming that the difference in chemical potentials is sufficiently small that the energy dependencies of the transmission and reflection coefficients may be neglected). As we have mentioned, a fraction R_{ii} of the current is reflected back into lead i. Similarly, lead j injects carriers into lead i as $(2e/h)T_{ji}\Delta\mu_j$. The net current flowing in lead i is therefore the net difference between the current injected from lead i to that injected back into lead i due to reflection and transmission from all the other leads, and we can write this current as

$$I_i = \frac{2e}{h}\left[(1 - R_{ii})\Delta\mu_i - \sum_{j \neq i} T_{ij}\Delta\mu_j\right]. \tag{6.17}$$

Conservation of the particle flux requires that

$$(1 - R_{ii}) = T_{ii} = \sum_{i \neq j} T_{ji} = \sum_{j \neq i} T_{ij}. \tag{6.18}$$

Therefore the reference potential drops out of the problem and we have

$$I_i = \frac{2e}{h}\left[(1 - R_{ii})\mu_i - \sum_{j \neq i} T_{ij}\mu_j\right]. \tag{6.19}$$

The generalization of equation (6.19) to the multi-channel case is obtained by simply considering that in each lead there are N_i channels at the Fermi energy μ_i. A generalized scattering matrix may be defined which connects the different leads and the different channels in each lead to one another. The elements of this scattering matrix are labeled $t_{ij,mn}$, where i and j label the leads and m and n label the channels in the two leads. The probability for a carrier incident in lead j in channel n to be scattered into channel m of lead i is given by $T_{ij,mn} = |t_{ij,mn}|^2$. Likewise, the probability of being reflected within lead i from channel n into channel m is given by $R_{ii} = |t_{ii,mn}|^2$. The reciprocity of the generalized scattering matrix with respect to magnetic flux now becomes $t_{ij,mn}(\Phi) = t_{ji,nm}(-\Phi)$. The reservoirs are assumed to feed all the channels equally up to the respective Fermi energy of a given lead. If we define the reduced transmission and reflection coefficients as

$$T_{ij} = \sum_{m,n} T_{ij,mn}, \quad R_{ii} = \sum_{m,n} R_{ii,mn}, \tag{6.20}$$

and the total current in lead i becomes

$$I_i = \frac{2e}{h}\left[(N_i - R_{ii})\Delta\mu_i - \sum_{j \neq i} T_{ij}\Delta\mu_j\right], \tag{6.21}$$

which differs from equation (6.19) by the presence of the number of modes entering the lead and the reflection and transmission terms are now sums over modes themselves.

Each of the chemical potentials may be associated with a local voltage $\mu_i = eV_i$, so that equation (6.21) may be written in matrix form as

$$\mathbf{I} = \mathbf{GV}. \tag{6.22}$$

Here, $\mathbf{I}$ and $\mathbf{V}$ are column matrices and $\mathbf{G}$ is the square conductance matrix. The size of the conductance matrix is set by the number of probes that exist in the sample, and the elements in the matrix maintain the same symmetry given by equation (6.16). Thus the conductance matrix is equal to its transpose under reversal of the magnetic flux. Of course, such relations are expected in linear response theory for a system governed by a local conductivity tensor satisfying the Onsager–Casimir symmetry relation.

The resistance matrix may be defined from the inverse of the conductance matrix, and this matrix maintains the same symmetry properties as those discussed above. This result may be used to establish an important property that assures that the reciprocity relation is maintained. We define the sample specific resistance term $R_{ij,kl}$ as the resistance obtained by passing current through leads i and j while measuring the voltage at leads k and l, or

$$R_{ij,kl} = \frac{V_k - V_l}{I_{ij}}. \tag{6.23}$$

Here, the two voltages are related to specific rows of the resistance matrix, while the only non-zero current entries are $I_i = I$ and $I_j = -I$. Thus, we may write equation (6.23) as

$$R_{ij,kl} = \frac{(R_{ki} - R_{kj})I - (R_{li} - R_{lj})I}{I} = R_{ki} + R_{lj} - R_{kj} - R_{li}. \tag{6.24}$$

Now, if we reverse the magnetic field (and flux), then we obtain

$$R_{ij,kl}(\Phi) = R_{ik} + R_{jl} - R_{jk} - R_{il} = R_{kl,ij}(-\Phi). \tag{6.25}$$

This well-known experimental result basically states that in the absence of a magnetic field, the resistance measured by passing current through one pair of contacts and measuring the voltage between the other two is identical to that measured if the voltage and current contacts are swapped. In the presence of a magnetic field, the more general result (6.25) requires that this be carried out while simultaneously reversing the flux through the sample. These are the requirements of the reciprocity theorem in circuits.

6.3.1 Two-terminal conductance

Let us first consider the simplest case, where current is injected through lead 1 and drained through lead 2. These same two terminals will be used to measure the voltage. By conservation of particle flux, we must have

$$N_1 - R_1 = T_{21}$$
$$N_2 - R_2 = T_{12}.$$

(6.26)

Using these two relations, we can now write the currents as

$$I = I_1 = \frac{2e}{h} T_{12}(\mu_1 - \mu_2)$$

$$-I = I_2 = \frac{2e}{h} T_{21}(\mu_2 - \mu_1).$$

(6.27)

These two equations imply that $T = T_{12} = T_{21}$, and that this transmission must be symmetric in the magnetic flux because it is diagonal. The measured voltage is the difference in the chemical potentials, and we achieve the Landauer formula

$$G = \frac{eI}{\mu_1 - \mu_2} = \frac{2e^2}{h} T.$$

(6.28)

6.3.2 Three-terminal conductance

The situation with three probes is a little more complicated and instructive. The situation we consider is shown in figure 6.6, and we assume that no current flows through terminal 3. Thus, this latter terminal corresponds to an ideal voltage probe. First consider the resistance $R_{12,13}$, which signifies the ratio of the voltage measured between contacts 1 and 3 for a certain current I flowing from contact 1 into contact 2. Using equation (6.21), the condition of zero current flowing in contact 3 gives

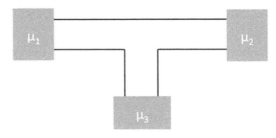

Figure 6.6. A hypothetical three-terminal sample.

$$0 = \frac{2e}{h}[(N_3 - R_{33})\mu_3 - T_{31}\mu_1 - T_{32}\mu_2]. \tag{6.29}$$

Since particle conservation implies that $N_3 - R_{33} = T_{13} + T_{23}$, we may write this equation as

$$\mu_3 = \frac{T_{31}\mu_1 + T_{32}\mu_2}{T_{13} + T_{23}}. \tag{6.30}$$

We now turn to probes 1 and 2, where we may write these equations using equation (6.21) as

$$I = I_1 = \frac{2e}{h}[(N_1 - R_{11})\mu_1 - T_{12}\mu_2 - T_{13}\mu_3]$$
$$-I = I_2 = \frac{2e}{h}[(N_2 - R_{22})\mu_2 - T_{21}\mu_1 - T_{23}\mu_3]. \tag{6.31}$$

If we rearrange equation (6.30), we can replace μ_2 in the second of these equations, and use the continuity of the flux entering probe 2 to give

$$-I = \frac{2e}{h}\left[\frac{(T_{13} + T_{23})(T_{12} + T_{32}) - T_{23}T_{32}}{T_{32}}\mu_3 - \frac{(T_{12} + T_{32})T_{31} + T_{21}T_{32}}{T_{32}}\mu_1\right]. \tag{6.32}$$

In the absence of a magnetic field, the various T are symmetrical and the numerators are the same for both terms. (In the presence of a magnetic field, μ_2 is influenced by the Hall voltage and more care is required in establishing the various resistances.) Then, we find that

$$R_{12,13} = \frac{\mu_1 - \mu_3}{eI} = \frac{hT_{32}}{2e^2 D}, \tag{6.33}$$

where

$$D = T_{13}(T_{12} + T_{32}) + T_{23}T_{12} = T_{31}(T_{21} + T_{23}) + T_{23}T_{21}. \tag{6.34}$$

In a similar manner, we could solve for μ_1 in equation (6.30) and insert this into the first of equation (6.31). Then, using the conservation of flux through probe 1, we could then find that

$$R_{12,32} = \frac{\mu_3 - \mu_2}{eI} = \frac{hT_{31}}{2e^2 D}. \tag{6.35}$$

Finally, using the sum of the two resistances in equations (6.33) and (6.35), we find the total resistance of the device to be

$$R_{12,12} = \frac{\mu_1 - \mu_2}{eI} = \frac{h(T_{32} + T_{31})}{2e^2 D}. \tag{6.36}$$

The net two-terminal conductance of the device with three probes will be given by the inverse of this last equation.

6.3.3 The quantum Hall device

By starting from a picture of edge states as ballistic one-dimensional channels, it is possible to account for the main features of the quantum Hall effect, when we use the Büttiker–Landauer approach described above. We begin this discussion by assuming that the value of the magnetic field is such that ν spin-split Landau levels are occupied and that, within the flat interior of the sample, the Fermi level lies between the νth and $(\nu + 1)$st Landau level. We consider a conductor with a Hall bar geometry, and with several Ohmic contacts to its 2DEG, as we illustrate in figure 6.7. The problem of solving for the Hall resistance of this system is one that involves calculating the different electrochemical potentials at which the contacts sit, under conditions of a fixed applied current (I). We are guided in this analysis by several key assumptions, the first of which is that all edge states leaving any given contact are assumed to be fully equilibrated with it (that is, it is assumed that the electrochemical potential of these edge states is the same as that of the contact that they are leaving).

The other important assumption is that, whenever any contact of the Hall bar is configured as a voltage probe, it then draws no electrical current. It should be recognized that this statement applies to the net current flowing through the voltage probe. This current is comprised of two contributions, the first which is due to incoming edge states that are equilibrated at the electrochemical potential of the neighboring contact (downstream in the sense of the edge-state flow), while the other is carried by edge states leaving the probe. As is common in experiments, we consider that contacts 2, 3, 5, and 6 in figure 6.7 are voltage probes, while contacts 1 and 4 are used to source and sink the current, respectively. We can then write the following expressions for the current drawn by the voltage probes:

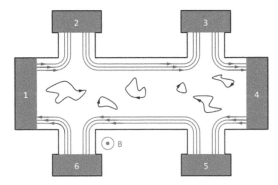

Figure 6.7. A descriptive schematic of a typical Hall bar sample. The edge states are indicated by the blue trajectories. Typical closed cyclotron orbits within the bulk of the sample are indicated in black.

$$I_2 = 0 = \frac{ev}{h}(\mu_2 - \mu_1)$$

$$I_3 = 0 = \frac{ev}{h}(\mu_3 - \mu_2)$$

$$I_5 = 0 = \frac{ev}{h}(\mu_5 - \mu_4)$$

$$I_6 = 0 = \frac{ev}{h}(\mu_6 - \mu_5).$$

(6.37)

As indicated in the first line, we have made the tacit assumption that $T_{21} = T_{32} = T_{43} = T_{54} = T_{65} = T_{16} = v$, and the reverse transmissions are zero. Hence, these equations lead us to the result that

$$V_3 = V_2 = V_1$$
$$V_6 = V_5 = V_4.$$

(6.38)

The two current probes of course do draw current through the device, as these two probes are the actual source and sink of the current. Thus, these two equations can be written as

$$I_1 = I = \frac{2ev}{h}(\mu_1 - \mu_6)$$

$$I_4 = -I = \frac{2ev}{h}(\mu_4 - \mu_3).$$

(6.39)

With these results, we can now calculate the Hall resistance, which can be measured either between probes 1 and 6 or between 2 and 5. With the first pair, we have

$$R_H = R_{14,62} = \frac{V_2 - V_6}{I} = \frac{h}{ve^2}\frac{V_2 - V_6}{V_1 - V_6} = \frac{h}{ve^2},$$

(6.40)

where we have used equation (6.38) to set $V_1 = V_2$. This last equation is, of course, just equation (6.15).

The other unique feature of the quantum Hall effect is the vanishing of the longitudinal resistance for the same ranges of magnetic field where the quantized plateaus are observed. This feature can also be explained by the picture of non-dissipative transport via adiabatic edge states. For the geometry of figure 6.7, the longitudinal resistance (R_{xx}) may be determined by measuring the voltage drop between probes 2 and 3, or 5 and 6. R_{xx} vanishes for either configuration, however, just as is found in experiments, as the voltage is the same at each of the probes selected.

6.3.4 Selective population of edge states

The edge-state picture of the quantized resistances in the quantum Hall effect illustrates clearly that such quantization arises from the perfect transmission of carriers from one ideal contact to another due to the suppression of back-scattering. The resistance is essentially given in terms of a fundamental constant times the number of transmitted edge states at the Fermi energy. A number of experiments have probed this picture of high-field magnetotransport through independent control of the number of edge states that are transmitted through a potential barrier [11, 12]. Such experiments typically feature a geometry similar to that shown in figure 6.8, in which the key innovation is that a gate is now introduced to generate a potential barrier across the primary current path. In practice, the barrier may be realized by using a split-gate quantum point contact (QPC), as was discussed in chapter 2. To understand the resulting behavior in this system, it is important that we appreciate that the edge states corresponding to different Landau levels propagate along the edges of the sample while following equipotential paths with distinctly different guiding-center energies. That is, the edge states arising from the highest occupied Landau level (or its spin-split counterpart) will lie the furthest from the edge of the sample. Edge states from lower-lying Landau levels will be closer to the edge. Thus a gate can selectively reflect some Landau levels while allowing others to propagate.

With these considerations, the problem becomes one of calculating the conductance of the gated device, using similar arguments to those employed in our analysis of the quantum Hall effect. To further simplify the analysis we are free to define the voltage of probe 4 to be equal to zero ($V_4 = 0$), so that the voltages of the remaining probes are then all referenced with respect to this. The most important difference with our earlier analysis is that now the edge states entering probes 3 and 6 originate from different probes, whereas before the absence of any barrier meant that they always came from the same probe. Here, we assume that ν_R edge states are reflected at the barrier, and replace all of the transmission coefficients by the number of transmitted and reflected edge states. Then, the various probe currents become

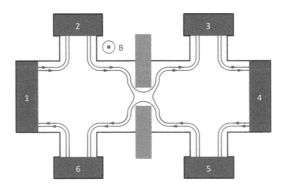

Figure 6.8. A quantum Hall bar with a gate (orange) placed over a portion of the device to allow controlled reflection of some edge states. Here, we indicate two edge states in the device, of which only one is allowed through the gated region.

$$I_1 = I = \frac{e}{h}(\nu\mu_1 - \nu\mu_6)$$

$$I_2 = 0 = \frac{e}{h}(\nu\mu_2 - \nu\mu_1)$$

$$I_3 = 0 = \frac{e}{h}[\nu\mu_3 - (\nu - \nu_R)\mu_2 - \nu_R\mu_5]$$

$$I_4 = -I = \frac{e}{h}(\nu\mu_4 - \nu\mu_3) \tag{6.41}$$

$$I_5 = 0 = \frac{e}{h}(\nu\mu_5 - \nu\mu_4)$$

$$I_6 = 0 = \frac{e}{h}[\nu\mu_6 - (\nu - \nu_R)\mu_5 - \nu_R\mu_2].$$

By simple inspection of these equations, we note that the previous voltage equality among the non-current-carrying probes on the two sides of the bar is upset. To simplify the analysis, we take $V_4 = 0$ to be our reference potential. Hence, we now have

$$V_5 = V_4 = 0$$
$$V_2 = V_1$$
$$V_3 = \left(1 - \frac{\nu_R}{\nu}\right)V_2 \tag{6.42}$$
$$V_6 = \frac{\nu_R}{\nu}.$$

With these voltage expressions, we can now compute the longitudinal resistance to be [12]

$$R_{xx} = R_{14,23} = \frac{V_2 - V_3}{I} = \frac{h}{\nu e^2}\frac{V_2 - V_3}{V_1 - V_6} = \frac{h}{e^2}\left(\frac{1}{\nu - \nu_R} - \frac{1}{\nu}\right). \tag{6.43}$$

If we have four edge states in the un-gated region, and we reflect two of them, then the longitudinal resistance is $h/4e^2$ according to this equation. Similarly, if we have two edge states in the un-gated region, and reflect one of them, the longitudinal resistance should be $h/2e^2$ according to this equation. In figure 6.9, we show results from Haug *et al* [12] that demonstrate this result and show that the plateaus obtained are quite flat and precise in their nature. It is another result that shows the clear nature of the quantization with the edge states.

6.3.5 Nature of the edge states

More information can be obtained about the nature of the edge states by probing them with a SGM. In this experiment, the AFM tip is metalized and with a bias voltage applied can be used as a scanning gate to locally probe the conductance and the carrier density. The sample uses the ideas of figure 6.8 except that in-plane gates (figure 2.1) are used so that no gate metal can interfere with the AFM tip. The sample was fabricated in an InAlAs/InGaAs/InAlAs quantum well heterostructure

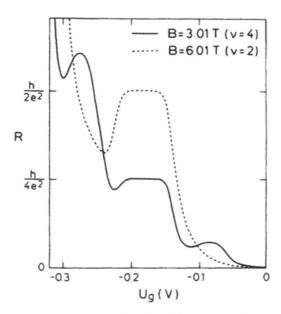

Figure 6.9. Resistance across a gated region as a function of the gate bias. The gate width is 10 μm, and the temperature is 0.55 K. Two different values of magnetic field are used, which lead to bulk filling factors of 2 and 4 as indicated. (Reprinted with permission from [12]. Copyright 1988 the American Physical Society.)

with the InGaAs quantum well lying 45 nm below the surface [13]. Shubnikov–de Haas measurements showed that there were two subbands occupied with densities of 7.2×10^{11} cm^{-2} and 2.1×10^{11} cm^{-2}. With the gate regions biased, no edge states of the second subband were observed to pass through the QPC, so that these were apparently totally reflected. Studies of the relative effects of tip bias and gate bias showed that the capacitances differed by about a factor of three, so that quantitative measurements of the conductance, and its change, could easily be made. With a magnetic field of 6.6 T applied to the sample, the quantum Hall effect was clearly established with approximately five edge states established in the lowest subband in the bulk of the sample. With a gate bias of −7.0 V, these separate so that only one edge state propagates through the QPC. Using equation (6.43), this leads to a conductance through the QPC of approximately $1.25e^2/h$. At less negative gate bias, a second edge state is observed to transit the QPC, and equation (6.43) would suggest a rise in conductance to approximately $3.3e^2/h$, although a somewhat smaller value is found in this case. In figure 6.10, we plot a series of line scans which depict the local conductance as the tip is scanned across the QPC in a direction normal to the current flow for a tip bias of −1.0 V. It is clear that at the most negative bias used, the tip bias leads to a vanishing conductance as it is sufficient to deplete the entire area. The existence of the propagating edge state is clear in these scans as indicated by the plateaus that exist near $1.25e^2/h$. The interesting feature is, in fact, the existence of these plateaus that suggest that the potential drop and density of the edge state are not the simple ideas appearing in

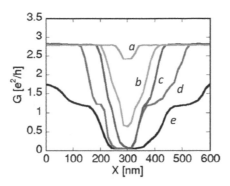

Figure 6.10. Conductance profiles determined by a scanning gate with a tip bias of −1.0 V at a temperature of 280 mK. The letters a–e depict curves for in-plane gate biases of −3.8, −4.2, −5.0, −5.8, and −7.4 V, respectively. (Adapted with permission from [13]. Copyright 2005 the American Physical Society.)

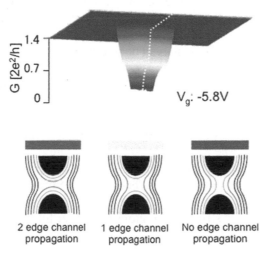

Figure 6.11. The top panel is a three-dimensional representation of curve *d* of figure 6.10. The bottom panel shows three representations of the edge states with color bars indicating to which portion of the three-dimensional image they refer.

figure 6.3. Instead, something more complicated is occurring, to which we shall return shortly.

We can understand what we are seeing in figure 6.10 with a different approach. In the top panel of figure 6.11, we plot a three-dimensional image of the conductance for −5.8 V on the in-plane gates (curve *d* of figure 6.10). Below this panel are three schematics of the QPC and the edge states with a color bar indicated the portion of the three-dimensional plot. In the bluish region, the QPC is sufficiently open that two edge states propagate through the QPC with a conductance as discussed above. In the yellowish region, however, the QPC has closed down from the tip bias so that only a single edge state propagates through it. Finally, the reddish portions of the three-dimensional image are where the tip has closed the QPC so that all edge states are reflected. The fact that the conductance shows a plateau in the state in which only one edge state is propagated is a reflection of the fact that the local potential should also show such a plateau, which is different than indicated in figure 6.4.

The above results suggest that a proper understanding of the microscopic structure of the edge states is vital to the quantitative analysis of many experiments. This understanding has been put forward by Chlovskii *et al*, who carried out a self-consistent determination of the electrostatics of the edge states associated with a potential boundary, such as that formed by a gate [14, 15]. They showed that the position of the ξth spin-split edge state from the gate edge is approximately

$$x = d\left[1 + \frac{\nu^2 + \xi^2}{\nu^2 - \xi^2}\right], \qquad (6.44)$$

where ν is the bulk filling factor and $2d$ is the depletion width around the gate at zero magnetic field. More importantly, the space occupied by the level as it passes through the Fermi energy is not the small amount indicated in figure 6.4, but is a larger space as required by the electrostatics. If the density changed instantaneously as indicated by this latter figure, this would require a massive change in the potential to support this discontinuity in density. The screening properties of the 2DEG in the presence of a magnetic field requires that self-sufficient evaluation of the potential as the magnetic field is varied. This causes the self-consistent potential to develop a series of broad terraces that are separated in energy by the same energy as that of the bulk levels. The electrons sit in the spaces of these terraces and provide compressible regions of the electron gas in the sense that they correspond to spatial regions where many electrons are available near the Fermi energy. These regions are characterized by a metallic conductivity and contribute the screening which, in turn, leads to the plateaus in potential. The spatial regions between the edge states now correspond to incompressible regions, as they have a lack of available states close to the Fermi level. These incompressible stripes are regions of integer filling factor. These plateaus in potential are precisely the constant conduction stripes that appear in figures 6.10 and 6.10.

6.4 The fractional quantum Hall effect

As remarkable as the quantum Hall effect is, it was discovered in a Si MOSFET device. As interest moved to heterostructure material systems, where the quality of the material was much higher, and consequently the mobility was much higher, a new structure was discovered within the quantum Hall effect. This second quantum Hall effect is the fractional quantum Hall effect, or FQHE. In general, one expects the Hall resistance to show the simple plateaus predicted by equation (6.15) and the longitudinal resistivity (or conductivity) to show a set of zeroes at the plateaus. As ν is an integer for these plateaus, this has come to be called the integer quantum Hall effect, or IQHE. Experimentally, in a high quality material such as that of a GaAs/AlGaAs heterostructure, the behavior is qualitatively different from this simple picture. The first measurements were made by Störmer and Tsui [16], in which they saw particularly strong plateaus at values of 1/3 and 2/3 for ν. As the quality of the 2DEG material has improved over time, the richness of the structure in the FQHE has also increased [17], with the presence of better defined zeroes at more fractions. As can be seen from figure 6.12, a quite rich structure exists in the longitudinal resistance that is accompanied by the appearance of further Hall plateaus as one enters the regime of fractional filling factor. Crucially, however, these fractions are found to correspond to very specific combinations of integer numerators and denominators.

The deep minimum which Störmer and Tsui observed at $\nu = 1/3$, which is much stronger than any of the other fractional states [16], led to the conclusion by Laughlin that the electrons had to be condensing into a new collective ground state [18, 19]. That is, while the IQHE could be understood within a single-particle picture, the FQHE must be a complicated many-body state. Laughlin predicted that the new collective state was a quantum fluid for which the elementary excitations, quasi-electrons and quasi-holes, were fractionally charged [18]. Moreover, Laughlin proposed a ground state wave function that possessed angular momentum, with the eigenvalue of $1/\nu$. Thus, the $\nu - 1/3$ state had an angular momentum of $3\hbar$. (For this discovery, Laughlin shared the 1998 Nobel Prize in physics with Störmer and Tsui.) This leads one to conclude that the quantity ν has a deeper meaning than simply

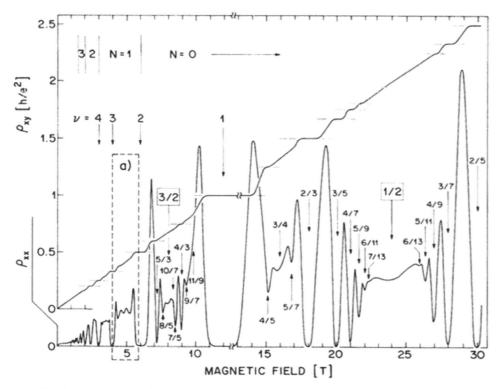

Figure 6.12. An overview of the longitudinal resistivity ρ_{xx} and the Hall resistivity ρ_{xy} for a very high mobility heterostructure at 150 mK. The use of a hybrid magnet required composition of this figure from four different field traces (breaks at 12 T are noted). (Reprinted with permission from [17]. Copyright 1987 the American Physical Society.)

being an integer that counts filled edge states. If we return to our derivation of the Hall resistance, we can rewrite the electron density as

$$n_s = \frac{B}{eR_H} = \frac{\nu e B}{h}.$$ (6.45)

Let us now multiply both sides by the area A of the sample, to obtain the total number of electrons in the sample, as

$$N = n_s A = \frac{\nu e B A}{h} = \nu \frac{\Phi}{\Phi_0},$$ (6.46)

where Φ is the total flux passing through the sample and Φ_0 is the flux quantum h/e. Hence it is clear that ν is the number of electrons per flux quantum in the system. For $\nu = 1$, we have one electron for each flux quantum. Again, for $\nu = 1/3$, we have one electron for every three flux quanta. This then suggests that in the many-body state for this latter fractional plateau, each electron is likely to be bound to three flux quanta.

An important consequence of the collective origin of the FQHE is the general incompressibility of the electronic state, just as in the integer FQE (IQHE). While such incompressibility is a simple consequence of the Pauli principle for the IQHE, its existence at the fractional fillings must be a result of the repulsive interactions between the electrons [20]. Haldane noted that by extending Laughlin's ideas into a spherical geometry, one could obtain a translationally invariant version which was readily extended to an entire hierarchy of fractional states [21]. The entire hierarchy of fractional (and integer) states can be written as [22, 23], in its most common form,

$$\nu = \frac{p}{2ps + 1}.$$ (6.47)

States with $s = 0$ correspond to the IQHE, while states for $s > 0$ give rise to the set of FQHE states. It was also suggested that an excitation gap existed between the ground state of the FQHE and any excited states, and that the kinetic energy needed to bridge this gap was quenched by the high magnetic field [24]. The cause of such a gap is presumably due to the many-body correlations arising from the Coulomb interaction among the electrons, as well as with the flux quanta. An important further recognition was that the arguments above could be understood on topo-logical grounds, and that the flux was exceedingly important in this context [25].

6.5 Composite fermions

In the above discussion, it is clear that only fractional factors in which the denominator was an odd integer appeared in the experimental data. However, in 1987, experimental evidence was first presented in which an even-denominator plateau was seen to begin to form [17], and there is evidence for this in figure 6.12. In this work, in fact, evidence appeared for the $\nu = 1/2$, $3/2$, and $5/2$ plateaus. Observation of the latter two plateaus was particularly significant, since they correspond to the observation of fractional Hall quantization under conditions where more than one spin-resolved Landau level is occupied. Moreover, it was particularly surprising that the $\nu = 5/2$ state was the better resolved of the three, showing the clearest minima in the longitudinal resistivity. Expansion of figure 6.12 around the $\nu = 5/2$ state is shown in figure 6.13. It was also found that tilting the magnetic field led to a rapid collapse of these even-denominator plateaus, an effect which is not seen in the odd-denominator fractional states [26]. While there is no rational reason to exclude the even-denominator fractional states, it does not sit well with the description introduced above.

Shortly after the experiments, Haldane and Rezayi [27] showed theoretically that a new incompressible quantum-liquid state of electrons existed which gave half-integral quantum Hall effect quantization for a non-polarized spin-singlet ground state. Moore and Read [28] then pointed out that the attachment of two quanta of a

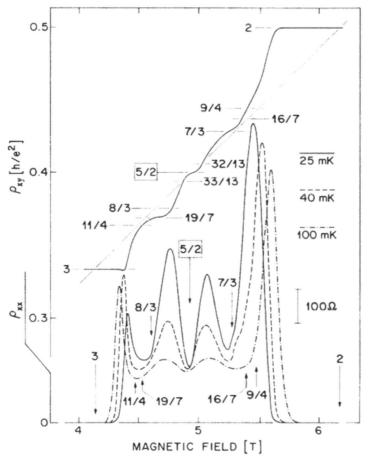

Figure 6.13. Expansion of figure 6.12 around the $\nu = 5/2$ state, showing the detailed number of different plateaus that are found. This is the box marked 'a' in the previous figure. A range of temperatures from 25 to 100 mK are shown. The filling factors are indicated in the longitudinal resistance while quantum numbers p/q ($=\nu$) are shown in the Hall resistance. (Reprinted with permission from [17]. Copyright 1987 the American Physical Society.)

fictitious flux to each electron would lead to an acceptable order parameter for this spin-singlet state. This so-called fictitious magnetic field comes from a Chern–Simmons gauge transformation, which introduces a gauge magnetic field [29]. The important idea here is that the electron, plus the two flux quanta, form what is known as a *composite fermion*. In this new quasi-particle, the flux quanta are actually connected with quantized vortices [30]. Here, the gauge field is just exactly strong enough to cancel the external field at $\nu = 1/2$. That is, when one rewrites the total Hamiltonian, with the gauge transformation, in terms of these composite particles, there is no magnetic field remaining at the value of the applied field for this filling factor. Thus, the composite fermions represent a system of spinless fermions in a (net) zero magnetic field. An additional astonishing fact is that, if the density of electrons n_s is held fixed, then the magnetic field corresponding to $\nu = p/(2p + 1)$ ($p = 1$ for the $\nu = 1/3$ plateau) satisfies

$$\Delta B = B - B_{\frac{1}{2}} = \frac{hn_s}{ep},$$

(6.48)

where

$$B_{\frac{1}{2}} = \frac{2hn_s}{e}$$

(6.49)

is the magnetic field at which $\nu = 1/2$. That is, the fractional plateaus correspond to the integer plateaus for the composite fermions [22]! Experimental studies of the composite fermion quickly established the reality of the composite fermion by measurements of the magnetotransport.

When we say that the composite fermion is a quasi-particle, this means that it behaves like e.g. an electron with a different mass from that at low magnetic field. Values for this mass have been inferred from Shubnikov–de Haas studies with a range of values of 0.53–$0.92m_0$ [31], $0.43m_0$ with a magnetic field dependence [32], $0.91m_0$ right at the $\nu = 1/2$ plateau, with indications of interactions between the composite fermions [33]. More recent measurements have focused upon using cyclotron resonance, with surface acoustic waves, to determine the effective mass [34] and this has shown both a density dependence and a magnetic field dependence [35]. In figure 6.14, the cyclotron resonance frequency is shown as a function of magnetic field around the $\nu = 1/2$ plateau for two different wave numbers of the rf and for two different densities. At the same cyclotron resonance frequency, there is a dependence upon the wave number as well as the density. A given frequency occurs at a different magnetic field for different wave numbers with surface acoustic waves. In figure 6.15, the dispersion of the cyclotron frequency and the inferred composite fermion mass are plotted, using a scaling that seems warranted by the experiments. The value of mass extrapolated to zero wave number appears to scale as $1.6(n_s a_B^2)^{1/2}$ away from the $\nu = 1/2$ plateau, where a_B is the Bohr radius (about 10 nm in GaAs).

As we have progressed from the IQHE to the FQHE to the composite fermion, the physics has gotten more complicated. But, this progression may not be complete.

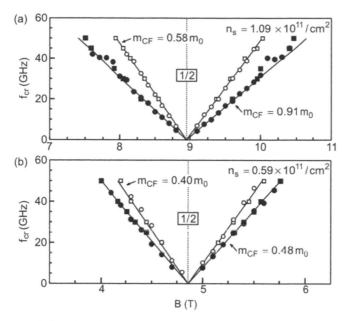

Figure 6.14. The magnetic field dependence of the composite fermion cyclotron resonance mode for two values of the wave vector, $k_{SAW} = 10.5 \times 10^7$ m^{-1} (open symbols) and 3.9×10^7 m^{-1} (solid symbols), and two electron densities: (a) $n_s = 1.09 \times 10^{11}$ cm^{-2} and (b) $n_s = 0.59 \times 10^{11}$ cm^{-2}. (Reprinted with permission from [35]. Copyright 2007 the American Physical Society.)

There are indications that they may well be a cascade of more phase transitions near the $\nu = 5/2$ plateau [36], and that might suggest that there is more exciting physics still to be discovered.

Problems

1. Consider a 2DEG at zero temperature. Construct plots that show the variation of the different Landau level energies, as well as the associated variation of the Fermi level, as a function of magnetic field. Compute these plots for a 2DEG of density 2×10^{11} cm^{-2} and 5×10^{11} cm^{-2}. Explain the difference between the resulting plots.

2. Consider figure 6.13 from [29]. (a) Use the Hall resistance data to determine the 2DEG density. (b) Use the Shubnikov–de Haas oscillations to construct a so-called Landau plot, and thus determine the electron density. (c) Confirm that the plateau in the Hall resistance near 4.2 T is consistent with the electron density determined for the sample. (d) The Shubnikov–de Haas oscillations show an additional splitting at high magnetic fields. Explain the physical origin of this effect.

3. When discussing the Hall effect in a 2DEG, the two-dimensional nature of current flow leads to the expression $J = \sigma E$, where σ is the conductivity tensor (6.11). This matrix, and the associated resistivity matrix given by the inverse of (6.11). Consider a heterostructure in which two subbands are occupied

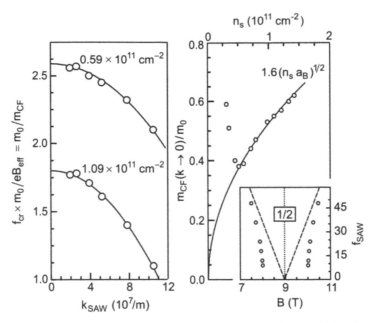

Figure 6.15. Left: k dependence of the cyclotron resonance mode energy of composite fermions (expressed in terms of inverse cyclotron mass) for $n_s = 1.09 \times 10^{11}$ cm^{-2} and $n_s = 0.59 \times 10^{11}$ cm^{-2}. Right: The composite-fermion cyclotron mass in the limit $k \to 0$ as a function of density. The inset plots the B-field location of the cyclotron-resonance mode versus the SAW frequency. The dashed line marks the commensurability condition where the SAW wavelength equals the composite- fermion cyclotron diameter. (Reprinted with permission from [35]. Copyright 2007 the American Physical Society.)

(with densities n_1 and n_2 and mobilities μ_1 and μ_2). In this case, the total current is obtained by adding the appropriate conductivity matrix elements, before inverting to get the total resistivity tensor. (a) By defining an effective number density n_{eff} and relating this to the measured Hall resistance show that, at low magnetic fields (where $\omega_c \tau < 1$), $n_{\text{eff}} = (n_1\mu_1 + n_2\mu_2)^2/(n_1\mu_1^2 + n_2\mu_2^2)$ (Hint: use the identity $\omega_c \tau = \mu B$). (b) Show that the longitudinal resistivity develops a parabolic dependence on magnetic field in the presence of the two-channel transport.

4. let us consider a device such as that in figure 6.9, but where the bottom two probes are removed (they do not exist). (a) Write expressions for the current flowing through the four different contacts, in terms of their associated voltages. (b) Derive an expression for the resistance that would be measured by passing current between contacts 1 and 4 and measuring the voltage drop between 2 and 3. (c) What is the measured resistance when all of the edge states in the 2DEG are transmitted through the constriction? Also, what is the value of the resistance when a fixed number of edge states pass through the QPC and the Landau index $N_L \to \infty$ (i.e. $B \to 0$)? What do these observations tell us about the physical origin of the quantized conductance in one-dimensional conductors?

References

[1] von Klitzing K, Dorda G and Pepper M 1980 *Phys. Rev. Lett.* **45** 494

[2] Ando T, Fowler A and Stern F 1982 *Rev. Mod. Phys.* **54** 437

[3] Ferry D K, Goodnick S M and Bird J P 2009 *Transport in Nanostructures* 2nd edn (Cambridge: Cambridge University Press) ch 4

[4] Seeger K 1973 *Semiconductor Physics* (Berlin: Springer)

[5] Englert T, Tsui D C, Gossard A C and Uihlein C 1982 *Surf. Sci.* **113** 293

[6] Mohr P J and Taylor B N 2000 *Rev. Mod. Phys.* **72** 351

[7] Laughlin R B 1981 *Phys. Rev.* B **23** 5632

[8] Janssen T J B M, Fletcher N E, Goebel R, Williams J M, Tzalenchuk A, Yakimova R, Kubatkin S, Lara-Avila S and Falko V I 2011 *New J. Phys.* **13** 093026

[9] Büttiker M 1986 *Phys. Rev. Lett.* **57** 1761

[10] Büttiker M 1988 *IBM J. Res. Dev.* **32** 317

[11] Komiyama S, Hirai H, Sasa S and Hiyamizu S 1989 *Phys. Rev.* B **39** 8066

[12] Haug R J, MacDonald A H, Streda P and von Klitzing K 1988 *Phys. Rev. Lett.* **61** 2797

[13] Aoki N, da Cunha C R and Akis R *et al* 2005 *Phys. Rev.* B **72** 155327

[14] Chklovskii D B, Shklovskii B and Glazman L I 1992 *Phys. Rev.* B **46** 4026

[15] Chklovskii D B, Matveev K A and Shklovskii B 1993 *Phys. Rev.* B **47** 12605

[16] Tsui D C, Störmer H L and Gossard A C 1982 *Phys. Rev. Lett.* **48** 1559

[17] Willett R, Eisenstein J P and Störmer H L *et al* 1987 *Phys. Rev. Lett.* **59** 1776

[18] Laughlin R B 1983 *Phys. Rev.* B **27** 3383

[19] Laughlin R B 1983 *Phys. Rev. Lett.* **50** 1395

[20] Haldane F D M 1983 *Phys. Rev. Lett.* **51** 605

[21] Haldane F D M and Rezayi E H 1985 *Phys. Rev. Lett.* **54** 237

[22] Jain J K 1989 *Phys. Rev. Lett.* **63** 199

[23] Murthy G and Shankar R 2003 *Rev. Mod. Phys.* **75** 1101

[24] Girvin S M, MacDonald A H and Platzman P M 1985 *Phys. Rev. Lett.* **54** 581

[25] Avron J E and Seiler R 1985 *Phys. Rev. Lett.* **54** 237

[26] Eisenstein J P, Willett R, Störmer H L, Tsui D C, Gossard A C and English J H 1988 *Phys. Rev. Lett.* **61** 997

[27] Haldane F D M and Rezayi E H 1988 *Phys. Rev. Lett.* **60** 1886

[28] Moore G and Read N 1991 *Nucl. Phys.* B **360** 362

[29] Harris J J 1989 *Rep. Prog. Phys.* **52** 1217

[30] Heinonen O 1998 *Composite Fermions* (Singapore: World Scientific)

[31] Du R R, Störmer H L, Tsui D C, Pfeiffer L N and West K W 1993 *Phys. Rev. Lett.* **70** 2944

[32] Du R R, Yeh A S, Störmer H L, Tsui D C, Pfeiffer L N and West K W 1995 *Phys. Rev. Lett.* **75** 3926

[33] Kukushkin I V, Smet J H, von Klitzing K and Eberl K 2000 *Phys. Rev. Lett.* **85** 3688

[34] Kukushkin I V, Smet J H, von Klitzing K and Wegscheider W 2000 *Physica* E **20** 96

[35] Kukushkin I V, Smet J H, Schuh D, Wegscheider W and von Klitzing K 2007 *Phys. Rev. Lett.* **98** 066403

[36] Falson J, Tabrea D, Zhang D, Sodemann I, Kozuka Y, Tsukazaki A, Kawasaki M, von Klitzing K and Smet J H 2018 *Sci. Adv.* **4** eaat8742

IOP Publishing

Transport in Semiconductor Mesoscopic Devices
(Second Edition)

David K Ferry

Chapter 7

Spin

In the previous chapter, we encountered spin-splitting in the measurements of transport at high magnetic fields. Such spin-splitting is a result of the Zeeman effect [1], in which the energy of a free carrier is modified by the magnetic field interacting with the spin. Normally, the effect is most familiar with optical spectroscopy of atoms or of impurity levels in a solid. For the free carriers in a semiconductor, however, the Zeeman effect leads to just two levels, given as the additional energy

$$E_Z = g\mu_B \mathbf{S} \cdot \mathbf{B} = \pm\frac{1}{2}g\mu_B B_z, \tag{7.1}$$

where the magnetic field is oriented in the z-direction. Here, μ_B ($= e\hbar/2m_0$) is the Bohr magneton, 57.94 μVT^{-1}. The factor g is referred to as the Landé g-factor, which has a value of 2 for a truly free electrons. It differs greatly from this value in semiconductors, and can even be negative (~-0.43 in GaAs at low temperatures [2]). This negative value has the effect of reversing the ordering of the two spin-split energy levels. As was discussed in the previous chapter, the Zeeman effect leads to splitting of the Landau levels and can be seen in the Shubnikov–de Haas oscillations at high magnetic fields. This splitting of the spin degeneracy of the Landau levels led to the IQHE [3].

While the Zeeman effect is the best known of the spin effects on transport, there are other effects which have become better known since the intense interest in spin-based semiconductor devices arose a few decades ago. This interest was spawned by the idea of a spin-based transistor [4], but has grown over the possibility of spin-based logic gates which will not be subject to the capacitance limitations of charge-based switching circuits. Many of these new concepts are dependent upon the propagation of spin channels, and the use of the spin orientation as a logic variable, and this has fostered the term *spintronics* [5].

doi:10.1088/978-0-7503-3139-5ch7

7.1 The spin Hall effect

The spin of an electron can be manipulated in a large number of ways, but in order to take advantage of current semiconductor processing technology, it would be preferable to find a purely electrical means of achieving this. For this reason, a great deal of attention has centered on the spin Hall effect in semiconductors. The idea was apparently first suggested by Dyakonov and Perel [6] and later, and independently by Hirsch [7]. The basic idea was that, in the presence of scatterers, it was possible for spins of one orientation to be scattered in a different direction than spins of the opposite orientation. This would lead different spins to accumulate on opposite sides of the sample, a result of the presence of anisotropic scattering in the presence of the spin–orbit interaction [8]. Thus, a transverse spin current arises in response to a longitudinal charge current, without the need for magnetic materials or externally applied magnetic fields [9, 10]. The spin–orbit interaction is a common part of the energy bands of a semiconductor, where it splits the otherwise triply degenerate top of the valence band at the Γ point. But, there are other forms of the spin–orbit interaction that are of interest in situations in which symmetries are broken in the semiconductor device. In the spin Hall effect, we achieve edge states, as in the quantum Hall effect, but here these edge states are spin polarized. The impurity driven spin separation is known as the extrinsic spin Hall effect, and there can be an intrinsic spin Hall effect directly from the spin–orbit interaction when asymmetries exist in the device.

7.1.1 The spin–orbit interaction

The quantum structure of atoms can lead to the angular momentum of the electrons mixing with the spin angular momentum of these particles. Since the energy bands are composed of both the *s*- and *p*-orbitals of the individual atoms in many of the semiconductors, it has been found that the spin–orbit interaction affects band structure calculations. The spin–orbit interaction is a relativistic effect in which the angular motion of the electron interacts with the gradient of the confining potential of the atom to produce an effective magnetic field. This field couples to the spin in a manner similar to the Zeeman effect. Early papers, which used first-principles calculations of the band structure, clearly demonstrated that the spin–orbit interaction was important for the detailed properties of the bands [11, 12]. Not the least of these effects is the splitting of the three-fold bands at the top of the valence band, producing the so-called split-off band. This latter band lies from a few meV to a significant fraction of an eV below the top of the valence band in various semiconductors.

The first inclusion of the spin–orbit interactions in modern pseudopotential calculations for the band structure of semiconductors is thought to be due to Bloom and Bergstresser [13], who extended the interaction Hamiltonian of Weisz [14] to compound semiconductors. We can illustrate how the spin–orbit interaction modifies the transport by considering a simple Hamiltonian for the electrons in a periodic potential $V(\mathbf{r})$ as [15]

$$H(\mathbf{p}, \mathbf{r}) = \frac{p^2}{2m} + V(\mathbf{r}) + \frac{\hbar}{4m^2c^2}(\boldsymbol{\sigma} \times \nabla V) \cdot \mathbf{p}, \tag{7.2}$$

where $\mathbf{p}$ is the normal momentum operator and the last term represents the spin–orbit interaction. The quantity $\boldsymbol{\sigma}$ is a vector whose components are the normal Pauli spin matrices, as

$$\boldsymbol{\sigma} = \sigma_x \mathbf{a}_x + \sigma_y \mathbf{a}_y + \sigma_z \mathbf{a}_z, \tag{7.3}$$

where (see appendix D)

$$\sigma_x = \begin{bmatrix} 0 & 1 \\ 1 & 0 \end{bmatrix}, \ \sigma_y = \begin{bmatrix} 0 & -i \\ i & 0 \end{bmatrix}, \ \sigma_z = \begin{bmatrix} 1 & 0 \\ 0 & -1 \end{bmatrix}. \tag{7.4}$$

It is clear from these matrices that the wave function is now more complicated with a spin component, to which we return below. The direct effect of the spin–orbit interaction may be qualitatively understood as an extra energy cost for the alignment of the intrinsic magnetic moment of the electron with the magnetic field that arises from its own orbital motion. As a result, this term leads to a modification of the momentum of the carriers, through a gauge transformation of the wave function much like the Peierls' modification from the presence of the magnetic field.

7.1.2 Bulk inversion asymmetry

Bulk inversion asymmetry arises in crystals which lack an inversion symmetry, such as the zinc-blende materials. In these crystals, the basis pair at each lattice site is composed of two dissimilar atoms, such as In and P, or Ga and Sb. Because of this, the crystal has lower symmetry than, e.g., the diamond lattice, where the basis pair is two Si atoms. Without this inversion symmetry, one still can have symmetry of the energy bands $E(\mathbf{k}) = E(-\mathbf{k})$, but the periodic part of the Bloch functions no longer satisfies $u_k(\mathbf{r}) = u_k(-\mathbf{r})$. As a result of this, the normal two-fold spin degeneracy is no longer required throughout the Brillouin zone [16]. The importance of this interaction is already recognized by people who study the electronic band structure. The inclusion of this term via the spin–orbit interaction leads to the warped surface of the valence bands [17]. For the conduction band, the perturbing Hamiltonian can be written as

$$H_{BIA} = \eta \Big[\{k_x, k_y^2 - k_z^2\}\sigma_x + \{k_y, k_z^2 - k_x^2\}\sigma_y + \{k_z, k_x^2 - k_y^2\}\sigma_z \Big], \tag{7.5}$$

where k_x, k_y, and k_z are aligned along the [100], [010], and [001] axes, respectively, and the σ_i are the Pauli spin matrices introduced in equation (7.4). The terms in curly brackets are modified anti-commutation relations given by

$$\{A, B\} = \frac{1}{2}(AB + BA). \tag{7.6}$$

The parameter η comes from the spin–orbit interaction and is given by [18]

$$\eta = \frac{4i}{3} P P' Q \left[\frac{1}{(E_G + \Delta)(\Gamma_0 - \Delta_c)} - \frac{1}{E_G \Gamma_0} \right], \tag{7.7}$$

where E_G and Δ are the primary energy gap and the spin–orbit splitting of the valence band in a material in which the minimum of the conduction band and the maximum of the valence band both occur at the Γ point in the Brillouin zone. The quantities P, P' and Q are matrix elements in the spin–orbit interaction along the line of

$$P = \frac{i\hbar}{m_0} \langle S | p_z | Z \rangle, \tag{7.8}$$

where S and Z are s-symmetry and p-symmetry wave functions and p_z is the momentum operator. Finally, in equation (7.7), Γ_0 is the splitting of the two lowest conduction bands at the zone center and Δ_c is the spin–orbit splitting of the lowest conduction band at the zone center. This interaction is stronger in materials with small band gaps, as may be inferred from equation (7.7). Note that equation (7.5) is cubic in the magnitude of the wave vector and is often referred to as the k^3 term.

While the above expressions apply to bulk semiconductors, much of the interest in recent years has been directed at Q2D systems in which the carriers are confined in a quantum well such as exists at the interface between AlGaAs and GaAs. Often, this structure is then patterned to create a quantum wire. For example, a common configuration is with growth along the [001] axis, so that there is no net momentum in the z-direction, and $\langle k_z \rangle = 0$, while $\langle k_z^2 \rangle \neq 0$ is a representation of the quantization energy in the z-direction. Then, equation (7.5) can be written as

$$H_{BIA} = \eta \left[\langle k_z^2 \rangle (k_y \sigma_y - k_x \sigma_x) + k_x k_y (k_y \sigma_x - k_x \sigma_y) \right]. \tag{7.9}$$

This is an important result. The prefactor of the first term in the square brackets is constant, and depends upon the material and the details of the quantum well. This average over the z-momentum corresponds to the different subbands in the quantum well, so that only a single value will result when the carriers are only in the lowest subband. But, this structure has now split equation (7.5) into a k-linear term and a k^3 term.

To explore equation (7.9) a little closer, let us chose a set of spinors to represent the spin-up and -down states as follows:

$$|+\rangle = |\uparrow\rangle = \begin{bmatrix} 1 \\ 0 \end{bmatrix}, \quad |-\rangle = |\downarrow\rangle = \begin{bmatrix} 0 \\ 1 \end{bmatrix}, \tag{7.10}$$

as in appendix D. Then, the linear first term in equation (7.9) gives rise to an energy splitting according to

$$\Delta E_1 \sim -\eta \langle k_z^2 \rangle (k_x \pm ik_y) \tag{7.11}$$

in the rotating coordinates discussed in appendix D. Now, the spin 'up' state rotates around the z-axis in a right-hand sense (with the thumb in the z-direction) with the spin polarization tangential to the constant energy circle in two dimensions. On the other hand, the spin 'down' state rotates in the opposite direction, but with the spin polarization still tangential to the energy circle (we will illustrate this rotation below).

If we ignore the cubic terms in momentum (the last term in equation (7.9)) for the moment and solve for the eigenvalues of the two spin states, we have

$$H = \begin{bmatrix} \dfrac{\hbar^2 k^2}{2m^*} & -\eta\langle k_z^2\rangle(k_x + ik_y) \\ -\eta\langle k_z^2\rangle(k_x - ik_y) & \dfrac{\hbar^2 k^2}{2m^*} \end{bmatrix}, \tag{7.12}$$

where we have assumed that the normal energy bands are parabolic for convenience. Then, we find that the energy levels of the two states are given as

$$E = \frac{\hbar^2 k^2}{2m^*} \pm \eta\langle k_z^2\rangle k. \tag{7.13}$$

Thus, we find that not only is the energy splitting linear in k, but it is also isotropic in the two-dimensional momentum space. The resulting energy bands for the two states are composed of two inter-penetrating paraboloids, and a constant energy surface is composed of two concentric circles (see figure 7.1 below). The inner circle represents the positive sign in the above equation while the outer circle corresponds to the negative sign. The eigenfunctions are no longer pure spin states, but are an admixture given by

$$\varphi_z^{(\pm)} = \frac{1}{\sqrt{2}}\begin{bmatrix} 1 \\ e^{\pm i\vartheta} \end{bmatrix}, \tag{7.14}$$

where ϑ is the angle that k makes with the [100] axis of the underlying crystal in the heterostructure quantum well. This angle is the angle defined by the polar coordinates in the two-dimensional momentum space. Hence, the root with the upper sign in equation (7.13), which we think of as being mostly spin 'up', has the spin polarization tangential to the inner circle. Correspondingly, the root with the lower sign in equation (7.13), which we think of as mostly spin 'down', has the spin polarization tangential to the outer circle.

If we now add in the cubic terms in equation (7.9), the energy levels of equation (7.13) are modified to be

$$E = \frac{\hbar^2 k^2}{2m^*} \pm \eta\langle k_z^2\rangle k \left[1 + \left(\frac{k^4}{\langle k_z^2\rangle^2} - 4\frac{k^2}{\langle k_z^2\rangle} \right)\sin^2\vartheta\cos^2\vartheta \right]^{1/2}. \tag{7.15}$$

This is a much more complicated momentum and angle dependence, but suggests that the constant energy circles are now warped in the same manner as the valence band. The transport is no longer isotropic in the transport plane. Similarly, the phase

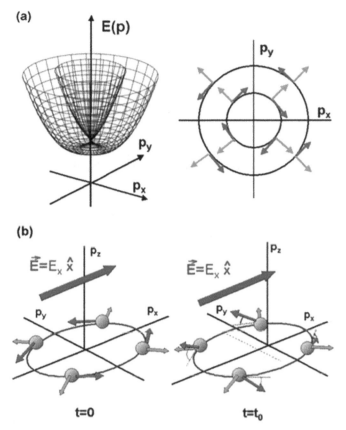

Figure 7.1. (a) Energy structure for the Rashba spin–orbit coupled system. The momentum is shown by green arrows and the spin polarization is shown by the red arrows. (b) When an electric field is applied for some time, the spin is rotated out of the plane according to the direction of the transverse momentum. (Reprinted with permission from [10]. Copyright 2004 the American Physical Society.)

on the down spin contribution to the eigenfunction equation (7.14) is no longer simply the angle, but acquires a 'wobble' as it rotates around the circle.

7.1.3 Structural inversion asymmetry

The spin Hall effect most commonly originates from the Rashba form of spin–orbit coupling [19], which is present in a 2DEG formed in an asymmetric semiconductor quantum well. Such an asymmetric quantum well is the quasi-triangular well found at the interface between AlGaAs and GaAs or at the interface of a Si MOSFET shown in figure 1.2, although the spin–orbit interaction is small in Si (but still very important). This is known as structural inversion asymmetry. In either of the quantum wells mentioned above, the structure is asymmetric around the hetero-junction (or oxide) interface. As a result, there is a relatively strong electric field in the quantum well, and motion normal to this can induce an effective magnetic field. This is the structural inversion asymmetry known as the Rashba effect [19]. Just as in the bulk inversion asymmetry, the electric field in the quantum well can lead to spin splitting without any applied magnetic field, due to the spin–orbit interaction. If we take the z-axis as normal to the heterojunction interface, then the spin–orbit interaction equation (7.2) can be written as

$$H_{SIA} = r\boldsymbol{\sigma} \cdot (\mathbf{k} \times \nabla V) \rightarrow [\boldsymbol{\alpha} \cdot (\boldsymbol{\sigma} \times \mathbf{k})]_z, \tag{7.16}$$

where

$$r = \frac{P^2}{3}\left[\frac{1}{E_G^2} - \frac{1}{(E_G + \Delta)^2}\right] + \frac{P'^2}{3}\left[\frac{1}{\Gamma_0^2} - \frac{1}{(\Gamma_0 + \Delta_c)^2}\right] \tag{7.17}$$

arises from the spin–orbit interaction in the creation of the band structure, and the symbols have the same meanings as in equation (7.7). The parameter is composed of the constants and the electric field, and for the configuration discussed, we have

$$H_{SIA} = \alpha_z(k_y\sigma_x - k_x\sigma_y), \tag{7.18}$$

where

$$\alpha_z = \frac{1}{4}\left(\frac{\hbar}{m_0c}\right)^2 \frac{\partial V}{\partial z}. \tag{7.19}$$

If we continue to use the basis set defined in equation (7.10), then the Rashba contribution to the energy is simply

$$E_R = \mp i\alpha_z(k_x \pm ik_y). \tag{7.20}$$

While the spin states are split in energy, this does not simply add to the bulk inversion asymmetry. First, the two spin states are orthogonal to each other, and then they are phase shifted (with opposite phase shift) relative to the previous results. It is easier to understand the effect of this Rashba term if we diagonalize the Hamiltonian for the two spin states with the Rashba contribution. We can write the Hamiltonian, for parabolic bands using equation (7.20) as

$$H = \begin{bmatrix} \dfrac{\hbar^2 k^2}{2m^*} & \alpha_z(k_x + ik_y) \\[2mm] \alpha_z(k_x - ik_y) & \dfrac{\hbar^2 k^2}{2m^*} \end{bmatrix}. \tag{7.21}$$

When this Hamiltonian is compared to equation (7.12), the two off-diagonal terms are shifted not only by the minus sign in front, but also by a phase factor of $-\pi/2$ within the term in parentheses in the upper right term and $\pi/2$ in the lower left term. Nevertheless, we can still diagonalize the Hamiltonian to give the new energies

$$E = \frac{\hbar^2 k^2}{2m^*} \pm \alpha_z k. \tag{7.22}$$

As in the case of the linear term in the bulk inversion asymmetry, the energy splitting is linear in k, and also isotropic with respect to the direction of **k**. Thus, the energy bands also are composed of two inter-penetrating paraboloids, and a constant energy surface is composed of two concentric circles (see figure 7.1). The inner circle represents the positive sign in equation (7.22), while the outer circle corresponds to the negative sign. The two eigenfunctions are given by

$$\varphi_z^{(\pm)} = \frac{1}{\sqrt{2}} \begin{bmatrix} 1 \\ e^{i(\vartheta \mp \pi/2)} \end{bmatrix}, \tag{7.23}$$

where ϑ is the angle that k makes with the [100] axis of the underlying crystal, within the heterostructure quantum well described in the previous section. The form of equation (7.23) clearly shows the phase shift relative to the bulk inversion asymmetry wave function. The spin direction remains tangential to the two circles, but pointed in the negative angular direction for the inner circle and in the positive angular direction for the outer circle. When both spin processes are present, the spin behavior becomes quite anisotropic in the transport plane [20]. However, the Dresselhaus bulk inversion asymmetry is generally believed to be much weaker than the Rashba terms discussed here. The strength of the Rashba effect can be modified by an electrostatic gate applied to the heterostructure, as it modifies the potential gradient term in equation (7.19).

7.1.4 Berry phase

In chapter 3, we discussed the Aharonov–Bohm effect in which trajectories passing through the top or the bottom of a mesoscopic ring would gain a phase difference when the ring was penetrated by a magnetic field. Here the difference in phase was given by (3.4) as

$$\delta\phi = \frac{e}{\hbar} \int_0^{2\pi} \mathbf{A} \cdot \mathbf{a}_\vartheta r d\varphi = \frac{e}{\hbar} \int \mathbf{B} \cdot \mathbf{a}_z dS = 2\pi \frac{\Phi}{\Phi_0}. \tag{3.4}$$

Now, this phase will occur whether the two paths cover the two separate paths or a single trajectory moves completely around the ring. Now, there is a complimentary quantization that arises from the Schrödinger equation. We can get at this by assuming a complex wave function and using it in the Schrödinger equation [21–23]. The form of the wave function is given as

$$\psi(x, t) = A(x, t)\exp(iS(x, t)/\hbar). \tag{7.24}$$

The quantity S is known as the *action* in classical mechanics, and leads to a continuity equation from which we may conclude that the velocity is related to this as

$$v = \frac{1}{m} \nabla S. \tag{7.25}$$

Obviously, then ∇S represents the momentum associated with the probability flow. In essence, this is a treatment of the quantum wave function as a form of hydrodynamics. But there is more to say about S. Examination of equation (7.24) tells us that S is defined modulo h (Planck's constant). Since the momentum is normally described by ∇S, the periodic nature of S leads to a quantization condition

$$\oint \boldsymbol{p} \cdot d\boldsymbol{r} = nh, \tag{7.26}$$

where n is an integer. Thus, S is not determined entirely by the local dynamics, as would be the classical case, but must also satisfy a topological condition which follows from its origin as the phase of an independent complex field—the wave function [24]. Another way of expressing this is that, if there are any vortices that form in the quantum flow, their angular momentum must be quantized as well.

Equation (7.26) was originally proposed by Einstein [25], and further developed by Brillouin [26] and Keller [27]. It is usually referred to as EBK quantization due to these authors. But, it is important to understand how this simple quantization relates to the topological nature of the wave function. The Aharonov–Bohm effect does not require quantization of the phase, and experiments have shown a smooth phase variation as the magnetic field is varied. This means that the form (7.26) is not the ultimate understanding. This importance was highlighted in discussions of the geometrical phase associated with the wave function by Berry [28]. We consider motion around a closed path according to equation (3.4) which has a period of T around this path such that $\boldsymbol{r}(T) = \boldsymbol{r}(0)$. The state of the system then evolves according to the Schrödinger equation. While the energy may be slowly varying in space and time (despite the constant energy assumption used previously), quite generally there is no firm relation between the phases at different points in space. If the system evolves adiabatically, then the states will evolve in a manner in which the wave function can be written as

$$\psi(t) \sim e^{-\frac{i}{\hbar} \int_0^t E \, dt'} e^{i\gamma(t)} \psi(\boldsymbol{r}(t)), \tag{7.27}$$

which differs from equation (7.24) only by the second exponential. Berry points out that this phase factor, $\gamma(t)$, is non-integrable and cannot be written as a function of the position variable, as it is usually not single-valued as one moves around the circuit. This phase factor is determined by the requirement that equation (7.27) satisfies the Schrödinger equation, and direct substitution leads to the relationship

$$\frac{\partial \gamma(t)}{\partial t} = i \langle \psi(\boldsymbol{r}(t)) | \nabla \psi(\boldsymbol{r}(t)) \rangle \cdot \frac{\partial \boldsymbol{r}(t)}{\partial t}. \tag{7.28}$$

Integrating this around the closed contour discussed above gives

$$\gamma(C) = i \oint_C \langle \psi(\boldsymbol{r}(t)) | \nabla \psi(\boldsymbol{r}(t)) \rangle \cdot d\boldsymbol{r}. \tag{7.29}$$

If we now recognize the momentum operator, this becomes

$$\gamma(C) = -\frac{1}{\hbar} \oint \langle p \rangle \cdot d\mathbf{r}. \tag{7.30}$$

This phase is often referred to as the Berry phase, with the previous expectation of equation (7.28) denoted as the Berry connection. Comparing this to equation (7.26) tells us that the left-hand side of equation (7.30) may often be an integer.

We can go a little further and relate this to the Aharonov–Bohm effect discussed above. Let us replace the momentum in equation (7.26) by the adjusted momentum due to the Peierls' substitution. Then, equation (7.26) becomes

$$\oint p \cdot d\mathbf{r} - e \oint A \cdot d\mathbf{r} = nh. \tag{7.31}$$

Let us now take the second term on the left and rewrite it using Stoke's theorem as

$$e \oint A \cdot d\mathbf{r} = e \int (\nabla \times A) \cdot n d\Omega = eB\Omega = \frac{h\Phi}{\Phi_0}, \tag{7.32}$$

where Ω is the area enclosed by the contour discussed above (the ring in the AB effect), and the other symbols have their normal meanings. Now, we see that the Aharonov–Bohm effect is an example of the Berry phase and the topology of the ring. As the flux enclosed by the ring (Φ) changes, this requires the momentum in the ring to also change in order to maintain the constant right-hand side of equation (7.31). This change in momentum is sensed outside the ring as a change in conductance that gives rise to the oscillations characteristic of the Aharonov–Bohm effect.

A great deal of attention has been given to the Berry phase in quasi-two-dimensional systems, such as mesoscopic systems. This particularly true in demonstrating the topological stability of the quantum Hall effect, as illustrated by Thouless and Kohmoto [29, 30]. These papers demonstrated that the Hall conductivity of a 2DEG in a magnetic field perpendicular to the two-dimensional transport plane is proportional to the Berry phase through the vector potential introduced above and seen by the nth subband eigenstate φ_n. Thus, the Hall conductivity arises from equation (7.32) as

$$\sigma_H = \frac{e^2}{h} \sum_n \frac{1}{2\pi i} \oint d\mathbf{k} \cdot \mathbf{A}_n(\mathbf{k}). \tag{7.33}$$

Different subbands (or Landau levels) may see different values of the vector potential as they often have different enclosed magnetic fields due to their radii. It was later shown by Chang and Niu that the velocity operator (7.25) for a semi-classical wave packet can acquire an anomalous phase from the Berry curvature

$$\Omega_n(k) = i \sum_{n'=1 \neq n}^{q} \left[\left\langle \left| \frac{\partial u_n}{\partial k_1} \right| u_{n'} \right\rangle \left\langle u_{n'} \left| \frac{\partial u_n}{\partial k_2} \right| \right\rangle - \text{c. c} \right], \tag{7.34}$$

where u_n is the cell-periodic part of the Bloch function and q is the highest occupied subband [31]. With the spin Hall effect, the spin–orbit coupling can lead to crossing/anti-crossings of various energy levels, and the anti-crossing effect is enhanced by the Berry curvature [32], which may be rewritten as

$$\Omega_n(k) = i \sum_{n'=1 \neq n}^{q} \left[\frac{\langle u_n | \frac{\partial H}{\partial k_1} | u_{n'} \rangle \langle u_{n'} | \frac{\partial H}{\partial k_2} | u_n \rangle - c.c}{(E_n - E_{n'})^2} \right]. \tag{7.35}$$

At those values where the anti-crossings occur, the denominator becomes small, and the Berry curvature is greatly enhanced. We have observed this effect in subband anti-crossings in the spin Hall effect in nanowires [33].

This can be even more pronounced in the transition-metal di-chalcogenides (TMDCs). These materials were discussed in section 4.4, as layered compounds. The Brillouin zone is similar to that of graphene and these monolayer materials have a direct bandgap at the K and K' points. But, the lack of inversion symmetry leads to some interesting spin effects. In the TMDCs, in the presence of the spin–orbit interaction, the Hamiltonian can be written as [34]

$$H = at(\tau k_x \sigma_x + k_y \sigma_y) \pm \frac{\Delta}{2} - \frac{\lambda \tau}{2}(\sigma_z - 1)s_z, \tag{7.36}$$

where the σ's are the Pauli matrices for the pseudo-spin basis functions of the valleys, τ is the valley pseudo-spin index, a is the lattice constant, t is the nearest neighbor hopping energy, Δ is the energy gap, 2λ is the spin splitting at the valence band top, and s_z is the Pauli matrix for spin. Hence, each of the two bands is also spin split. Here, the energy is measured from the mid-gap point. Generally, it is the spin–orbit interaction that produces the splitting of the doubly degenerate valence and conduction bands. Each of the two bands (conduction and valence) has a Kramers degeneracy due to the spin of the carriers. When the spin–orbit interaction is included, the Kramers doublets are split. Interestingly, the spin orientation is different at K and K'. The splitting of the two valence bands is about 150 meV in MoS_2 and about 430 meV in WS_2 [35]. While the conduction band splitting is small in MoS_2, it is about 30 meV in WS_2 and can be as large as 50 meV in many of the other TMDCs. Thus, it is conceivable that only a single spin state is occupied in each valley, whether we are considering electrons or holes.

While we think of the spin as being either 'up' or 'down', the spins actually lie in the plane of the monolayer material [36]. A constant energy circle is found around each valley (see figure 7.1) with the momentum normal to this circle. However, the spin direction is tangent to the circle, and in opposite directions in the two valleys, as shown in figure 7.1. As a result, if one integrates the spin momentum around the constant energy circle, a Berry phase is produced and this leads to a Berry curvature Ω [37]. This potential acts like a pseudo-magnetic field and therefore has to point in opposite directions for the two valleys in order to maintain time reversal symmetry (that is, the pseudo-magnetic field is oriented in the z-direction for the K valleys and is oriented in the $-z$-direction for the K' valleys, where z is normal to the monolayer). The Berry curvature reacts with a longitudinal electric field to produce a transverse momentum. This gives rise to what would normally be referred to as the spin Hall effect when a longitudinal electric field is applied to the crystal [6, 7, 37].

As remarked, the spin is coupled to the valleys due to the difference in the orientation of the spin splitting [34, 38]. This produces a valley-spin Hall effect, which again arises from the fact that, in the presence of the longitudinal electric field, the Berry curvature leads to a transverse velocity $\mathbf{E} \times \mathbf{\Omega}$. Hence, the carriers in the two valleys will be driven to opposite sides of the ribbon width. The importance of this lies in the recognition that spin separation may lead to useful information processing devices [39] and there have been many attempts to make useful devices using this concept [40]. Xiao *et al* [34] have evaluated the Berry curvature for the TMDC conduction band in terms of the material parameters as

$$\Omega_c(k) = -\tau \frac{2a^2t^2\Delta}{[\Delta^2 + 4a^2t^2k^2]^{3/2}}, \tag{7.37}$$

where k is the wave number (measured from K or K') and the other parameters are as defined above. The values of these parameters are taken from [34], where these authors give a value for the Berry curvature for WS$_2$ of 9.57×10^{-16} cm^2. This curvature is normal to the monolayer of the TMDC, and as discussed above, is oppositely directed in the K and K' valleys.

7.1.5 Studies of the spin Hall effect

One of the more remarkable features of the Rashba spin–orbit term (the structural inversion asymmetry terms) is that this effect gives rise to the intrinsic spin Hall effect in a nanowire, in which the longitudinal (charge) current along the nanowire gives rise to a transverse spin current. In this situation, one spin state will move to one side of the nanowire, while the other spin state moves to the opposite side. There can be an intrinsic spin Hall effect that does not rely upon the presence of any impurities with their spin scattering [10]. We can illustrate this with a consideration of the energy surfaces. In figure 7.1(a), the nested double parabola discussed in the previous sections is sketched, and this leads to the constant energy surfaces being two nested circles in momentum space, as shown in the right panel of figure 7.1(a). In this latter case, the direction of the momentum in the Rashba spin–orbit coupled system are indicated by the green arrows, and the two eigen-spinor polarizations are shown by the red arrows. If we apply an electric field along the x-axis, the current-carrying direction, this produces a force on the spin, as indicated in figure 7.1(b). After some time (but small compared to the scattering time), the electrons experience an effective torque that rotates the spin polarization. This torque rotates the polarization upward, and giving a positive polarization in the z-direction for positive y-momentum, and giving a negative polarization in the z-direction for negative y-momentum. This difference drives spin-up electrons to one side of the sample and spin-down electrons to the other side, yielding a net spin current in the device. These authors suggested that this intrinsic spin Hall effect should have a universal value of [10]

$$\sigma_{sH} = -\frac{J_{s,y}}{E_x} = \frac{e}{8\pi}. \tag{7.38}$$

Other studies showed that, in the infinite two-dimensional sample limit, arbitrarily small disorder introduces a vertex correction that exactly cancels out the transverse spin current [41]. However, in finite systems such as quantum wires, the spin Hall effect survives in the presence of disorder and manifests itself as an accumulation of oppositely polarized spins on opposite sides of the wire [42], and that the observed value of the spin Hall effect could be larger or smaller than that of equation (7.38), depending upon the magnitude of the spin–orbit coupling, the Fermi energy, and the degree of disorder [43]. This has led to the proposal for a variety of devices that utilize branched, Q1D structures to generate and detect spin-polarized currents through purely electrical measurements [44], and experiments have been performed to try to measure these effects [39]. We will return to this in section 7.3. In addition to Rashba spin–orbit coupling, a term due to the bulk inversion asymmetry of the host semiconductor crystal, the Dresselhaus spin–orbit coupling [16] discussed above, can also yield a spin Hall current.

What was probably the first experimental effort was carried out by Awschalom *et al* [45]. Here, experiments were based on an *n*-GaAs layer 2 μm thick, grown on top of an AlGaAs thin layer itself grown on a semi-insulating GaAs substrate. The *n*-layer was doped to 3×10^{16} cm^{-3} in order to achieve long spin lifetimes. Measurements were made using Kerr rotation of an optically polarized beam. The magneto-optical Kerr effect is a rotation of the optical polarization due to an electric field in the quantum well, which modifies the dielectric function. This electric field can be produced by the optical beam itself, but the effect is modified by the presence of a spin polarization. Then, the Kerr rotation can be used to measure the spin polarization in the sample. When an electric field is applied along the length of the sample, the spin Hall effect can be developed as discussed above. A magnetic field is applied transverse to the electric field and in the plane of the sample. The Kerr signal occurs for small magnetic fields (of order a few mT around zero magnetic field). It was clear in the measurements that opposite spin directions were found at the two longitudinal edges of the sample. The effect diminishes as the magnetic field is increased due to spin precession (see appendix E). Further information on this spatial imaging of the spin separation was given in a subsequent paper [46].

A somewhat different approach has been presented by Wunderlich *et al* [47]. In this experiment, a *p*-type layer is produced in an AlGaAs/GaAs heterostructure. This structure is then patterned into a ribbon so that *n*-type layers can be placed in close proximity. These latter layers then produce two light-emitting diodes at the edges of the ribbon. The spin Hall effect is especially strong in the hole layer, and this guides the spin polarized carriers to the diodes, but each diode sees a different spin polarization. This is used to create optical circular polarization from the diodes

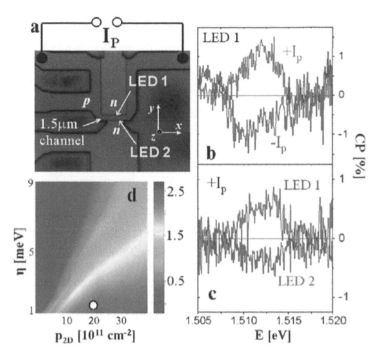

Figure 7.2. The intrinsic spin Hall effect in a GaAs structure. (a) A micrograph of the device structure. The two LEDs are sensitive to the spin polarization of the holes at the two sides of the p-layer. (b) Reversing the current in the p-layer reverses the polarization of the LED emission. (c) The two LEDs sense the opposite spins on either side of the p-layer. (d) The quasi-particle lifetime as a function of the hole density. The color scale corresponds to the ratio of the measured spin Hall effect to the universal value of (7.24). (Reprinted with permission from [47]. Copyright 2005 the American Physical Society.)

due to the spin polarization of the injected holes. Because of the optical selection rules, the presence of a particular spin polarization will create a corresponding circular polarization in the light, and this can be measured. These results are shown in figure 7.2. The device itself is depicted in panel (a), and the two diodes are labeled as LED 1 and LED 2. As indicated in panel (b), the optical circular polarization is reversed when the current through the *p*-layer is reversed. This indicates that the opposite spin is pushed toward that LED when the current is reversed, as expected for the spin Hall effect. Then, the two outputs from the two LEDs are shown in panel (c), and it is clear that they are affected by opposite spins. Finally, the group investigated whether the spin Hall effect has a universal value, as given by equation (7.38). This is shown in panel (d) as a color plot of the quasi-particle lifetime as a function of the hole density. The color denotes the magnitude of the spin Hall effect relative to the 'universal' value given in equation (7.38). The measured strength is the white dot at the lower center of the image, and suggests that this measured value is roughly twice the value expected if it were universal. They term this a 'strong intrinsic spin Hall effect.

7.2 Spin injection

As semiconductor devices have grown continually smaller, it has become apparent that this trend cannot continue indefinitely. In fact, the nominal 'size' in the 2020 generation of integrated circuits from ARM is only 7 nm, which is roughly 30 times the spacing of the Si atoms. Even though this 'size' is not a real physical dimension any more, it is clear that there is a limit that is rapidly approaching. As a new path to continued growth in the capabilities of integrated circuits, many researchers have taken to exploring alternative approaches to computer logic devices, by changing from the charge basis of our traditional transistors. One possible approach is to use spin as the variable, and this possibility burst into the integrated circuit world with the proposal for the 'spin FET' [4]. In this proposal, the base material for a HEMT-like device was InAs, which has a large spin–orbit splitting and so should have a reasonably long spin lifetime (the time before the spin polarization randomizes). The source and drain contacts, however, were made of iron, a magnetic material. Thus, a spin polarized contact would inject a polarized spin into the InAs channel. Of course, this spin would precess, and the amount of precession would vary with the electric field provided by the gate bias. Hence, the ability of the polarized spin carrier to leave the drain contact would depend upon the polarization when this contact was reached. As a result, the drain current could be switched by the gate bias, as is normally expected. The key issue, of course, was the ability to successfully inject a single spin state from the source contact.

Of course, the idea of spin injection did not begin with the above device proposal. The concept of spin injection into semiconductors seems to have originated with Aronov [48, 49]. One of the earliest experiments used a permalloy contact on aluminum to inject from the polarized permalloy into the nonmagnetic aluminum, where the spin polarization could be analyzed and the spin lifetime determined [50]. In fact, it was later suggested to extend this structure to produce a spin bipolar transistor [51]. The quest for the spin FET then led to the first experiments to successfully inject spin polarized carriers into a semiconductor, in this case CdTe [52] and InAs [53–55], but it was generally found that one needed to make a tunnel barrier between the magnetic metal and the semiconductor to avoid conductivity mismatch in the two materials. In fact, an actual spin FET has been realized using the InAs channel [56], which was further reviewed by Johnson in [40]. In figure 7.3, we illustrate the device and the measurements. The structure is shown in panels (b) and (c) and constitutes an InAs quantum well grown on GaAs with GaAlSb barriers. The InAs channel was patterned to a width of about 0.9 μm. The magnetic metal layers are placed over this structure and spaced some 10.6 μm. A picture of the sample is shown in panel (a), and shows that six separate channels are connected in parallel to create the device, although other devices with a single channel of width 15 μm were also measured, this device is primarily discussed. While the device is symmetrical, metal F1 was typically used as the injector and metal F2 was used as the detector, which acts as a spin sensitive potentiometer (or resistance). The detected signal is shown in panel (d) of the figure for the single-channel device, and the difference between the up sweep and the down sweep is typical for spin

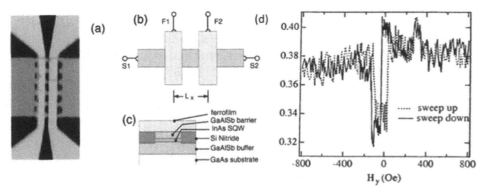

Figure 7.3. (a) Micrograph of the structure showing six parallel channels. (b), (c) Structure of the heterostructure spin system used to propagate spin from one contact to a second contact utilizing an InAs quantum well. (d) Measurement of the spin propagation detected at F2 for a single channel of 15 μm width (see the text for the discussion of the hysteretic curves). Other details are given in the text. (Reprinted with permission from [40]. Copyright (2005) American Chemical Society.)

measurements. It is clear that the spin has been successfully injected and then detected at a distance away, with the detection being sensitive in this case to the magnetic field.

To understand the signal in figure 7.3(d), we have to consider the experiment in a little more detail. When the electrodes F1 and F2 are both ferromagnets, then the measurement is a voltage that is linearly proportional to the current being fed through the sample, and this is recorded as a resistance. This voltage (resistance) is relatively high when the two contact ferromagnets have parallel polarization and relatively low when their polarizations are opposite. In general, the InAs quantum well is asymmetrical in the growth direction (normal to the heterostructure interfaces) and this leads to an electric field in the z-direction (the growth direction), which produces a spin polarization in the conduction band according to equation (7.22). Normally, the two concentric circles (figure 7.1(a)) have the same Fermi energy, but when spins of one polarization are injected into the semiconductor, the electrochemical potentials of the two spin subbands are altered [57]. When contact F1 provides spin-down electrons, these propagate ballistically to the other contact, where the spin-down electrochemical potential is raised, and the spin-up electrochemical potential is lowered. When the magnetizations of the two contacts are parallel, this potential separation is measured and appears as the high resistance regions of the curves in figure 7.3(b). When the two magnetizations are anti-parallel, a smaller voltage corresponding closer to the equilibrium state is measured, and this is reflected in the dips in the curves near $B = 0$. The two contacts have slightly different coercive forces, $H_{F1} \neq H_{F2}$, which leads to the fact that one contact will switch its magnetic state before the other one, and this gives the observed curves.

Spin propagation in Si is more difficult due to the very small spin–orbit interaction that exists in this material. However, spin injection has been seen in Si, and detected as either light emission [56] or by the spin-valve effect of the InAs quantum well discussed in the previous paragraph [58]. In the Si device, an Al–Al$_2$O$_3$–Al tunnel junction is used to inject unpolarized electrons into a layer of

CoFe. There is an exponentially different mean free path for the two different spin orientations in the polarized CoFe film, so that a dominant single-spin polarization is injected over the barrier into the Si film, where it transports through the layer. The spin polarization is then measured by a NiFe/Cu layer and the last Si layer. If the polarization of the NiFe layer is compatible, the spins can propagate through to the last Si layer. When the polarizations of the two magnetic layers are the same, the detected current will be higher than when the two polarizations are opposite. This detected current is changed as the magnetization of the magnetic layers is reversed.

7.3 Spin currents in nanowires

As mentioned in the previous section, there is an interest in spin as a logical variable for use in future nanoelectronic devices. There is also some interest in using spin as a quantum bit (qubit) in quantum computing since the natural two state nature of spin, and its precession on the Bloch sphere are essentially the same as that conceived for an analog qubit. One suggestion for a semiconductor qubit actually uses wave propagation in a pair of parallel quantum wires [59, 60]. Such a qubit would depend upon the interaction of the carrier waves in the two wires, especially at a designated interaction region. It has been shown that the wave definitely can be switched from one wire to the other with an applied magnetic field [61], which supports the idea of using spin as the important variable.

The ability to obtain spin separation in a quantum wire certainly depends upon the Rashba effect and the spin–orbit interaction. The induced magnetic field can lead to spin separation through a process which is essentially that of the spin Hall effect discussed above. In figure 7.4, the spin resolved squared magnitude of the wave function in a 100 nm wide quantum wire in the lowest subband of an InAs quantum well heterostructure is shown as a function of the wave momentum down the wire [62]. Here, the Rashba coefficient (7.19) is 20 meV-nm. The yellow color represents the spin along the z-direction, while the red corresponds to the opposite spin direction. It can be seen that there is a change of the spin wave functions as a

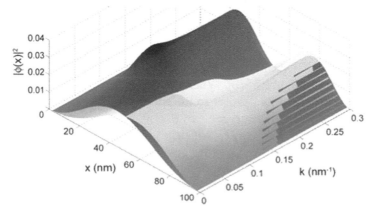

Figure 7.4. The spin resolved square magnitude of the wave function in an InAs nanowire as a function of the momentum along the wire. (Reprinted with permission from [62].)

function of the momentum, and this arises from the fact that the various subbands cross and interact in the effective magnetic field, so that a wave can hybridize between the various subbands. Nevertheless, the ability to spatially separate the two spins across the nanowire leads to the ability to make a spin filter by splitting the nanowire in a Y-type branch [44], as shown in figure 7.5(a) (a spin branch cascade is actually shown in this figure). It is conceived that the individual nanowires would be InAs with a width of 100 nm, and the branches would have a radius of curvature of 100 nm. In figure 7.5(b), we show the spin resolved square magnitude of the wave functions in the various channels of the structure at very low temperatures. The color indications are the same as above, as are the spin–orbit parameters, but the doping is assumed to be 3×10^{10} cm^{-2}, so that only the lowest subband is occupied, although there are of course two spin states. In the input wire, the spin Hall effect leads to the two spin states being spatially separated. When the first branch is reached, the opposite spins move to opposite sides of the branch and into the different arms. Because the electrons in the output wires are polarized out of the plane of the heterostructure and therefore undergo precession as they move down the wires. This precession leads to the wave function moving from one side of the wire to the other, a wobbling motion that has often been called zitterbewegung.

The efficiency of the Y-branch switch is best characterized by the degree of spin polarization of the electrons in the output arms. We find that this polarization can rise to a value of almost 60%. While small, this is considerably larger than that usually achieved by spin injection.

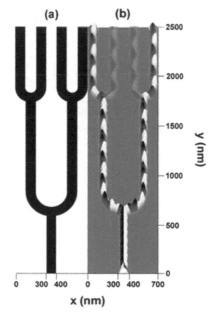

Figure 7.5. (a) A two-stage cascade of Y-branch switches which can be used for spin filtering. (b) The spin resolved squared magnitude of the wave function for the device in (a). (Reprinted with permission from [62].)

To make it more useful, one can develop a cascade of such Y-branches [63], as shown above in figure 7.5. As indicated in this figure, the use of a second set of Y-branch switches really does not add much under the proper conditions, as the results of the first Y-branch is sufficient to isolate the spin states into only one arm of the second set of switches. However, under proper conditions, the spin state can be switched from one arm to the other in the second set of gates, which can be achieved by changing the propagation length of the intermediate wires. Alternatively, the two outputs of the second set of Y-branch can be used as a spin polarization detector. The precession can also be reduced by the use of an in-plane magnetic field so that a true spin filter can then be created [62].

Attempts to experimentally measure such spin filters have had reasonable success [64, 65]. One such example is shown in figure 7.6(a) [65]. The device has been fabricated from an InAs heterostructure by electron-beam lithography and reactive-ion etching [66]. The common horizontal wire(in the figure) is oriented along the [1, 1, 0] direction of the crystal, as this direction exhibits the highest mobility. The contacts are the typical AuGe contacts used for many III–V materials. The top gate, used to adjust the carrier concentration in the InAs quantum well, is separated from the semiconductors by a polymer layer about 0.24 μm thick. At low temperatures, it is hoped that the transport through the structure will be ballistic in nature. The central wire, discussed above, is 1 μm long and has a width of 150 nm. Current is injected at port A and the current that flows to the other arms is measured. The corresponding voltages allow a determination of the effective conductance at each of the other ports. In figure 7.6(b), the conductance from the source to output arms C (blue) and D (green, as indicated in the left panel) are shown for different conductances through the central arm. One can see that, as the magnetic field is varied, the currents in these two arms oscillate, indicating that they are measuring a

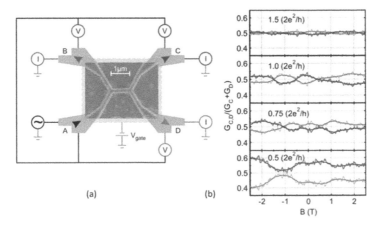

(a) (b)

Figure 7.6. (a) Micrograph of an InAs heterostructure with patterned Y-branch switches at left and right sides. The color code and terminals of measured conductances are indicated in the figure. (b) Measured conductances at terminals C (blue) and D (green) are plotted for various values of the conductance in the central arm. Both the raw data and the smoothed data (lines) are shown. (Reprinted with permission from [65]. Copyright 2013 AIP Publishing LLC.)

spin polarization of the output signals. These oscillations in the output of this second filter are interpreted as being a result of spin polarization in the first stage (left-hand side) of the Y-branch filter.

7.4 Spin qubits

As we mentioned above, the spin–orbit interaction is relatively small in Si, but it is large enough to consider making spin-based qubits in this material. The idea for a qubit based upon Si effectively dates from the suggestion of Kane to use a ^{31}P dopant in isotopically pure ^{28}Si [67]. In this approach, it was suggested to use the nuclear spin of the P atom, which is 1/2, and the fact that the particular isotope of Si does not have a nuclear spin. Then, the electron wave function for the positively charged donors extends extensively through the conduction band and can mediate the interaction between the nuclear spins on neighboring P atoms. Such a qubit is interesting because it has the promise of a long spin-relaxation time (discussed in the next section), and is scalable [68]. Getting the P atoms into the right position thus constitutes a serious problem, but the latter authors have demonstrated that it is possible to create an atomically precise linear array of single P bearing molecules on a Si surface, and that these P atoms can act as quantum qubits.

The process of getting a P atom to a precise location turns out not to be so difficult. One method is to hydrogenate the dangling bonds on the Si surface, then to remove particular hydrogen atoms using a scanning tunneling microscope (in ultra-high vacuum) [69]. Removing the hydrogen allows the molecule to be chemically attach to the dangling Si bond at that particular site [70]. In the case mentioned above, the molecules from phosphine gas are used to deposit the P atom at the desired position, then the difficult task is to deposit isotopically pure Si while maintaining the P atom at the desired position [67]. Another approach to locating the P atom is to pursue single atom implantation [71–73]. This latter technology was developed some years ago, and qubits based upon P atoms have been fabricated with both approaches.

The operation of a single donor qubit under control of surface gates and measured with a single-electron transistor was evaluated by Dehollian et al [74]. In figure 7.7, we give some information about the operation of this donor qubit formed with the P atom. Here, they use gate set tomography (GST), which is a tool for characterizing logic operations in the qubit. The qubit itself is formed by the spin states of an electron bound to a ^{31}P atom, implanted into the isotopically pure silicon substrate. The spin states are split by an applied magnetic field, and switching is achieved via coupling of the electron spin to the nuclear spin, resulting in a two spin, four level system. Qubit preparation is handled by tunneling to/from a single-electron transistor, which is also used for sensing the state. The aluminum gates are placed on top of a SiO_2 layer. In addition, an electron spin resonance signal is used for qubit manipulation.

It has also been suggested to use the electron spin rather than the nuclear spin in a semiconductor nanostructure [75]. One reason to use the electron spin lies in the fact that the spin resonance transition for the electron can be rather effectively tuned by

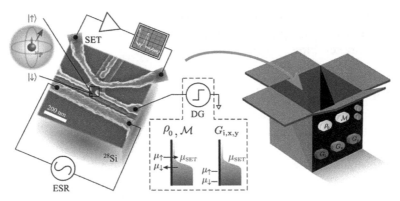

Figure 7.7. Diagram of qubit device and the gate set tomography (GST) model of a qubit. SEM image of the on-chip gate structure is shown to the left. The aluminum gates have been false colored for clarity. Depicted in red are the source-drain n^+ regions which couple the single-electron transistor to the current measurement electronics. For initialization and measurement, the donor gates are pulsed such that the various Fermi energies are $\mu_\uparrow > \mu_{SET} > \mu_\downarrow$, inducing spin-dependent tunneling between the donor and the SET. When applying a gate sequence, the DG are pulsed to higher voltage to prevent the donor electron from tunneling to the SET. The inset diagram (upper left)—zoomed from the approximate donor location—represents the Bloch sphere of the qubit, consisting of the spin of an electron confined by an implanted ^{31}P donor, with its nuclear spin frozen in an eigenstate. The model treats the qubit as a black box with buttons (right) which allow one to initialize (ρ_0), apply each gate in the gate set ($G_{i,x,y}$) and measure (M) in the observable basis ($|\uparrow\rangle$ or $|\downarrow\rangle$). Reproduced under the Creative Commons Attribution 3.0 licence from Dehollian *et al* [74].

electrostatic gates, which are a common part of the single-electron quantum dot structures (discussed more in section 8.2). The use of the electron spin has been reviewed recently [76]. It has also been suggested [77] to move to Ge for better tenability via the bias induced Stark effect, but most work remains in the Si system. While much of the effort on single impurity qubits has focused on donors, there has also been some work directed at the use of acceptors as well. There is a feeling that the dopants for this type of qubit need to be near an interface in order to interact well with surface gates. Hence, the use of acceptors near the surface of both Ge an Si have been studied as well [78, 79].

The case for nearly all of the above dopant-based qubits is that the dynamic variable is the spin itself. It is possible to consider a hybrid qubit in which the dynamical variable depends upon both spin and charge. The hybrid qubit requires neither nuclear-state preparation nor micro-magnets for control, and becomes considerably more amenable to systems [80]. Such a hybrid qubit has been studied by these latter authors in a traditional double dot system, similar to those in single-electron transistors (again, discussed more in section 8.2). Three electrons are used in the system, with the gates tuned so that there are two electrons in the left dot and one in the right dot to define the (2,1) state. Changing the voltage on gate L to raise the energy difference between the dots favors the opposite situation and the state (1,2). The singlet state in the right dot, state (2,1), is taken to be the $|0\rangle$ state. The triplet state in the right dot, the (1,2) state, is taken to be the logical $|1\rangle$. The presence of the extra electron means that the hybrid states of the dots are not pure singlet or triplet

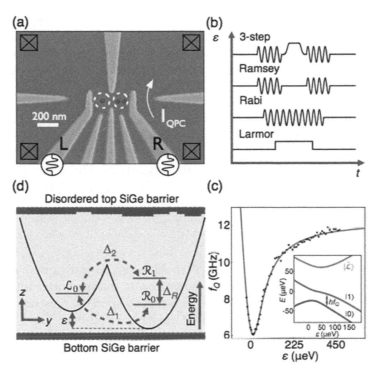

Figure 7.8. Experimental and theoretical qubit set-up, and resulting energy dispersions. (a) Scanning electron microscope image of a device nominally identical to the one used in the experiment. The gate voltages applied to the various leads are tuned to form two quantum dots, located approximately within the dashed circles, where the red spots represent electrons in a (1,2) charge configuration (one level left dot and two levels right dot). (b) Schematic of the four pulse sequence employed in the experiments to obtain the qubit parameters. (c) The experimentally measured dispersion of the quantum-dot hybrid qubit. The red line is a least squares fit to the data. The inset is the three energy eigenstates determined for the qubit. (d) A schematic cartoon illustrating the model for the qubit. Reprinted with permission from [82], copyright 2018 by the American Physical Society.

and fast electric field techniques can be used to manipulate the qubit in either X or Z rotations on the Bloch sphere (discussed in appendix E). The use of the three electron system tends to extend the coherence [81]. The singlet-triplet qubit is attractive for many reasons, particularly for its high fidelity, at least for single qubit operations. In figure 7.8, we show such a hybrid qubit, with its excitation and control characteristics [82]. The double-dot system is excited with microwave pulses; four different pulse sequences are used. These set of pulse sequences allow one to map out the energy dispersion of the qubit. Also in the figure, Δ_1 and Δ_2 refer to the tunnel couplings between the various charge states and Δ_R is the energy splitting between the two basis states for the qubit, which are the singlet and triplet states in the right quantum dot. Tuning the dispersion was found to be critical to finding the exact parameters for best qubit operation.

For two qubit operations leading to entanglement, hybrid qubits can be problematic in a sense that there are a range of operations over which the interaction can occur and this leads to a preferred set for good performance [83], just as was found for the double-quantum dot qubit in the previous paragraph [82]. In addition, the interaction between a hybrid qubit (discussed above) and other types of qubits has been studied, since such an interaction can usually be tuned between different operating modes via the gates [84].

Pure spin qubits formed of quantum dots in the Si/SiO$_2$ system have been formed in a manner to utilize from 1 to 3 electrons in the dots [85]. Then, pulsed electron spin resonance is used to exercise coherent control over the qubit. Quantum dots have also been created in the Si/SiGe heterostructure system. High fidelity gating of a single hybrid qubit has been achieved in this system [86]. The optimal geometry for the gates for such a spin qubit in this heterostructure (or in GaAs) has been studied for a single qubit [87] and for a qubit linear array [88].

7.5 Spin relaxation

The decay, or decoherence, of a spin polarization can be instigated by a magnetic impurity, or by a normal impurity in a material with the spin–orbit interaction. As our various spin applications discussed above depend upon the presence of this latter interaction, we can ignore the discussion of magnetic impurities. The key issue then is the change of polarization that can occur during the scattering of the carrier by the impurity. In the Elliott–Yafet mechanism [89, 90], a spin flip can occur during the scattering of the carrier by the impurity. In the presence of spin–orbit scattering, the electron state is not a pure single polarization, as there is always a small admixture of the other spin in the wave function. Hence, under the scattering process, there is a possibility that the overlap of the initial and final state wave functions will lead to a flip of the spin. Thus, there is a random spin rotation that results from the scattering and this gradually breaks up the spin coherence. But, the spin generally is not affected between the scattering processes, so that the spin lifetime is proportional to the momentum relaxation time. This is not limited to just 'impurities, but can also arise from boundary scattering, interface scattering, and phonon scattering [89]. A more complicated interaction can also arise from the phonon effect on the spin–orbit interaction of the lattice ions [91]. Yafet showed that the spin relaxation rate (the inverse of the lifetime) basically follows the temperature dependence of the resistivity of the sample. In a heavily doped semiconductor in which there is little freeze-out of the carriers, this would lead to an independence of temperature at low temperatures, as observed experimentally [92]. On the other hand, the resistivity is strongly affected by the carrier density, which is given by the ionized impurity density. At higher temperatures, the resistivity decreases as the carrier density increases (for materials in which the donors or acceptors are not fully ionized), and hence the spin lifetime will also decrease with temperature. Figure 7.9 shows the spin lifetime in Ge, which was Sb-doped at 5×10^{17} cm^{-3} [93]. The lifetime is estimated from a fit to Hanle effect data. Also shown in the figure is a result from a theoretical work using the Elliott–Yafet mechanism [94]. While the quantitative fit differs from the theory, the qualitative dependence on temperature is as expected for this mechanism.

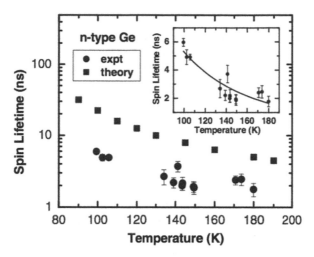

Figure 7.9. Spin lifetime in Ge as a function of temperature. Here, the dominant interaction is the Elliott–Yafet mechanism. (Reprinted with permission from [93]. Copyright 2012 AIP Publishing LLC.)

A second important mechanism for spin relaxation is the Dyakonov–Perel spin relaxation mechanism [95, 96], which occurs in materials which lack an inversion symmetry (due to either a lack of bulk inversion symmetry or structural inversion symmetry). This mechanism depends upon the presence of the spin–orbit interaction in the material to lead to an effective magnetic field which affects the spin during its transport. In addition, the spin–orbit interaction lifts the degeneracy of the two spin states, as discussed above in section 7.1. This induced magnetic field leads to spin precession during the motion of the carrier, and the impurity induces a change in the polarization which randomizes the precession of the spin. Additionally, the magnetic field itself can cause spin flip through the spin–orbit interaction. The randomization of the precession caused by the impurities leads to spin relaxation, but Dynakonov and Perel showed that the lifting of the spin degeneracy was sufficient to cause spin relaxation. From section 7.1, we recognize that the interaction term, given by the appropriate term added to the Hamiltonian for the spin–orbit interaction, is larger for larger momentum. Since each carrier has a different momentum, the precession will be different for each electron, leading to the dephasing within the ensemble. In the Dyakonov–Perel mechanism, the precession angle around any particular reference axis diffuses so that the square of the precession increases with time roughly as $(t/\tau)(\omega\tau)^2$, where τ is the momentum relaxation time, and ω is a typical precession frequency (see appendix E) [97]. The spin relaxation time is then defined to be the time when this precession angle becomes of order unity, so that it is defined as $\tau_s = 1/\omega^2\tau$. Hence, as the mobility and the conductivity rise (for constant carrier density), τ is getting larger, and the spin relaxation time is getting smaller. One form of the spin relaxation time for the Dyakonov–Perel mechanism is given as [98]

$$\tau_s = \frac{105\hbar^6}{64\eta^2(E_F m^*)^3 \tau},$$

(7.39)

where η is the spin–orbit coupling parameter in the bulk inversion asymmetry case.

In heavily doped materials, particularly when they are p-type materials, another mechanism that can lead to spin relaxation is the Bir–Aronov–Pikus mechanism [99]. This process depends upon the electron and hole interaction, and the spins of the interacting particles. The spin–orbit interaction is important for both electrons and holes and the exchange interaction between the electron and hole can create an effective magnetic field via this process. As before, this magnetic field leads to spin precession of the electron. However, the spin of the hole varies much faster than the precession of the electron, and this leads to a fluctuating effective magnetic field. This, in turn, introduces a fluctuation in the precession of the electron spin. This process is relevant in those semiconductors where the spin–orbit interaction introduces a significant overlap between the electron and hole wave functions. Surprisingly, it is found that the structure of the scattering rate for the electron–hole exchange interaction is qualitatively the same as found for the carrier–phonon scattering [100]. This mechanism for spin relaxation has been shown to be important in Mn-doped GaAs, as the Mn is a acceptor dopant as well as being important in producing magnetism in this material [101].

Problems

1. Consider two electrons in a state in which the radius (around some orbit center) is normalized to unity, and the angular wave function is $\varphi(\vartheta) = f(\cos\vartheta)$. If these two electrons have opposite spins and are located at $\vartheta = 0$ and π, discuss the anti-symmetry properties of these electrons. Can you infer the nature of the function which describes these angular variations?

2. In a particular semiconductor with effective mass of $0.04\, m_0$, it is found that the spin-split Landau levels are such that the spin-splitting (the shift upward of the upper spin state) is exactly enough that all the levels are equally spaced at a magnetic field of 10 T. What is the value of g at this magnetic field?

3. It has been reported that, in some semiconductors at high magnetic field, the lower spin state of one Landau level becomes degenerate with the upper spin state of the next lower Landau level. What must the value of g be for this condition to occur?

4. The Kane perturbation model is perhaps the simplest approach to including both the $\mathbf{k} \cdot \mathbf{p}$ and spin–orbit interactions. How do the parameters in this theory compare to those that appear in equations (7.7), (7.8), and (7.17)?

Appendix D Spin angular momentum

As is commonly known, electrons can have two spin orientations. Typically, these are called spin up and spin down. But, this is for an arbitrary orientation of the electron's own magnetic moment. Recognition of the two spin states arises from the introduction of the Pauli exclusion principle whenever we consider a fully quantized state. These two spin states must have opposite polarization, hence the designations as up and down. In fact, the spin can be oriented into any desired direction by the application of external forces, such as a magnetic field. The most common orientation, arising in the Zeeman effect, is along the z-axis which is achieved via an external magnetic field oriented in this direction. But, other directions are useful in a variety of applications such as the use of spin orientation in a qubit (quantum bit) for quantum computation. In this appendix, we wish to spend a little time talking about the values and orientations of the spin and get to the Pauli spin matrices.

Most people get a smattering of angular momentum in their undergraduate courses in classical mechanics or atomic physics. For sure, the angular momentum of an electron orbiting in a centrally symmetric potential, such as the Coulomb potential around an atom, possesses a quantized value for the spin angular momentum. In this atomic case, both the total angular momentum L^2 and the z-directed angular momentum L_z can be made to commute. The fact that they could both be made to commute with the Hamiltonian tells us that they can both be diagonalized along with the Hamiltonian and therefore could both be measured at the same time as the total energy. In the present case, the only angular momentum to be considered is the spin angular momentum, which we take to be oriented in the z-direction, just as in the atomic case. We expect that, just as in the atomic case, the total spin S^2 will continue to commute with the z-component of spin S_z, just as in the atomic case. This will be useful in what follows.

Let us begin by recalling that, in any quantum confinement problem such as the atom or a quantum well, the total wave function is a sum over a set of eigenfunctions $\psi_i(\mathbf{r})$. For each of these functions, which are defined by a set of quantized values due to the confinement, one has not considered the spin. If we now want to also include the spin, then we need an additional part of each eigenstate wave function. Typically, this is a multiplicative term describing the spin state. Traditionally, this is a two component wave function called a spinor. From the Zeeman effect, we know that one typically denotes the extra energy for the spin-up state by the value 1/2, and the value of the spin-down state by the value −1/2. We use this to denote the two possible states and their spinors as

$$\varphi\left(\frac{1}{2}\right) = \begin{bmatrix} 1 \\ 0 \end{bmatrix}, \ \varphi\left(-\frac{1}{2}\right) = \begin{bmatrix} 0 \\ 1 \end{bmatrix}, \tag{D.1}$$

where the first equation refers to the *up* state and the second equation refers to the *down* state. Thus, the eigenvalues correspond to those adopted in the Zeeman effect, as mentioned above.

Because the spin angular momentum has been taken to be oriented along the z-axis, we expect the spin matrix for the z-component of angular momentum must be diagonal for two reasons. First, it must commute with the total spin and with the Hamiltonian, and, second, it must produce the eigenvalues found from the Zeeman effect. Thus, we simply state that

$$S_z = \frac{\hbar}{2} \begin{bmatrix} 1 & 0 \\ 0 & -1 \end{bmatrix}, \tag{D.2}$$

and this gives

$$S_z \times \varphi\left(\frac{1}{2}\right) = \frac{\hbar}{2}\begin{bmatrix} 1 & 0 \\ 0 & -1 \end{bmatrix}\begin{bmatrix} 1 \\ 0 \end{bmatrix} = \frac{\hbar}{2}\varphi\left(\frac{1}{2}\right)$$

$$S_z \times \varphi\left(-\frac{1}{2}\right) = \frac{\hbar}{2}\begin{bmatrix} 1 & 0 \\ 0 & -1 \end{bmatrix}\begin{bmatrix} 0 \\ 1 \end{bmatrix} = -\frac{\hbar}{2}\varphi\left(-\frac{1}{2}\right)$$

(D.3)

as expected. We know from the study of normal angular momentum that the value of L^2 is given as $l(l+1)\hbar^2$. Thus, we expect a similar result for the spin angular momentum, and using $s = 1/2$, we get $S^2 = s(s+1)\hbar^2 = 3\hbar^2/4$. More strictly, this value is for the square of the magnitude of the total spin angular momentum. Again, the matrix representation of this total spin angular momentum must be diagonal, and this is given by

$$|S|^2 = \frac{3\hbar^2}{4}\begin{bmatrix} 1 & 0 \\ 0 & 1 \end{bmatrix}.$$

(D.4)

To find the other components of the spin angular momentum, and the other spin matrices, we introduce the rotating coordinates, as

$$S_+ = S_x + iS_y, \qquad S_- = S_x - iS_y.$$

(D.5)

Then, we can write the square magnitude of the total spin angular momentum as

$$S^2 = S_x^2 + S_y^2 + S_z^2 = S_+S_- + S_z^2 - i[S_x, S_y].$$

(D.6)

The commutator relations for these spin components are given as

$$[S_x, S_y] = i\hbar S_z, \quad [S_y, S_z] = i\hbar S_x, \quad [S_z, S_x] = i\hbar S_y.$$

(D.7)

Then, (D.6) becomes

$$S^2 = S_x^2 + S_y^2 + S_z^2 = S_+S_- + S_z^2 + i\hbar S_z.$$

(D.8)

Similarly, if we reverse the two rotating terms we get

$$S^2 = S_x^2 + S_y^2 + S_z^2 = S_-S_+ + S_z^2 - i\hbar S_z.$$

(D.9)

We can combine these last two equations, and use the results of equations (D.2) and (D.4) to yield

$$S^2 - S_z^2 = \frac{1}{2}(S_+S_- + S_-S_+) = \frac{\hbar^2}{2}.$$

(D.10)

The operators S_+ and S_- act as creation and annihilation operators for the spin angular momentum. That is, operating on a spinor with the first of these operators will raise the angular momentum, which can only occur if it acts on the spin-down state and produces the spin-up state, or

$$S_+\varphi\left(-\frac{1}{2}\right) = \hbar\varphi\left(\frac{1}{2}\right) \rightarrow S_+ = \hbar\begin{bmatrix} 0 & 1 \\ 0 & 0 \end{bmatrix}.$$

(D.11)

Similarly, acting with the operator S_- removes a quantum of angular momentum and lowers the spin angular. This can occur only if the operator acts upon the spin-up state and produces the spin-down state, or

$$S_-\varphi\left(\frac{1}{2}\right) = \hbar\varphi\left(-\frac{1}{2}\right) \rightarrow S_- = \hbar\begin{bmatrix} 0 & 0 \\ 1 & 0 \end{bmatrix}. \tag{D.12}$$

We can now easily invert equation (D.5) to give

$$\begin{aligned} S_x &= \frac{1}{2}(S_+ + S_-) = \frac{\hbar}{2}\begin{bmatrix} 0 & 1 \\ 1 & 0 \end{bmatrix} \\ S_y &= \frac{1}{2}(S_+ - S_-) = \frac{\hbar}{2}\begin{bmatrix} 0 & -i \\ i & 0 \end{bmatrix} \end{aligned}. \tag{D.13}$$

The Pauli spin matrices are just the matrices that appear in the definitions of the components of the spin angular momentum, as given by equation (7.4) above.

Appendix E The Bloch sphere

As we discussed in the previous appendix, the spin of the electron is characterized by its eigenstate, for which the wave function is a two component spinor. In a computer, the bit of information can also be expressed as a two component spinor, where the state $|0\rangle$ might correspond, for example, to the spin-down state and the state $|1\rangle$ would then correspond to the spin-up state. It makes natural sense then to spend a little more time with the idea of two level systems [102]. To simplify the notation slightly, let us use the generalized spinors

$$\alpha = \begin{bmatrix} 1 \\ 0 \end{bmatrix}, \quad \beta = \begin{bmatrix} 0 \\ 1 \end{bmatrix}. \tag{E.1}$$

The equation of motion for these spinors is, as usual, given by the Schrödinger equation in which the Hamiltonian is a 2×2 matrix. If there are no external forces acting upon our simple two level system, then the Hamiltonian is independent of time and the wave functions evolve as

$$\psi(t) = e^{-iHt/\hbar}\psi(0), \tag{E.2}$$

where ψ corresponds to either of the spinors α or β. In general, we diagonalize the Hamiltonian to provide the eigenvalues, which we take to be E_α and E_β. The total wave function can be written as a sum of the two spinors as

$$\psi = c_1\alpha + c_2\beta, \quad |c_1|^2 + |c_2|^2 = 1. \tag{E.3}$$

We can use the above properties of the Hamiltonian and the spinors to determine a number of intuitive properties. For example, if we want to determine how β evolves into α, we need only examine the total Hamiltonian, as

$$\beta^\dagger e^{-iHt/\hbar}\alpha = [0 \quad 1]e^{-iHt/\hbar}\begin{bmatrix} 1 \\ 0 \end{bmatrix}$$
$$= \frac{H_{21}}{E_\beta - E_\alpha}(e^{-iE_\alpha t/\hbar} - e^{-iE_\beta t/\hbar}), \tag{E.4}$$

where

$$H_{21} = \langle\beta|H|\alpha\rangle = [0 \quad 1]H\begin{bmatrix} 1 \\ 0 \end{bmatrix}. \tag{E.5}$$

If we examine the magnitude squared value of this term, we discover that it oscillates, and that the occupation oscillates from one state to the other with a frequency given by the difference in the eigen-energies of these two states.

The oscillation that occurs above is suggestive, as it points out that we can define the total state equation (E.3) itself as a single spinor whose components are the complex coefficients given in this equation. What this means is that the two spinors α and β define a two-dimensional space for our wave function. Where the quantum well was an infinite-dimensional space of eigenfunctions, the problem here is limited to just these two states. Thus, a spinor whose coefficients are the coefficients in equation (E.3) is a vector in this two component space, as

$$\psi = \begin{bmatrix} c_1 \\ c_2 \end{bmatrix} = \begin{bmatrix} |c_1|e^{i\vartheta_1} \\ |c_2|e^{i\vartheta_2} \end{bmatrix}. \tag{E.6}$$

But, this description is not unique since the squares of the magnitude must sum to unity. Thus, the two phases can be rather arbitrary, and we can only specify in detail the relative phase $\vartheta_1 - \vartheta_2$. So, if we write $\vartheta_2 = \vartheta_1 + \varphi$, we cannot then tell the difference between the states ψ and $e^{i\vartheta_1}\psi$, which means we have to somehow believe that these are the same state. In a sense, this all arises from the periodicities of the angles and that is a property we will use below.

A useful quantity to use in dealing with two level systems is the corresponding density matrix, defined via

$$\rho = \psi\psi^\dagger = \begin{bmatrix} c_1 \\ c_2 \end{bmatrix}[c_1^* \quad c_2^*] = \begin{bmatrix} |c_1|^2 & c_1 c_2^* \\ c_2 c_1^* & |c_2|^2 \end{bmatrix}. \tag{E.7}$$

An important property of this matrix is that its trace is unity, as required by equation (E.3). Now, we come to an important point. A fundamental property of any 2×2 matrix, whose trace is unity, is that it can be written in terms of the Pauli spin matrices and a quantity known as the polarization as

$$\rho = \frac{1}{2}(I + \mathbf{P} \cdot \boldsymbol{\sigma}), \tag{E.8}$$

where the individual Pauli spin matrices are given in equations (D.2) and (D.13). Comparing this last result with equation (E.7), we can identify the various component connections through

$$|c_1|^2 = \frac{1}{2}(1 + P_z)$$

$$|c_2|^2 = \frac{1}{2}(1 - P_z) \qquad . \qquad \text{(E.9)}$$

$$c_1 c_2^* = (c_2 c_1^*)^* = \frac{1}{2}(P_x - iP_y)$$

We can now invert these equations to identify the polarization components themselves as

$$P_x = 2Re(c_1 c_2^*)$$
$$P_y = -2Im(c_1 c_2^*). \qquad \text{(E.10)}$$
$$P_z = |c_1|^2 - |c_2|^2$$

It is easy to see that the trace of the density matrix is unity. Not only is this required by equation (E.3), but this trace is required to be the sum of the eigenvalues of the two wave function components. Since these are 0 and 1, we also satisfy this requirement. This now implies that one of the eigenvectors of this density matrix is ψ itself. The other eigenvector must be orthogonal to ψ so that their inner product yields 0, but this other eigenvector is still arbitrary at this point.

An important property arises from the expectation values of the various spinors. We examine this with the x-spinor as

$$\langle \sigma_x \rangle = Tr\{\rho\sigma_x\} = \frac{1}{2}(Tr\{\sigma_x\} + \mathbf{P} \cdot Tr\{\boldsymbol{\sigma}\sigma_x\})$$
$$= \frac{1}{2}P_x Tr\{\sigma_x^2\} = P_x \qquad . \qquad \text{(E.11)}$$

This can be repeated for each of the other components, which yields the important result that

$$\mathbf{P} = \langle \boldsymbol{\sigma} \rangle = Tr\{\rho\boldsymbol{\sigma}\}. \qquad \text{(E.12)}$$

That is, the polarization of our two level system is defined by the expectation value of the spin vector. The direction of the spin is uniquely connected to the polarization of the system. Another important point is that since the wave function ψ is an eigenfunction of the density matrix, it is also an eigenfunction of the last term in the density matrix

$$\mathbf{P} \cdot \boldsymbol{\sigma}\psi = \psi. \qquad \text{(E.13)}$$

To examine the nature of the polarization itself, let us make some angular definitions of the various components of the wave function equation (E.6), as

$$c_1 = e^{i\gamma}\cos(\vartheta/2), \quad c_2 = e^{i(\gamma+\varphi)}\sin(\vartheta/2), \qquad \text{(E.14)}$$

so that

$$P_x = \cos(\varphi) \sin(\vartheta)$$
$$P_y = \sin(\varphi) \sin(\vartheta) \; .$$
$$P_z = \cos(\vartheta)$$

(E.15)

We recognize these angles as the angles in a spherical coordinate system which relate to the normal rectangular coordinates. The angle ϑ is the polar angle and the angle φ is the azimuthal angle. The spherical system in which we utilize a spherical shell of unity radius is known as the Bloch sphere, and is shown in figure E1. From the definition of our spinors and the above coefficients, we recognize that our states may be defined as

$$\beta = |0\rangle, \; \vartheta = 0, \mathbf{P} = \mathbf{a}_z$$
$$\alpha = |1\rangle, \; \vartheta = \pi, \mathbf{P} = -\mathbf{a}_z$$

(E.16)

Hence, moving around within the state ψ means that we move around on the surface of the Bloch sphere by varying the two angles.

One of the properties of the spin that we are familiar with is the fact that it precesses under a variety of different forces. We can investigate that by introducing a simple form for the Hamiltonian. This form is suggested by the density matrix itself, and we will denote it as

$$H = \frac{1}{2}(Q_0 I + \mathbf{Q} \cdot \boldsymbol{\sigma}).$$

(E.17)

The time rate of change of the polarization is then given by the well-known relation in quantum mechanics

$$\frac{d\mathbf{P}}{dt} = -\frac{i}{\hbar} \langle [\boldsymbol{\sigma}, H] \rangle.$$

(E.18)

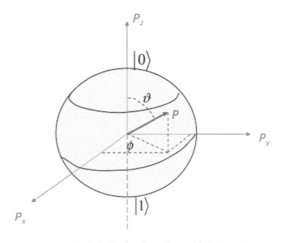

Figure E1. The Bloch sphere is a spherical shell of unit radius, which describes our unique two level system.

The only component of the Hamiltonian that is important now is the second term in equation (E.17), as the first term commutes with everything. Then, we need to evaluate the quantity

$$\boldsymbol{\sigma}(\mathbf{Q} \cdot \boldsymbol{\sigma}) - (\mathbf{Q} \cdot \boldsymbol{\sigma})\boldsymbol{\sigma} = \mathbf{Q} \times (\boldsymbol{\sigma} \times \boldsymbol{\sigma}). \tag{E.19}$$

If $\boldsymbol{\sigma}$ were a simple vector, the last term in parentheses would vanish, but this is not a simple vector. Rather, we find that

$$(\boldsymbol{\sigma} \times \boldsymbol{\sigma}) = \begin{vmatrix} \mathbf{a}_x & \mathbf{a}_y & \mathbf{a}_z \\ \sigma_x & \sigma_y & \sigma_z \\ \sigma_x & \sigma_y & \sigma_z \end{vmatrix} = 2i\boldsymbol{\sigma}, \tag{E.20}$$

which is surprising but leads to

$$\frac{d\mathbf{P}}{dt} = \frac{1}{\hbar}\mathbf{Q} \times \langle \boldsymbol{\sigma} \rangle = \frac{1}{\hbar}\mathbf{Q} \times \mathbf{P}. \tag{E.21}$$

So, the precession of the spin polarization arises from any vector term in the Hamiltonian which is not parallel to the polarization itself. It is easy to show that this motion does not change the amplitude of the polarization, which is of course required for the spin itself. If the vector $\mathbf{Q}$ is a constant amplitude vector, then it defines a precession energy corresponding to a precession frequency $\omega_Q = Q/\hbar$.

References

[1] Zeeman P 1897 *Phil. Mag. Ser.* 5 **43** 226
[2] Oestreich M and Rühle W W 1995 *Phys. Rev. Lett.* **74** 2315
[3] von Klitzing K, Dorda G and Pepper M 1980 *Phys. Rev. Lett.* **45** 494
[4] Datta S and Das B 1990 *Appl. Phys. Lett.* **58** 665
[5] Žutic I, Fabian J and das Sarma S 2004 *Rev. Mod. Phys.* **76** 323
[6] Dyakonov M I and Perel V I 1971 *JETP Lett.* **13** 467
[7] Hirsch J E 1999 *Phys. Rev. Lett.* **83** 1834
[8] Zhang S 2000 *Phys. Rev. Lett.* **85** 393
[9] Murakami S, Nagaosa N and Zhang S 2003 *Science* **301** 1348
[10] Sinova J, Culcer D, Niu Q, Sinitsyn N A, Jungwirth T and MacDonald A H 2004 *Phys. Rev. Lett.* **92** 126603
[11] Falicov L M and Cohen M H 1963 *Phys. Rev.* **130** 92
[12] Liu L 1962 *Phys. Rev.* **126** 1317
[13] Bloom S and Bergstresser T K 1968 *Solid State Commun.* **6** 465
[14] Weisz G 1966 *Phys. Rev.* **149** 504
[15] Taylor P L and Heinonen O 2002 *A Quantum Approach to Condensed Matter Physics* (Cambridge: Cambridge University Press)
[16] Dresselhaus G 1955 *Phys. Rev.* **149** 580
[17] Ferry D K 2020 *Semiconductors: Bonds and Bands* 2nd edn (Bristol: IOPP) section 2.5.2
[18] Sakurai J J 1967 *Advanced Quantum Mechanics* (Reading, MA: Addison-Wesley) 85–7
[19] Bychov Y A and Rashba E I 1984 *J. Phys. C: Solid State Phys.* **17** 6039
[20] Cummings A W, Akis R and Ferry D K 2011 *J. Phys.: Condens. Matter* **23** 465301
[21] Madelug E 1926 *Z. Phys.* **40** 322

[22] Kennard E H 1928 *Phys. Rev.* **31** 876

[23] Bohm D 1952 *Phys. Rev.* **85** 166

[24] Ferry D K 1998 *VLSI Des.* **8** 165

[25] Einstein A 1917 *Verh. Deutsche Phys. Gesell.* **19** 82

[26] Brillouin L 1926 *J. Phys. Radium* **7** 353

[27] Keller J B 1958 *Ann. Phys.* **4** 180

[28] Berry M V 1989 *Proc. Roy. Soc. London* **424** 219

[29] Thouless D J, Kohmoto M, Nightingale M P and den Nijs M 1982 *Phys. Rev. Lett.* **49** 405

[30] Kohmoto M 1985 *Ann. Phys.* **160** 343

[31] Chang M-C and Niu Q 1996 *Phys. Rev.* B **53** 7010

[32] Berry M V 1984 *Proc. R. Soc. London* A **392** 45

[33] Cummings A W, Akis R and Ferry D K 2009 *J. Phys.: Condens. Matter* **21** 055502

[34] Xiao D, Liu G-B, Feng W, Xu X and Yao W 2012 *Phys. Rev. Lett.* **108** 196802

[35] Kormányos A, Burkard G, Gmitra M, Fabian J, Zólyomi V, Drummond N D and Fal'ko V 2015 *2D Mater.* **2** 022001

[36] Lebègue S and Eriksson O 2009 *Phys. Rev.* B **79** 115409

[37] Jungwirth T, Niu Q and MacDonald A H 2002 *Phys. Rev. Lett.* **88** 207208

[38] Xiao D, Yao W and Niu Q 2007 *Phys. Rev. Lett.* **99** 236809

[39] Jacob J, Meier G, Peters S, Matsuyama T, Merkt U, Cummings A W, Akis R and Ferry D K 2009 *J. Appl. Phys.* **105** 093714

[40] Johnson M 2005 *J. Phys. Chem.* B **109** 14278

[41] Inoue J, Bauer G E W and Molenkamp L W 2004 *Phys. Rev.* B **70** 041303

[42] Nikolic B K, Souma S, Zarbo L B and Sinova J 2005 *Phys. Rev. Lett.* **95** 046601

[43] Sheng L, Sheng D N and Ting C S 2005 *Phys. Rev. Lett.* **94** 016602

[44] Cummings A W, Akis R and Ferry D K 2006 *Appl. Phys. Lett.* **89** 172115

[45] Kato Y K, Myers R C, Gossard A C and Awschalom D D 2004 *Science* **306** 1910

[46] Shih V, Myers R C, Kato Y K, Lau W H, Gossard A C and Awschalom D D 2005 *Nat. Phys.* **1** 31

[47] Wunderlich J, Kaestner B, Sinova J and Jungwirth T 2005 *Phys. Rev. Lett.* **94** 047204

[48] Aronov A G 1976 *Teor. Fiz.* **71** 370

[49] Aronov A G and Pikus G E 1976 *Sov. Phys. Semicond.* **10** 698

[50] Johnson M and Silsbee R H 1988 *Phys. Rev.* B **37** 5326

[51] Johnson M 1995 *J. Magn. Magn. Mater.* **140–144** 21

[52] Oestrich M, Hübner J, Hägele D, Klar P J, Heimbrodt W and Rühle W W 1999 *Appl. Phys. Lett.* **74** 1251

[53] Schmidt G, Müller G, Molenkamp L W, Behet M, De Boeck J and Panissod P 1999 *J. Magn. Magn. Mater.* **198–199** 134

[54] Monzon F G and Roukes M L 1999 *J. Magn. Magn. Mater.* **198–199** 632

[55] Hammer P, Bennett B R, Yang M J and Johnson M 1999 *Phys. Rev. Lett.* **83** 203

[56] Jonker B T, Kioseoglou G, Hanbicki A T, Li C H and Thompson P E 2007 *Nat. Phys.* **3** 542

[57] Hammer P R and Johnson M 2002 *Phys. Rev. Lett.* **88** 066806

[58] Applebaum I, Huang B and Monsma D J 2007 *Nature* **447** 295

[59] Bertoni A, Bordone P, Brunetti R, Jacoboni C and Reggiani S 2000 *Phys. Rev. Lett.* **84** 5912

[60] Reggiani S, Bertoni A and Rudan M 2002 *Physica* B **314** 136

[61] Harris J, Akis R and Ferry D K 2001 *Appl. Phys. Lett.* **79** 2214
[62] Cummings A W 2009 *PhD Dissertation* Arizona State University, Tempe, AZ
[63] Cummings A W, Akis R, Ferry D K, Jacob J, Matsuyama T, Merkt U and Meier G 2008 *J. Appl. Phys.* **104** 066100
[64] Jacob J, Lehmann H, Merkt U, Mehl S and Hankiewicz E M 2012 *J. Appl. Phys.* **112** 013706
[65] Benter T, Lehmann H, Matsuyama T, Hansen W, Heyn C, Merkt U and Jacob J 2013 *Appl. Phys. Lett.* **102** 212405
[66] Richter A, Koch M, Matsuyama T, Heyn C and Merkt U 2000 *Appl. Phys. Lett.* **77** 3227
[67] Kane B E 1998 *Nature* **393** 133
[68] O'Brien J L, Schofield S R, Simmons M Y, Clark R G, Dzurak A S, Curson N J, Kane B E, McAlpine N S, Hawley M E and Brown G W 2001 *Phys. Rev.* B **64** 161401
[69] Lyding J W, Albein G C, Shen T-C, Wang C and Tucker J R 1994 *J. Vac. Sci. Technol.* B **12** 3735
[70] Hersam M C, Guisinger N P and Lyding J W 2000 *Nanotechnology* **11** 70
[71] Matsukawa T, Fukai T, Suzuki S, Hara K, Koh M and Ohdomari I 1997 *Appl. Surf. Sci.* **117-118** 677
[72] Matsukawa T, Shinada T, Fukai T and Ohdomari I 1998 *J. Vac. Sci. Technol.* B **16** 2479
[73] Jamieson D N, Yang C, Hearne S M, Pakes C I and Prawer S 2005 *Appl. Phys. Lett.* **86** 202101
[74] Dehollian J P, Muhonen J T, Blume-Kohut R and Rudinger K M 2016 *New J. Phys.* **18** 103018
[75] Loss D and DiVencenzo D 1998 *Phys. Rev.* A **57** 120
[76] Dempsey K J, Ciudad D and Marrows C H 2011 *Phil. Trans. R. Soc.* A **369** 3150
[77] Sigillito A J, Tyryshkin A M, Beeman J W, Haller E E, Itoh K M and Lyon S A 2016 *Phys. Rev.* B **94** 125201
[78] Abadillo-Uriel J C and Calderón M J 2016 *Nanotechnol.* **27** 024003
[79] Salfi J, Tong M, Rogge S and Culcer D 2016 *Nanotechnol.* **27** 244001
[80] Kim D *et al* 2014 *Nature* **511** 70
[81] Thorgrimsson B *et al* 2017 *npj Quantum Inform.* **3** 32
[82] Abadillo-Uriel J C *et al* 2018 *Phys. Rev.* B **98** 165438
[83] Wolfe M A, Calderon-Vargas F A and Kestner J P 2017 *Phys. Rev.* B **96** 201307
[84] Serina M, Kloeffel C and Loss D 2017 *Phys. Rev.* B **95** 245422
[85] Veldhorst M, Ruskov R, Yang C H, Hwang J C C, Hudson F E, Flatté M E, Tahan C, Itoh K M, Morello A and Dzurak A S 2015 *Phys. Rev.* B **92** 201401
[86] Kim D, Ward D R, Simmons C B, Savage D E, Lagally M G, Friesen M, Coppersmith S N and Eriksson M A 2016 *npj Quantum Inform.* **1** 15004
[87] Malkoc O, Stano P and Loss D 2016 *Phys. Rev.* B **93** 235413
[88] Zajac D M, Hazard T M, Mi X, Nielsen E and Petta J R 2016 *Phys. Rev. Appl.* **6** 054013
[89] Elliot R 1954 *Phys. Rev.* **96** 226
[90] Yafet Y 1963 *Solid State Phys.* **14** 1
[91] Overhauser A W 1953 *Phys. Rev.* **89** 689
[92] Ochiai Y and Matsuura E 1976 *Phys. Status Solidi* **38** 243
[93] Guite C and Venkataraman V 2012 *Appl. Phys. Lett.* **101** 252404
[94] Li P, Song Y and Dery H 2012 *Phys. Rev.* B **86** 085202
[95] Dyakanov M I and Perel V I 1971 *Sov. Phys. JETP* **38** 1053

[96] Dyakanov M I and Perel V I 1972 *Sov. Phys. Solid State* **13** 3023

[97] Fabian J and das Sarma S 1999 *J. Vac. Sci. Technol.* **B 86** 1708

[98] Buss J H, Rudolph J, Schupp T, As D J, Lischka K and Hägele D 2010 *Appl. Phys. Lett.* **97** 062101

[99] Bir G L, Aronov A G and Pikus G E 1976 *Sov. Phys. JETP* **42** 705

[100] Lechner C and Rössler U 2005 *Phys. Rev.* **B 72** 153317

[101] Akimov I A, Dzhloev R I, Korenev V L, Kusrayev Y G, Zhukov E A, Yakovlev D R and Bayer M 2009 *Phys. Rev.* **B 80** 081203

[102] Merzbacher E 1970 *Quantum Mechanics* 2nd edn (New York: Wiley) ch 13

IOP Publishing

Transport in Semiconductor Mesoscopic Devices
(Second Edition)

David K Ferry

Chapter 8

Tunnel devices

Tunneling is an interesting process. It appears in nearly all books on quantum mechanics, but it is often overlooked as an important topic. Yet, it is well-known that the nuclear fusion process that makes our Sun work, and is therefore important to life on Earth, just would not work without tunneling. So, it seems a little bizarre that it is not held in higher regard in parts of the physics community. Moreover, it has become a more important part of the world of devices as these devices have continued to become ever smaller as a result of the world following Moore's law. It is an essential part of every FET, as tunneling is a major source of gate leakage currents. Then, since Esaki's discovery of the tunnel diode, tunneling has been studied on its own for device applications. And, as we continue to investigate mesoscopic devices, tunneling is recognized as often being an integral part of the device under study. New devices, such as the resonant tunneling diode (RTD) and the single-electron transistor (SET) have furthered this interest and the importance of tunneling.

We want to start by looking at the simple quantum point contact (QPC), because a great many of the devices in which we will be interested in this chapter use the QPC as an input or output connection to the environment. If we look at the QPC conductance in figure 2.2, there are the obvious plateaus corresponding to the full channels being transmitted through the QPC. Then, there is a little rounding on the more negative voltage end of each plateau, which arises from the thermal broadening of the Fermi–Dirac distribution. But, the main rise in current between each plateau is due to tunneling of carriers through the barrier that they see in the QPC. In the case of Schottky-gate-induced formation of the QPC, the actual potential that forms between the two gates appears to be a saddle potential such as that shown schematically in figure 8.1. The transverse potential—the rising potential seen in the figure as one moves away from the actual saddle maximum—is actually a harmonic oscillator parabolic potential, as was discussed in section 2.4.1. In the current flow

doi:10.1088/978-0-7503-3139-5ch8

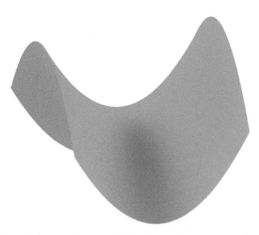

Figure 8.1. A schematic view of the saddle potential that forms in a quantum QPC due to surface Schottky barrier gates. The carrier flow direction is indicated by the purplish region coming in from the front.

direction (indicated by the purplish region), the potential decreases in a parabolic manner, and these opposing parabolic potentials create the saddle shape shown in the figure. The actual peak of the saddle comes from the bare potential plus the factor of $\hbar\omega/2$ that is apparent in the longitudinal energy bands shown in figure 2.3. As the carriers approach the QPC barrier, they cannot pass the barrier if the Fermi energy is below the saddle. This is represented by the purplish flow of carriers, which ends just below the saddle in figure 8.1. Consequently, if they are to pass the QPC, they must do so by tunneling. As the Fermi energy is increased, relative to the saddle potential, the tunneling probability increases and the resulting current increases. We can show this with a simple approach to the tunneling coefficient, which can be approximated as (see appendix F)

$$T \sim \exp(-\int \gamma(x)dx), \tag{8.1}$$

where γ is the decay constant of the quantum wave for energies below the height of the barrier [1]. The tunneling problem can be represented by the sketch of figure 8.2. Here, the parabolic potential in the current direction, which decreases away from the saddle as can be seen in figure 8.1, is indicated by the black line. The Fermi levels on the two sides of the barrier are shown in blue, with the dashed extension through the barrier. Hence the barrier has a spatially varying shape as indicated. The points $\pm x_0$ are the so-called turning points where the Fermi energy meets the potential. We can now write the integral in equation (8.1) as

$$\text{Int} \sim -\int_{-x_0}^{x_0} \sqrt{\frac{2m^*}{\hbar^2}(V_0 - \alpha x^2)}\, dx, \tag{8.2}$$

where α characterizes the downward curvature of the potential in the current direction, and has the units of eV cm^{-2}. If we measure V_0 from the Fermi energy, then the potential is zero at x_0. Hence, we can write the integral as

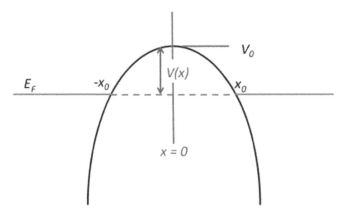

Figure 8.2. A schematic picture to help in evaluating the tunnel transmission through the QPC potential.

$$\text{Int} \sim -\frac{\pi}{2}\sqrt{\frac{2m^*V_0^2}{\alpha\hbar^2}}. \tag{8.3}$$

As the gate voltage is made more positive, the curvature of the potential decreases, but so does the potential amplitude, which leads to an overall increase in the tunneling current. If we adapt the Landauer formula and use equation (8.1) as the transmission, then the rising part of the steps in figure 2.2 is given approximately as

$$G \sim \frac{2e^2}{h}\exp\left(-\frac{\pi}{2}\sqrt{\frac{2m^*V_0^2}{\alpha\hbar^2}}\right), \tag{8.4}$$

but one must still determine experimentally just how one can determine V_0 and α. The latter can be determined by measuring the harmonic oscillator parameters as in figure 2.17. From this measurement, one can determine the energy level spacing, and this can be used with the gate voltage spacing of the steps to estimate the potential height as a function of gate voltage.

8.1 Coulomb blockade

When the dielectric material in a capacitor becomes too thin, then it is possible for the charge on the capacitor plate to tunnel through the insulator, and the capacitor becomes what is known as a leaky capacitor. Under some mesoscopic device concepts, this tunneling is a desirable effect, and can be used to create some interesting devices. In other cases, such as the gate oxide in a MOSFET, this is an undesirable effect. When, the capacitor is made with a small area, then another effect begins to occur. While the capacitor is small, we continue to describe it as a macroscopic capacitance associated with the system, although this description may not be fully valid. The change in electrostatic potential due to a change in the charge on an ideal conductor is associated with the linear relationship between the charge and the voltage

$$Q = CV, \tag{8.5}$$

where C is the capacitance, Q is the charge on the conductor, and V the electrostatic potential that exists between the two 'plates' of the capacitor. For an ideal metal, any charge added to it will rearrange itself such that the electric field inside the metal vanishes, and the surface of the metal becomes an equipotential surface. Therefore, the electrostatic potential associated with the metal relative to its reference is uniquely defined. In our capacitor, we consider two metal conductors connected by a dc voltage source. This leads to a charge $+Q$ on one conductor and a charge $-Q$ on the other. The capacitance of the two conductor system is then defined as $C = Q/V_{12}$. The electrostatic energy stored in the two conductor system is the work done in building up the charge Q on the two conductors and is given by

$$E = \frac{Q^2}{2C}. \tag{8.6}$$

In the case of very small capacitors, the charging energy given by this latter equation due to a single electron, $e^2/2C$, becomes comparable to the thermal energy, k_BT. The transfer of a single electron between conductors therefore results in a voltage change that is significant compared to the thermal voltage fluctuations and creates an energy barrier to the transfer of electrons. This barrier remains until the charging energy is overcome by sufficient bias. How small must the capacitor be for such effects to become important? If the energy stored in the capacitor is about the same as the thermal energy, then the capacitor has a value of 3×10^{-18} F, or 3 aF, at room temperature. Of course, at very low temperatures, the capacitance can be significantly larger.

Historically, Coulomb blockade effects were first predicted and observed in small metallic tunnel junction systems. As mentioned already, the conditions in metallic systems of high electron density, large effective mass, and short phase coherence length (compared to semiconductor systems) usually allow us to neglect size quantization effects. The dominant single-electron effect for small metal tunnel junctions is therefore the charging energy due to the transfer of individual electrons, $e^2/2C$. The effects of single-electron charging in the conductance properties of very thin metallic films was recognized in the early 1950s by Gorter [2] and Darmois [3]. It was found that these metal films formed arrays of small islands, and conduction occurs due to tunneling between these islands. Since the island size is small, the tunneling electron has to overcome an additional barrier due to the charging energy, which leads to an increase in resistance at low temperature. Such discontinuous metal films show an activated conductance, similar to an intrinsic semiconductor. Neugebauer and Webb [4] developed a theory of activated tunneling in which the activation energy resembles an energy gap and is therefore referred to as a *Coulomb gap*.

There have been many experimental studies of the transport properties of metal clusters or islands imbedded in an insulator that are then contacted by conducting electrodes. More interest in the area developed in studies of superconducting tunnel junctions [5], where the coulomb blockade interacted with the normal Josephson

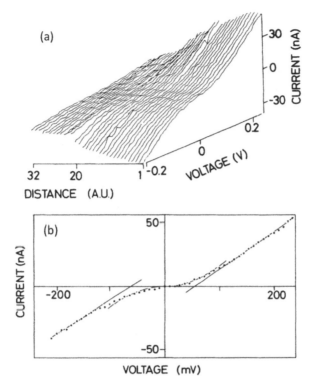

Figure 8.3. (a) The *I–V* characteristics of a tunnel junction between a STM tungsten tip and a stainless steel surface. The distance of the tip from the surface is reduced for each sweep (moving to the right in the figure). (b) A single curve from (a) plotted as the dots. The straight lines are guides to the eye, while the quadratic curve through the origin is a fit to theory. (Reprinted with permission from [8]. Copyright 1988 the American Physical Society.)

tunneling [6] (we will see this again when we return to superconducting qubits in section 8.4 below). Very soon, however, the coulomb blockade and the tunneling of single electrons were observed for normal metal systems [7, 8]. What is normally seen in these experiments is that no, or a very low, current flows until the applied voltage reaches $e/2C$, then the current begins to rise. So, this gives a plateau around zero bias, and the width of this plateau is typically e/C, corresponding to the energy in equation (8.6). We illustrate this with measurements made between a tungsten tip on an STM and a stainless steel metal substrate in figure 8.3 [8]. In panel (a), many *I–V* plots are shown corresponding to an increasing distance from the tip to the surface. In panel (b), one single trace is shown to illustrate the Coulomb blockade, which leads to the observed plateau. At higher values of the applied bias, the current approaches a linear dependence on the voltage. The plateau in this curve appears to be about 120 mV wide, which then corresponds to a capacitance of 1.3 aF. As the measurements were made at helium temperature, the charging energy (8.6) is certainly larger than the thermal energy. These measurements appear to be clear evidence for single-electron tunneling through their structure, by which only a single electron tunnels at any one point in time. These results correlate well with the theory [5].

8.2 Single-electron structures

A single capacitor is seldom used to make a structural device. Certainly, a single capacitor is used in dynamic random-access memory, but it is coupled there to a transistor which controls the charging and discharging of the capacitor. In meso-scopic devices, the idea of a single-electron device, or SET, has appeared, and this uses (at least) two capacitors connected in series. Between the two capacitors is a region that can accumulate charge. Typically, this small region is termed a quantum dot. In metals, it may contain more than a thousand electrons, but in semi-conductors the charge states may have their own quantization due to the small size of the dot and the number of electrons can be few, even down to zero.

8.2.1 A simple quantum-dot tunneling device

Let us consider the two-capacitor circuit shown in figure 8.4, in which the 'island' consists of a small quantum dot coupled weakly through thin insulators to metal leads as shown. Typically, the tunnel junction can be considered as a parallel combination of the tunneling resistance R and the actual capacitance C. In metal tunnel junctions, the tunneling barrier is typically very high and thin, while the density of states at the Fermi energy is very high. The tunneling resistance is therefore almost independent of the voltage drop across the junction, but of course is much larger than the $h/2e^2$ that corresponds to a single propagating mode. In the analysis that follows, we ignore this resistance, but will call upon it later. Electrons that tunnel through one junction or the other are therefore assumed to immediately relax due to carrier–carrier scattering, so that resonant tunneling through both barriers is simultaneously neglected. This assumption is made as we are interested in the charge that can accumulate in/on the quantum dot in the sequential tunneling approach. That is, charge may tunnel through only one of the two capacitors, and this will change the amount of charge on the dot. Tunneling represents the injection

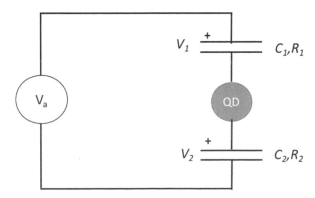

Figure 8.4. Circuit for two Coulomb blockaded capacitors and a central quantum dot on which to accumulate charge.

of single particles, which involve several characteristic time scales. The tunneling time (the time to tunnel from one side of the barrier to the other) is the shortest time (on the order of 10^{-14} s), whereas the actual time between tunneling events themselves is on the order of the current divided by e, which for typical currents in the nA range implies a mean time of several hundred picoseconds between events. The time for charge to rearrange itself on the electrodes due to the tunneling of a single electron will be something on the order of the dielectric relaxation time, which is also very short. Therefore, for purposes of analysis, we can consider that the junctions in the regime of interest behave as ideal capacitors through which charge is slowly leaked.

The net charge on the electrodes of the individual capacitors is given by equation (8.5), which can be rewritten for the present case as

$$Q_1 = C_1 V_1$$
$$Q_2 = C_2 V_2. \tag{8.7}$$

The net charge Q_{dot} on the island is the difference of these two charges. In the absence of tunneling, the difference in charge would be zero and the island neutral. Tunneling allows an integer number of excess electrons to accumulate on the island so that we find

$$Q_{\text{dot}} = -ne = Q_2 - Q_1. \tag{8.8}$$

Here, n is the net number of electrons, and can only exist if the charge is different on the two capacitors. This convention is chosen such that an increase in either n_1 or n_2, the number of electrons that define the charge on the two capacitors, corresponds to increasing either the junction charge Q_1 or Q_2, respectively, in equation (8.7). The sum of the two voltages across the two capacitors is just the applied voltage, V_a, so that, using equations (8.7) and (8.8), we may write the voltage drops across the two tunnel junctions as

$$V_1 = \frac{1}{C_1 + C_2}(C_2 V_a + ne)$$
$$V_2 = \frac{1}{C_1 + C_2}(C_1 V_a - ne) \tag{8.9}$$

In the following, we will write $C_T = C_1 + C_2$ to simplify the equations. The electrostatic energy stored in the two capacitors is given by

$$E = \frac{Q_1^2}{2C_1} + \frac{Q_2^2}{2C_2} = \frac{1}{2C_T}\left(C_1 C_2 V_a^2 + Q_{\text{dot}}^2\right). \tag{8.10}$$

However, this is not the total energy in the circuit. We must add to this the work done by the voltage source in transferring the charge to the two capacitors, which involves the various tunneling currents through them. This additional work is found by integrating the current over time, as

$$W_a = \int dt\, V_a I(t) = V_a \Delta Q. \tag{8.11}$$

Here, ΔQ is the total charge transferred from the voltage source, including the integer number of electrons that tunnel into the island and the continuous polarization charge that builds up in response to the change of electrostatic potential on the island. A change in the charge on the island due to one electron tunneling through capacitor 2 (so that $n'_2 = n_2 + 1$) changes the charge on the island to $Q' = Q + e$, and $n' = n - 1$. From equation (8.9), the voltage across junction 1 changes as $V'_1 = V_1 - e/C_T$. Therefore, from equation (8.7), a polarization charge flows in from the voltage source $\Delta Q = -eC_1/C_T$ to compensate. The total work done to pass in n_2 charges through junction 2 can then be written as

$$W_a(n_2) = -n_2 e V_a \frac{C_1}{C_T}. \tag{8.12}$$

By a similar approach, we can also find the work done in transferring n_1 charges through junction 1 to be

$$W_a(n_1) = -n_1 e V_a \frac{C_2}{C_T}. \tag{8.13}$$

With this, we can write the total energy in the system as

$$E(n_1, n_2) = \frac{1}{2C_T}\left(C_1 C_2 V_a^2 + Q_{\text{dot}}^2\right) + \frac{eV_a}{C_T}(C_1 n_2 + C_2 n_1). \tag{8.14}$$

The condition for Coulomb blockade is based on the change in this electrostatic energy with the tunneling of a particle through either junction. At zero temperature, the system has to evolve from a state of higher energy to one of lower energy. Therefore, tunneling transitions that take the system to a state of higher energy are not allowed, at least at zero temperature (at higher temperature, thermal fluctuations in energy on the order of $k_B T$ weaken this condition). At high enough temperature, the thermal fluctuations wash out the Coulomb blockade, and the capacitors become just leaky capacitors. We find the voltage at which the tunneling of single electrons can occur from the change in energy of the system when the charge on the dot changes by ± 1. let us assume first that the charge on C_2 changes by the addition or subtraction of a single electron, which leads to the change in energy of

$$\Delta E_2^{\pm} = E(n_1, n_2) - E(n_1, n_2 \pm 1)$$
$$= \frac{e}{C_T}\left[-\frac{e}{2} \pm (ne - V_a C_1)\right]. \tag{8.15}$$

Similarly, for a change in the charge on C_1 by the addition or subtraction of a single electron, the change in energy is

$$\Delta E_2^{\pm} = E(n_1, n_2) - E(n_1 \pm 1, n_2)$$
$$= \frac{e}{C_T}\left[-\frac{e}{2} \mp (ne + V_a C_1)\right] . \tag{8.16}$$

For all possible transitions on to or off of the island, the leading term involving the Coulomb energy of the island causes ΔE to be negative until the magnitude of V_a exceeds a threshold that depends on the lesser of the two capacitances. For $C_1 = C_2 = C$, the requirement becomes simply $|V_a| > e/2C$. Tunneling is prohibited and no current flows below this threshold, as evident in the I–V characteristics shown in figure 8.3 (although the figure is a single capacitor, the effect is the same). This region of Coulomb blockade is a direct result of the additional Coulomb energy, $e^2/2C_T$, which must be expended by an electron in order to tunnel on to or off of the island. The effect on the current voltage characteristics is a region of very low conductance around the origin. For large-area junctions where C_T is large, no regime of Coulomb blockade is observed, and current flows according to the tunnel resistance.

Figure 8.5(a) shows the equilibrium band diagram for a double-tunnel-junction system, illustrating the Coulomb blockade effect for equal capacitances. A Coulomb gap of width $e^2/2C_T$ has opened at the Fermi energy of the metal island, half of which appears above and half below the original Fermi energy, so that no states are available for electrons to tunnel into from the left and right electrodes. In essence, this energy gap is the charging energy required to put an electron onto the dot. For large capacitances, this energy is reduced toward zero. Similarly, electrons in the island cannot tunnel from the island to the metals, because there are no empty states in the metallic contacts. However, when a bias of $V_a \geqslant e/2C_T$ is applied, the bands shift to bring the dot level into alignment with one of the two contacts. Now, electrons can tunnel through the barriers, and with small capacitors, only a single electron at a time tunnels. When the first electron tunnels onto the dot, the energy will be shifted to show a Coulomb blockade to the second electron. This gap prohibits the tunneling until the voltage is raised to $3e/2C_T$, as is apparent from equations (8.15) and (8.16) with $n = \pm 1$ in this case.

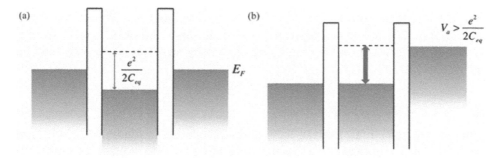

Figure 8.5. Band diagram of a two-capacitor single-electron circuit (a) in equilibrium, showing the Coulomb blockade, and (b) with a bias above the tunneling threshold so that the minimum energy can be supplied from the source.

An important aspect of the structure we have been discussing is the size of the quantum dot. As is apparent, the necessary voltage at which current begins to flow is dependent upon the size of the dot. In general, we can assume that the distance between the top electrode of C_1 and the bottom electrode of C_2 is held fixed in any experimental configuration. Of course, if an STM tip is used, this may not be the case, but we nevertheless assume that this will be the case for the studies we want to discuss now. Then, the capacitances will depend upon the size of the quantum dot. If the dot is made smaller, the capacitances will also become smaller due to the increase in the thickness of the insulating medium between the top electrode and the dot. This will lead to an increase in the voltage at which the current starts to rise. And the reverse will also be the case, as when we make the dot larger, the capacitances will also become larger, so that the required voltage is decreased. As a result, when a bias is applied that is larger than the turn-on voltages, the resulting tunnel current can be a measure of the size of the quantum dot. Thus, the tunnel current can be used to monitor this size, which may be used for an interesting quantitative sensing circuit. For example, if single molecules are used as the quantum dot, then these molecules can be interrogated by the tunnel current [9]. By bonding a pair of recognition molecules to the two metallic tips that represent the wires from the applied bias source, recognition can be done on a target molecule which has preferential hydrogen bonding to particular sites on the recognition molecules [10]. Moreover, one can identify particular DNA molecules by detecting the direct tunneling current through them, as the molecules are passed through a nanopore to which the tunneling electrodes are attached [11, 12]. That is, the different molecules crossing the DNA double helix are of different sizes, and hence different tunneling currents are associated with each of these molecules. Measuring the tunneling current as the DNA strand is pulled through the hole, between the electrodes, the sequence can be determined.

The above approach has since been extended to amino acids and peptides through the recognition tunneling method [13]. In this approach, the tunneling leads are modulated. Since the tunneling current depends exponentially upon both the voltage applied and the tunneling distance, the modulation leads to a series of current spikes. As an amino acid is moved into the gap between the electrodes, the amplitude of these spikes will vary according to the size of the molecule. These current spikes can then be used as a recognition signal to identify the particular amino acid that is in the tunneling gap. The process is explained further in the video of figure 8.6, from the ASU group [13]. The process of making the entire structure is further explained in [14].

8.2.2 The gated single-electron device

The single-electron circuit that we have been considering in figure 8.4 is interesting, but it was quickly realized that having an additional bias voltage to control the actual charge on the quantum dot would be beneficial to creating a SET. Consider the topologically equivalent SET circuit shown in figure 8.7. In this circuit, a separate voltage source, V_g, is coupled to the island through an ideal (infinite tunnel

Figure 8.6. The video describes the recognition of individual amino acids as they are pulled through the gap in a tunneling structure. Here, the acids play the role of the quantum dot in, e.g., figure 8.4. (The video is from Lindsay, included with his permission. Available at https://iopscience.iop.org/book/978-0-7503-3139-5.)

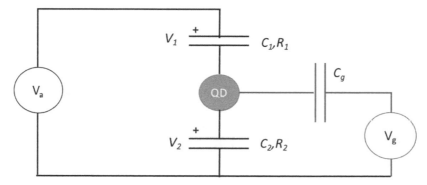

Figure 8.7. The simple circuit of figure 8.4 is now modified by the addition of a gate voltage source which is coupled to the quantum dot by a gate capacitor.

resistance) capacitor, C_g. This additional voltage modifies the charge balance on the island so that equation (8.7) requires an additional polarization charge that arises from this new bias source and its coupling capacitance. This new charge is

$$Q_g = C_g(V_g - V_2). \tag{8.17}$$

Now, the charge on the quantum dot is also affected, and

$$Q_{\text{dot}} = -ne = Q_2 - Q_1 - Q_g. \tag{8.18}$$

We can now combine the various equations above to give the new forms of the voltages across the two capacitors, which now include the effect of the gate and its capacitance. These equations can be written as

$$V_1 = \frac{1}{C'_T}[(C_2 + C_g)V_a - C_gV_g + ne]$$

$$V_2 = \frac{1}{C'_T}(C_1V_a + C_gV_g - ne) \qquad (8.19)$$

$$C'_T = C_T + C_g = C_1 + C_2 + C_g$$

As before, we can now write the energy due to the charge on the capacitors, equivalent to equation (8.10), as

$$E = \frac{1}{2C'_T}\left[C_gC_1(V_g - V_a)^2 + C_1C_2V_a^2 + C_gC_2V_g^2 + Q_{dot}^2\right]. \qquad (8.20)$$

The work performed by the voltage sources during the tunneling through junctions 1 and 2 now includes both the work done by the gate voltage and the additional charge flowing onto the gate capacitor electrodes. Equations (8.12) and (8.13) are now generalized to

$$W_a(n_2) = -\frac{n_2e}{C'_T}(V_aC_1 + V_gC_g)$$

$$W_a(n_1) = -\frac{n_1e}{C'_T}[V_aC_2 + C_g(V_a - V_g)] \qquad (8.21)$$

These different contributions to the total energy can now be combined as in equation (8.14). More important to us, however, are the relevant equations for the change of the charge on the dot by adding or subtracting a charge from one of the capacitors. These changes in energy can be written in analogy with equations (8.15) and (8.16) to be

$$\Delta E_2^{\pm} = \frac{e}{C'_T}\left[-\frac{e}{2} \pm (ne - C_1V_a - V_gC_g)\right], \qquad (8.22)$$

and

$$\Delta E_1^{\pm} = \frac{e}{C'_T}\left\{-\frac{e}{2} \mp [ne - V_gC_g + (C_2 + C_g)V_a]\right\}. \qquad (8.23)$$

The gate bias now allows us to change the charge on the island, and therefore to shift the region of Coulomb blockade. Thus, a stable region of Coulomb blockade may be realized for $n \neq 0$. As before, the condition for tunneling at low temperature is that $\Delta E_{1,2} > 0$ so that the system goes to a state of lower energy after tunneling. The conditions for forward and backward tunneling then become

$$-\frac{e}{2} \pm (ne - C_1V_a - V_gC_g) > 0$$

$$-\frac{e}{2} \mp [ne - V_gC_g + (C_2 + C_g)V_a] > 0 \qquad (8.24)$$

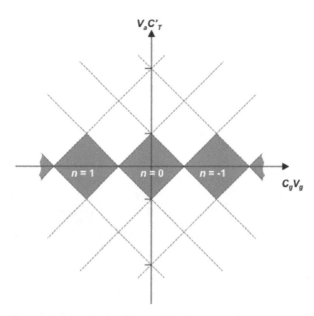

Figure 8.8. Stability diagram for the circuit of figure 8.7. The gate voltage can now be used to create new regions of Coulomb blockade (in blue). Each diamond is characterized by a different value of the charge on the quantum dot.

The four equation (8.24), for each value of n may be used to generate a stability plot in the V_a–V_g plane, which shows stable regions corresponding to each n for which no tunneling may occur. Such a diagram is shown in figure 8.8 for the case of $C_g = C_2 = C$, $C_1 = 2C$. The lines represent the boundaries for the onset of tunneling given by given by these equations for different values of n. The trapezoidal shaded areas correspond to regions where no solution satisfies the equations, and hence where Coulomb blockade exists. Each of the regions corresponds to a different integer number of electrons on the island, which is 'stable' in the sense that this charge state cannot change, at least at low temperature when thermal fluctuations are negligible. The gate voltage then allows us to tune between stable regimes, essentially adding or subtracting one electron at a time to the island.

It is possible to actually perform spectroscopy on the states of the quantum dot. In semiconductors, the lower density of states and small size of the quantum dot can lead to quantization of the wave function within the dot. Hence, the tunneling characteristics will be modified from the simple charging diagram due to this quantization. This has been probed using vertical GaAs/AlGaAs heterostructures in an elegant manner [15, 16]. In these structures, a vertical resonant tunneling structure is created. This involves two GaAlAs barriers with a GaAs quantum well in between them. GaAs provides the source and drain cladding layers on either side of the two GaAlAs layers, as shown in figure 8.9(a) [16]. A schematic of the potential across the resonant tunneling device, under bias, is shown in figure 8.9(b). The structure is then patterned into small vertical pillars, as shown in panel (c) of the figure, and gates can then be deposited around the periphery of the pillars as

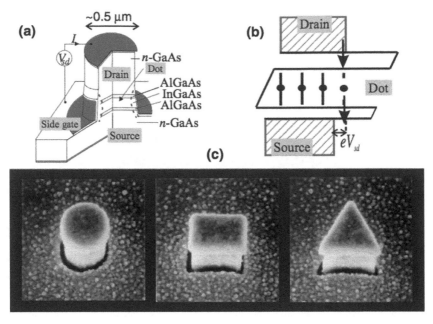

Figure 8.9. (a) Schematic of the vertical resonant single-electron device, showing the surrounding gate electrode. (b) Plot of the bias across the device. (c) Micrographs of various pillar shapes that have been made. (Reprinted with permission from [16]. Copyright 2001 IOP Publishing.)

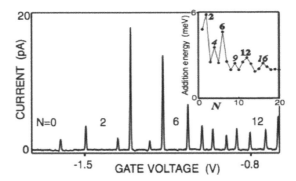

Figure 8.10. Current flowing through the circular vertical single-electron device as the gate voltage is varied. Each peak corresponds to the tunneling of a single electron. N indicates the number of electrons that reside on the quantum dot. (Reprinted with permission from [16]. Copyright 2001 IOP Publishing.)

indicated in panel (a). This gate allows one to completely deplete the quantum dot of electrons, and then allow individual electrons to tunnel into the dot by slowly increasing (lowering the negative bias of the Schottky barrier gate) the gate voltage. As each edge of the Coulomb diamond is found, a current peak will occur that signals the tunneling of a single electron. This is shown in figure 8.10. In this latter figure, each peak of the current corresponds to adding a single electron to the charge on the quantum dot. It can be seen that there are larger gaps that appear when $N = 2, 6, 12, \ldots$ (see the inset), which is taken to be indicative of the Darwin–Fock spectrum (see appendix F) for a two-dimensional harmonic oscillator, especially when a normal magnetic field is applied. In the lowest quantum level of the dot, only two electrons of opposite spin can be accommodated. In the next level, four electrons in which each pair has opposite spin can be accommodated, and so on up the ladder of states. The non-uniform spacing of the tunneling peaks arises from the Coulomb charging energy plus the quantization energy and the varying many-body energies that arise as various numbers of electrons occupy the quantum dot.

In the previous example, the current was observed as a series of spikes as one passed the crossing points for the Coulomb diamonds. This was due to strong quantization in the central quantum dot. If the quantum dot does not show quantization, a different behavior can be found in the structure, in which one can see what is known as a Coulomb staircase. Let us consider the MOS structure shown in figure 8.11 [17]. Here, the MOS device is fabricated on a p type substrate with an initial silicon-on-insulator thickness of 300 nm over a buried oxide of 375 nm thickness. As indicated in panel (a), three gates are formed in the structure. The two side gates are biased to deplete the electrons from the inversion layer and thus to create potential barriers as shown in panel (b). A third, top gate is used to control the potential in the quantum dot that is created between the two side gates. Here, the central quantum dot is relatively large and does not show quantized energy levels, especially as the device will be operating at room temperature, rather than at low temperatures. The transfer characteristics are plotted in figure 8.12, where both the drain current and the transconductance are plotted as a function of the bias on the third gate, which is the central top gate. Initially, this bias is sufficiently negative to deplete the quantum dot of electrons. As the bias is made more positive, current begins to flow (red curve), and we can see the characteristic staircase behavior. The transconductance (black data) is the derivative of the current with gate bias, and shows peaks at each of the transitions in the staircase. From the peak spacing in the transconductance, which is about 1 V, it is felt that the gate capacitance is quite small, and about 0.16 aF. Note that the data are obtained at 300 K, which is the

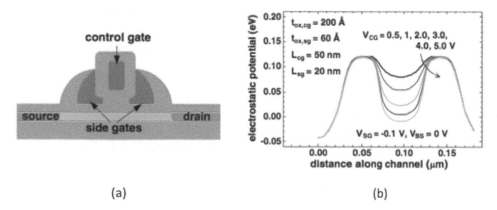

(a)

(b)

Figure 8.11. (a) Schematic depiction of the self-aligned double-gate single-electron device. (b) Simulation of the electrostatic potential along the channel for the device in (a). (Reprinted with permission from [17]. Copyright 2008 IOP Publishing. All rights reserved.)

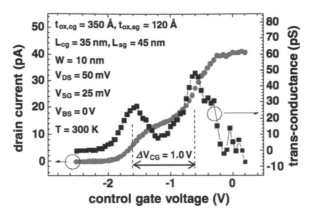

Figure 8.12. Transfer characteristics for the device shown in figure 8.11. The current (red) shows a Coulomb staircase rather than current spikes. The transconductance (black) oscillations indicate a gate capacitance of about 0.16 aF. (Reprinted with permission from [17]. Copyright 2008 IOP Publishing. All rights reserved.)

reason that the quantization effect is absent, and one therefore does not see current spikes as observed at low temperature. Hence, it is quite likely that the observed behavior is that due to single-electron tunneling along the source–drain direction.

One can carry the quantum dot size down to the ultimate of being a single atom, and this structure has been labeled as a single-atom transistor [18]. One of the earliest proposals to create a quantum computer in a condensed matter system was to use single-atom isotopes of ^{31}P implanted into a silicon crystal [19]. This idea would use the nuclear spin states as the qubit (discussed below in a later section). By hydrogenating the surface of silicon (100), all the dangling bonds are satisfied. A single hydrogen can be removed by using an STM to break the individual bond [20]. Then, the surface can be exposed to phosphine, and a single molecule will attach to the single bare dangling bond exposed by the hydrogen removal. Subsequent reaction leads to the phosphorus atom moving into the crystal, where it acts as a single donor atom [18]. The atomic potential of this atom, when placed between the metallic source and drain electrodes, plays the role of the quantum dot. This potential can be manipulated by the appropriate placement of one or more side gates, and thus provide the equivalent behavior of the system in figure 8.7.

8.2.3 Double dots

In the preceding sections, we established a correspondence between Coulomb-blockaded semiconductor quantum dots and an artificial atom through the spectrum of charge states that can exist in the dots. We want to extend this analogy to consider how a pair of dots may be coupled to each other in a simple manner such that they form what may be viewed as artificial molecules. As with the discussion of real molecules in nature, where new molecular orbitals are formed as a result of the wave function overlap between the component atoms, we will see here that such overlap can give rise to new electronic states in coupled quantum dots. The collective character of these states has the potential to lead to new classes of electronic devices, particularly for application to quantum computing, where one is interested in using the superposition states of quantum systems as the basis for computing. In discussions of such collective phenomena, however, there is an important need to distinguish between essentially classical collective effects that arise from Coulomb charging between otherwise isolated dots, and true quantum effects that are a consequence of controlled wave function coupling. In the following discussion, we first focus on the charging effects in coupled quantum dots. Here, we closely follow the treatment of van der Wiel *et al* [21].

Previously, we determined the conditions for which the Coulomb blockade existed by determining the total energy of the circuit and requiring that this energy be lowered as a result of single-electron tunneling. The extension of this concept to the problem of single-electron tunneling through a pair of sequentially coupled quantum dots is indicated in figure 8.13. In this new circuit, we take the number of electrons stored on dot 1 (2) to be N_1 (N_2) (note the difference from the notation n_1 and n_2 earlier, where the latter numbers denoted the number of electrons to have tunneled through junctions 1 and 2, respectively). The system now features three tunnel barriers (as opposed to two in the original problem). The new capacitor controls the electrostatic coupling between the two dots. As in the case of the SET, the electrostatic energy of each dot is regulated via an independent gate voltage (defined here to be V_{g_1} and V_{g_2}). Each of these gate voltages couples to the respective charge island via an associated capacitance (C_{g_1} and C_{g_2}). In order to focus on the key effects arising from the influence of the electrostatic coupling between the dots, we assume that the circuit model of figure 8.13 characterizes the system completely.

A classical analysis of the energetics of Coulomb-coupled quantum dots has been performed by several authors, in which the quantization of energy within the dots is ignored [22–24]. Following the derivation by van der Wiel *et al* [21], we begin by considering the case of linear transport through the double-dot system,

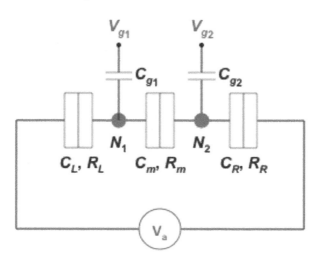

Figure 8.13. Simple circuit in which two quantum dots are coupled with individual gate voltages for each dot. The coupling between the dots is governed by the central capacitor C_m.

corresponding to $V_a = 0$. Under such conditions, it can be shown that the electrostatic energy of the coupled-dot system, containing N_1 and N_2 electrons on dots 1 and 2, respectively, may be written as

$$E(N_1, N_2) = \frac{e^2 N_1^2}{2}\left(\frac{C_2}{C_1 C_2 - C_m^2}\right) + \frac{e^2 N_2^2}{2}\left(\frac{C_1}{C_1 C_2 - C_m^2}\right)$$
$$+ \frac{e^2 N_1 N_2}{2}\left(\frac{C_m}{C_1 C_2 - C_m^2}\right) + f(V_{g1}, V_{g2}) \qquad (8.25)$$

The notation here differs from that used earlier. Note that, in figure 8.13, the main capacitors are now labels C_L and C_R. In equation (8.25), the capacitances C_1 and C_2 are the total capacitance seen by dots 1 and 2, respectively. These are given by

$$C_1 = C_L + C_{g1} + C_m$$
$$C_2 = C_R + C_{g2} + C_m \qquad (8.26)$$

The last term in equation (8.25) provides the energy changes that the two gate voltages provide, and may be written as [21]

$$f(V_{g1}, V_{g2}) = \left(eN_1 C_{g1} V_{g1} + \frac{C_{g1}^2 V_{g1}^2}{2}\right)\left(\frac{C_2}{C_1 C_2 - C_m^2}\right)$$
$$+ \left(eN_2 C_{g2} V_{g2} + \frac{C_{g2}^2 V_{g2}^2}{2}\right)\left(\frac{C_1}{C_1 C_2 - C_m^2}\right)$$
$$+ (eN_2 C_{g1} V_{g1} + eN_1 C_{g2} V_{g2} \qquad (8.27)$$
$$+ C_{g1} V_{g1} C_{g2} V_{g2})\left(\frac{C_m}{C_1 C_2 - C_m^2}\right)$$

We can examine the implications of these results by considering two extreme cases. In the limit of zero coupling between the dots, corresponding to $C_m \to 0$, only the first two terms on the right-hand side of equation (8.25) survive. This means that the total electrostatic energy of the system is simply that of the two isolated dots. The other limit arises in the situation where C_m is the largest capacitance in the system, in which case $C_{1(2)} \approx C_m$ and the problem essentially reduces to one involving the charging of a large single-dot formed by the two smaller ones. From this reasoning, one can construct the charge-stability diagram in the space of the two gate voltages for the first case ($C_m \to 0$) as shown in figure 8.14(a). Here, the charging regions simply are sets of squares in which the allowed number of electrons are shown schematically in this figure. Hence, a change in one gate voltage only affects the quantum dot to which it is attached. It does not affect the charge on the other quantum dot, as there is no connection to that dot. The gate capacitance and the appropriate other capacitance correspond just to the case of figure 8.4.

For non-zero coupling between the dots ($C_m \neq 0$) the contour becomes distorted, and the original four-fold intersections of the different charge squares in

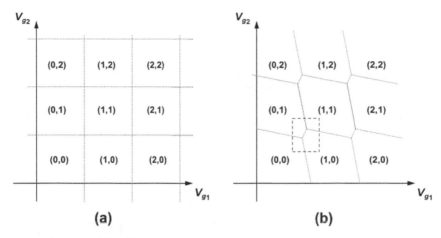

Figure 8.14. (a) Charge-stability diagram for the case when there is no coupling between the two quantum dots. (b) Charge-stability diagram for coupling between the two dots. The simple crossing evolves into a pair of triple points.

figure 8.14(a) develop instead into a pair of closely sited triple points, as shown in figure 8.14(b). The existence of these triple points is critical for transport, as they allow for the flow of current with a small source–drain bias present. This pair of points allows the electron number on both dots to fluctuate by a single electron. The origin of these triple points is explained in terms of the mutual capacitance between the two dots in the system. When this capacitance is non-zero, the charging of one of the dots by an additional single electron modifies the electrostatic energy of the other, and this effectively causes a repulsion of the resonance lines in the region where the resonance lines of both dots intersect. Consider the triple point highlighted in red in figure 8.14(b) which indicates the values of the gate voltages (V_{g_1} and V_{g_2}) for which the four charge populations (N_1, N_2) = (0,0), (1,0), (0,1), and (1,1) come together. The flow of current via single-electron tunneling is possible at this degeneracy point, via the charge sequence (0,0) → (1,0) → (0,1) → (0,0), and so on. This triple point therefore involves a single-electron transfer through the double-dot system.

As can be imagined current can flow freely between the pair of triple points. In figure 8.15, we show the stability diagram which plots the transconductance as a function of the two gate voltages for a double-dot system in a triple-layer graphene sheet [25]. The device structure is shown as the inset to the figure, and the different regimes are separated (and defined) through metal gates used to deplete the carriers under the gates. The gates were fabricated by standard electron-beam lithography; the actual gates were between the metallic Schottky barrier regions. Contacts to the individual regions were then formed as a Cr/Au double layer (10/50 nm, respectively). The triangular-shaped dots are estimated to have areas of 4×10^{-3} μm^2 for dot 1 (left-hand dot in the figure) and 5×10^{-3} μm^2 for dot 2 (right-hand dot in the figure). From the shape of the modified diamonds, the various capacitances have been estimated, using the approach in [21] (discussed below) to be $C_{g_1} = 2.2$ aF, $C_{g_2} =$

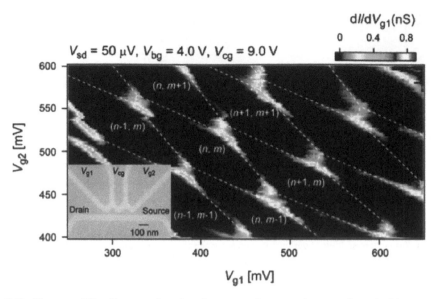

Figure 8.15. Charge-stability diagram, plotted as the measured transconductance, for a double-quantum-dot system on graphene. The gate patterns are shown in the inset to the figure. Current flows mainly between the paired triple points. (Reprinted with permission from [25]. Copyright 2009 American Chemical Society.)

2.9 aF, and $C_m = 10$ aF. The total dot capacitances are estimated to be 30 aF and 60 aF for C_1 and C_2, respectively. This gives a value for C_L of 17.8 aF and a value for C_R of 47.9 aF.

The shape of each of the Coulomb 'diamonds' that appears in figures 8.14(b) and 8.15 can be used to determine the values of the various capacitances that appear in figure 8.13. The basic procedure was described in [21]. The key is to measure the various voltages indicated in figure 8.16, where we show an expanded view of a single 'diamond', outlined in red. These can be simply related to the various quantities as

$$\Delta V_{g1} = \frac{e}{C_{g1}}, \quad \Delta V_{g2} = \frac{e}{C_{g2}},$$

$$\Delta V_{g1}^m = \Delta V_{g1}\frac{C_m}{C_2} = \frac{eC_m}{C_{g1}C_2}, \quad (8.28)$$

$$\Delta V_{g2}^m = \Delta V_{g2}\frac{C_m}{C_1} = \frac{eC_m}{C_{g2}C_1}.$$

Now, this gives four equations, but there are five capacitances that must be determined. To fully determine all the capacitances, one must go beyond the linear theory and apply an applied bias to the ends of the circuit. Of importance is the parameter $\alpha_{1(2)}$ that provides the scaling between effects induced from the gate and from the applied bias. These are determined by the capacitances as [21]

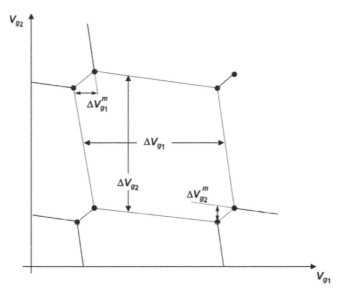

Figure 8.16. Illustration of a single-Coulomb 'diamond' to determine the parameters for the double quantum dot of figure 8.13.

$$\alpha_{1(2)}\Delta V_{g1(2)} = \frac{C_{g1(2)}}{C_{1(2)}}e\Delta V_{g1(2)} = |eV_a|. \tag{8.29}$$

With these parameters known, the full set of capacitors can be determined. In the work discussed in figure 8.15, these parameters were determined to be $\alpha_1 \sim 0.07$ and $\alpha_2 \sim 0.05$, both in meV/mV [25].

When the quantum dots are sufficiently small to lead to quantization, the regions around the triple points can be far more complicated, as excited states can influence these charging diagrams. Nonlinear effects can occur as well, as already mentioned. Discussion of the full analysis is beyond the level we consider here, but are discussed in [21].

8.3 Quantum dots and qubits

In order to understand how quantum dots can be envisioned as representing qubits, we have to first understand how the quantum bit, or qubit, differs from the normal computer binary bit. This difference is shown in figure 8.17. In panel (a), we show a two-dimensional space, in which the axes represent the binary 1 and the binary 0 for the nth bit of our computer. Now, with normal binary encoding, these two unit vectors are added to produce the 'state' of this bit according to

$$|n\rangle = a_n|0\rangle + b_n|1\rangle, \tag{8.30}$$

but this equation is subject to the conditions that either $a_n = 1$ and $b_n = 0$, or $a_n = 0$ and $b_n = 1$. These are the only two possibilities that are allowed. The entire computer runs on this approach. The standard storage is by bytes, which are eight bits, and

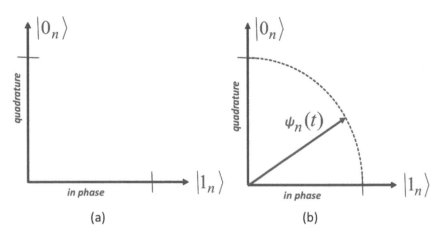

Figure 8.17. (a) The two-dimensional space for a standard computer binary bit. (b) The modification of this space for a qubit, where the wave function can have any analog, complex combination of the two states.

this can represent a number between 0 and 255. For example, the number 62 is represented by the many-bit state $|0111110\rangle$. As a further example, the standard compression of an image is via the jpeg standard, where a pixel, or picture element has its color encoded as three bytes, one byte each for red, blue, and green. This combination provides 16.78 million colors. Black is all zeroes and white is all ones. Any other color is somewhere between these limits.

Now, the desire to use quantum computation lies in the ability to store more information in each quantum bit, or qubit, and to use entanglement between the bits to enlarge the information content even more. For this purpose, we use the properties of the quantum wave function for each bit, as shown in figure 8.17(b). We still use the same basis set, as indicated in the figure, but now the coefficients only need to satisfy $|a_n|^2 + |b_n|^2 = 1$, which is required by the normalization of the wave function. Hence, the fact that the wave function is complex introduces the phase, and the vector orientation of the wave function can lead to a point that is anywhere on the unit circle (only the first quadrant is shown in the figure). Now, each qubit is a two level system, corresponding to the two axes, and the occupancy of each level is governed by the wave function. But, the two level system was discussed in appendix E in terms of the Bloch sphere. There, in figure E1, we see that the two axes are represented by the top of the sphere corresponding to the 'zero' state and the bottom of the sphere corresponding to the 'one' state. The actual qubit state can be any point on the surface of the sphere, and quantum computer 'operations' generally can be expressed as rotations in this three-dimensional space. These actual operations, and the concepts of quantum computing, are beyond the material we want to discuss here, but can be found in several textbooks [26, 27]. When we recall that the Bloch sphere was introduced to describe a particular two level system, that of spins, it becomes clear why there is a significant effort to utilize a spin qubit, as discussed above.

Let us examine how a pair of quantum dots can be formed to represent at least a binary bit. As we have discussed above, quantum dots can be formed in a very large number of ways. Here, we assume that the dots are formed in a GaAs/AlGaAs heterostructure, and their definition is by a set of Schottky barrier gates, as shown in figure 8.18(a). The black gates are the basic definition gates for a single large dot, which is then split into two smaller dots by the blue finger gates. The gates marked as V_2 and V_4 are used to fine tune the exact size of the individual dots. The potential profile is shown in figure 8.18(b), with the deep red color being above the Fermi energy, and the other colors representing various potential depths. The entire gate structure is considered to be about 1 µm high by 1.5 µm wide [28]. We will highlight the 19th eigenstate in the structure, as it shows switching from one dot to the other. Gate V_4 is maintained at a bias of −1 V, and the eigenstates are shown in figure 8.19. In panel (a), V_2 is also set to be −1 V, and the density is clearly in the left dot. On the other hand, when V_2 is reduced slightly, to −1.01 V, then the density switches to the right-hand dot, as shown in panel (b). It is clear that the small perturbation potential causes the switching between these two possible states, which can represent the two possible values for the bit.

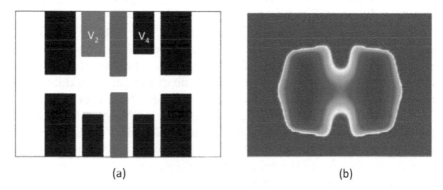

Figure 8.18. (a) Assumed gate pattern for the double-dot simulation. (b) Self-consistent potential profile for the dots. Reprinted from [28].

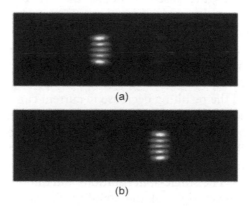

Figure 8.19. In panel (a) with $V_2 = -1.0$ V, the wave function is localized in the left dot. But, when $V_2 = -1.01$ V, panel (b) shows that the wave function is localized in the right dot. Reprinted from [28].

The potential above is a double-well potential, with each well appearing as a two-dimensional harmonic oscillator, although considerably deformed from the textbook examples, such as discussed in appendix C. If the two wells are exactly the same, then a given eigenstate in one well will have an exact equivalent in the other well. These two states will have the exact same energy, and this is not allowed by the exclusion principle. As a result, there will be a coupling between the wells, which can be controlled by the blue finger gates in figure 8.18(a). This coupling leads to an energy splitting corresponding to a lower-energy bonding interaction and a slightly higher-energy anti-bonding interaction. In each of these new states, the wave function has an equal fraction in each of the quantum dots. If we were to localize an electron in the left dot, it will oscillate back and forth between the dots. In order to localize the wave function in only one of the dots, they have to be not exactly equal. This inequality is provided by the discretization of the system in the simulations for figures 8.18 and 8.19. By slightly modifying the left potential, the localization switches to the other dot. In principle, the various potentials can be carefully adjusted so that the two states become merged, so that the wave function oscillates between the two dots, and it is thought that this oscillation can be used to create a qubit with these two dots.

The idea of using spin to create a qubit with quantum dots was apparently first proposed by Loss and DiVincenzo [29]. They developed a detailed scheme to achieve quantum computation with a pair of single-electron quantum dots. The qubit is realized with the spin of an excess electron in one of the quantum dots. Two-qubit quantum-gate operation is achieved by merely adjusting the barrier existing between the two dots (as, for example, adjusting the blue barriers in figure 8.18(a) above). If the potential barrier is high, the two qubits, one in each dot, do not interact. If the potential is lowered, then the two qubits are allowed to interact and the spins are affected by a coupling due to the spin–spin coupling energy. Many approaches burst upon the scene after this seminal paper.

We can illustrate the double-quantum-dot approach with some relatively recent work [30]. Here, the structure is fabricated on a GaAs/AlGaAs heterostructure with Schottky barrier metallic gates used to define the active region, which may be glimpsed below the magnetic material in figure 8.20(a). As above, this is a gate-defined double quantum dot, to which has been added a split Co micromagnet (the yellow regions in this figure). To the outside of the active double dot, a QPC has been added (indicated in the figure by the current path arrow). When the dot on the right has an electron (or more) in it, the Coulomb interaction causes the opening for the QPC to narrow, and the resulting decrease in current can signal the charge state. This sensor can be used to map out the stability regime for the double-dot system, shown in figure 8.20(b). Essentially, this is equivalent to the red square region in figure 8.14(b), except that the applied source–drain region opens the simple pair of triple points in the triangular structure seen here. Here, N_L and N_R denote the charge in the left and right dots. The single-spin rotations and qubit interactions are carried out in the (1,1) state (lower right of the stability diagram). To rotate each spin, electrically driven spin resonance is used, although the Co nanomagnets provide a spatially varying magnetic field due to the shape of these magnets. The spin

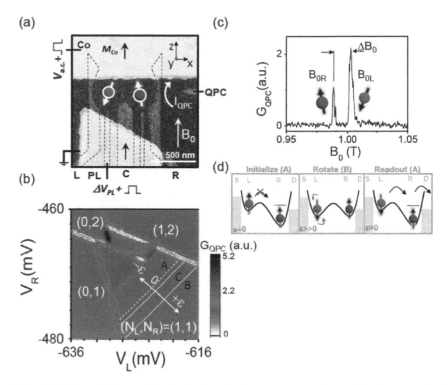

Figure 8.20. (a) A false color image of the device showing the two quantum dots and the electron spin. (b) The stability diagram around the area of interest. (c) Spin resonance signals for the left and right dots. (d) Measurement cycle for controlled spin rotations. (Reprinted with permission from [30]. Copyright 2011 the American Physical Society.)

resonance signal is seen in panel (c) of this figure. Panel (d) shows the shape of the double-well potential, which is controlled by the various gate electrodes (as depicted in figure 8.18(a)). During operation, a static magnetic field of 2 T in the left dot and 1.985 T in the right dot is applied, and the dot is excited with a microwave signal of 11.1 GHz. As indicated in the figure, various gate voltage pulses are also applied to sequence the interaction. For stage A in the figure, the two dots are set in the spin blockade phase, where the spins cannot interact with both spins either up or down. In stage B, the dots are isolated from one another and one of the spins is rotated by applying a pulsed microwave signal. Finally, at the second stage A, to the right, the spin can be read out with the QPC sensor in which the signal is proportional to the probability that the two spins are oppositely polarized. This demonstrates the ability to control the spin polarization. Finally, two-qubit gates are used to demonstrate operations via the interdot spin exchange operation, as suggested by Loss and DiVincenzo [29].

   If all of this seems to be a great deal of work to operate a pair of qubits, it is. Yet, this technology is still in its infancy, and experiments such as this demonstrate the feasibility of the scientific basis. In the experiments, the authors demonstrated that

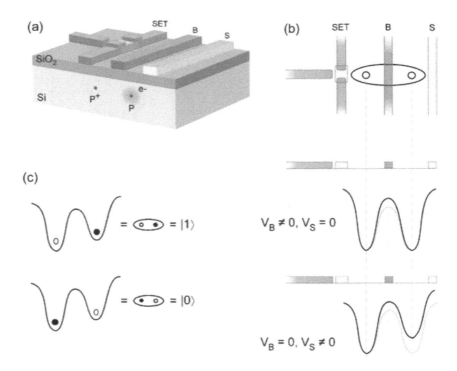

Figure 8.21. A qubit using a pair of phosphorus atoms, which have been implanted into the silicon substrate. (a) Schematic of the device showing the control gates and readout SET. (c) Two choices for the qubit state. (b) The barrier (B) and symmetry (S) gates which control the potential of the two quantum wells. (Reprinted with permission from [31]. Copyright 2004 the American Physical Society.)

an all electrical two-qubit gate could be realized and it could perform simple operations.

It is also possible to marry the previous double-dot qubit with the single-atom transistor discussed in the last section. A double-dot qubit has been realized using two single phosphorus atoms recently [31]. The double-well potential is created from the local potential well created by the individual P atoms. This qubit is a charge-based qubit in which the charge resides on one of the atoms, while the other is ionized, thus creating a P–P+ charge system. The qubit is manipulated by a set of gates and a SET, operated at a high ac frequency, is used for the readout. The structure is shown in figure 8.21. The device is created in a Si–SiO$_2$ with the pair of P atoms implanted into the Si substrate. In the figure, the structure, the control gates, and the SET are shown schematically. The control gates are used to move the charge from one P atom to the other, and this position can be sensed by the SET.

More recently, the use of a single phosphorus atom has been suggested to be adequate for a qubit which utilizes the spin of the electron on the atom [32]. The device is subject to an applied in-plane magnetic field of 1 T which provides well-defined spin-up and spin-down states of the electron. Transitions between these states are achieved through the use of a microwave spin resonance signal provided by an on-chip transmission line, and these transitions can be measured with a SET.

In the read/initialization stage, a spin-up electron will tunnel into the SET, to later be replaced by a spin-down electron. This causes a pulse of current through the SET, while a spin-down electron remains trapped on the donor. During the control stage, the electron spin states are placed well below the SET state, and cannot interact with it, while the spins are manipulated by the microwaves through electron spin resonance.

The case for nearly all of the above dopant-based qubits is that the dynamic variable is the spin itself, just as for the double dot qubits described earlier. It is possible to consider a hybrid qubit in which the dynamical variable depends upon both spin and charge. The hybrid qubit requires neither nuclear-state preparation nor micromagnets for control, and becomes considerably more amenable to systems [33]. This has been studied by these latter authors for the double quantum dot, which is defined lithographically by surface gates (as shown in panel (a) of figure 8.22). Here, a more traditional double dot system, similar to those in SETs, is used for the qubits. Three electrons are used in the system, with the gates tuned so that there are two electrons in the left dot and one in the right dot to define the (2,1) state. Changing the voltage on gate L to raise the energy difference between the dots favors the opposite situation and the state (1,2). The singlet state in the right dot, state (2,1), is taken to be the $|0\rangle$ state. The triplet state in the right dot, the (1,2) state, is taken to be the logical $|1\rangle$. The presence of the extra electron means that the hybrid states of the dots are not pure singlet or triplet and fast electric field techniques can be used to manipulate the qubit in either X or Z rotations on the Bloch sphere.

The coherence can be extended when the hybrid qubit is composed of three electrons in a double quantum dot [34]. Similar surface control gates have been used in the pure spin qubit case as well [35], where multiple dots are used to suppress charge noise. This was studied for a pair of singlet-triplet qubits, each of which was composed of two electrons in a double quantum dot. More recently, the importance of the spin–orbit interaction in these gate-defined dots has been studied [36, 37], and the impact of atomic-scale structure in the energy dispersion on decoherence was investigated [38]. In fact, figure 8.22 is from this last paper, and shows the double quantum dot qubit and its operation.

The singlet-triplet qubit is attractive for many reasons, as it seems to have a higher fidelity for single qubit operations. However, for two qubit operations leading to entanglement, it can be problematic in a sense that there are a range of operations over which the interaction can occur and this leads to a preferred set for good performance [39]. Spin qubits formed of quantum dots in the Si/SiO_2 system have been formed in a manner to utilize from 1 to 3 electrons in the dots [40] for this purpose. Then, pulsed electron spin resonance is used to exercise coherent control over the qubit. With these structures, these authors then demonstrated the achievement of valley splitting (normally, in the MOS system, the six ellipsoids of the conduction band break into a two-fold and a four-fold set due to the inversion potential, and one seeks to further split the lower lying two-fold set) [41]. The role of disorder on this valley splitting has also been studied [42].

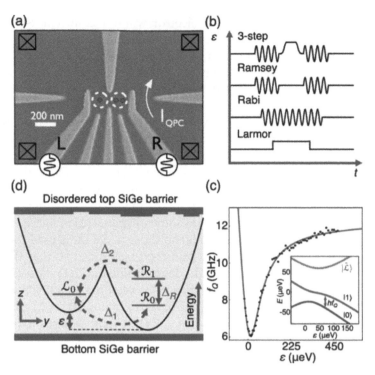

Figure 8.22. Experimental and theoretical setup, and resulting energy dispersions. (a) A scanning electron microscope image of a device nominally identical to the one used in the experiment. The gate voltages are tuned to form two quantum dots, located approximately within the dashed circles, where red dots represent electrons in a (1,2) charge configuration. (b) Schematics of the four pulse sequences employed in the experiments. The three-step sequence is used to obtain the qubit frequency data f_Q plotted in (c). The Ramsey pulse sequence is used to obtain the qubit frequencies and Ramsey decay rates. The Rabi and Larmor sequences are used to obtain Rabi fringes and f_Q. (c) The experimentally measured f_Q of a quantum-dot hybrid qubit as a function of detuning ε (black dots). The solid red line shows the results of a least-squares fit of the data to theory assuming ε-independent model parameters. Inset: The three energy eigenstates obtained by diagonalizing the theory Hamiltonian. (d) A schematic cartoon illustrating the theoretical model for both the quantum-dot hybrid qubit and the single- electron charge qubit, with the low-energy basis states $|L_0\rangle$, $|R_0\rangle$, and $|R_1\rangle$, as appropriate for the hybrid qubit. In our 2D tight-binding simulations, atomic-scale step disorder is introduced into the top interface as shown here. The lateral confinement potential is taken to be biquadratic, and the two dots are offset by energy ε. The interdot tunnel couplings are labeled $_1$ and $_2$, and we refer to $_R$ as the 'valley splitting,' although $|R_1\rangle$ may involve a valley-orbit excitation. (Reprinted with permission from [38]. Copyright 2004 the American Physical Society.)

8.4 The Josephson qubits

Kammerlingh Onnes, after liquefying helium at very low temperatures, spent most of his time examining the properties of various materials at this low temperature of 4.2 K. In studying Hg, he found a totally unexpected result, in that as he cooled Hg below the temperature of liquid helium, the resistance dropped dramatically. At 3 K, the resistance was less than 10^{-6} ohm. This phenomena is now known as *super-conductivity*, and has been observed in a significant fraction of the known elements.

Table 8.1. Superconductor transition temperatures.

Element	T_c (K)	Compound	T_c (K)
Pb	7.2	Nb_3Sn	18.05
V	3.72	V_3Ga	16.5
Ti	0.39	V_3Si	17.01
Nb	9.1	NbN	16.0
Ta	4.48	InSb	1.9
Hg	4.15	Nb_3Al	17.5
Zn	0.85	$YBa_2Cu_3O_7$	92
Sn	3.72	$HgBa_2Ca_2Cu_3O_8$	134

The onset of superconductivity is considered to be a phase transition, and the critical temperature below which it occurs is called the *transition temperature* T_c. If one plots the resistance as a function of temperature, there is a clear drop in the resistance by orders of magnitude at this transition temperature. Just above the transition temperature, there is a gradual drop in resistance, that gives a 'rounding' to the curve, and this is caused by thermal fluctuations at the transition temperature. These fluctuations are believed to be the fact that not all of the electrons are paired except at absolute zero temperature. Above T_c, on average none of the electrons are paired, so that the fraction that are paired is a function of temperature. In this region of enhanced conductivity above T_c, it is thought that small regions of the material are beginning to become superconducting, with the entire sample becoming so at further reductions of the temperature.

In general, metals which have very high conductivity do not become super-conductors. On the other hand, metals which are poor conductors, such as Pb, Ta, Nb, Hg, and so on, become quite good superconductors. Some of the transition temperatures are given in table 8.1. Also shown are a couple of members of a new class of materials known as high temperature superconductors, which have transition temperatures of tens of degrees Kelvin, up to ~200 K at high pressure. These latter materials tend to be cuprates and ceramics.

A magnetic field can be used to destroy the superconductivity. That is, there is a critical magnetic field H_c above which the superconductivity vanishes. Moreover, it is found that this critical field varies with temperature as

$$H_c = H_{c0}\left[1 - \left(\frac{T}{T_c}\right)^2\right],\tag{8.31}$$

where H_{c0} is the critical field at absolute zero of temperature. Comparing different material, it is found that in general a higher critical temperature will lead to a higher critical magnetic field. This critical magnetic field will also limit the amount of current that the superconductor can carry, since the current gives rise to a magnetic field by Ampere's law. However, there are other materials, such as the compounds in table 8.1, where the superconductivity does not end abruptly. In these materials, a lower critical magnetic field signals the beginning of the process and an upper critical field signals when superconductivity has finally gone away. These materials are sometimes called hard superconductors.

In superconductivity, it is found that the electrons would pair up as Cooper pairs [43]. Normally, electrons repel each other due to the Coulomb interaction between them. However, in metals there are typically something like a few times 10^{22} electrons cm^{-3}. Such a high electron concentration usually heavily screens the Coulomb interaction, so that it is a relatively weak interaction, especially if the electrons are not fairly close to one another. Cooper hypothesized that because of this weakness, electrons could actually interact with each other if there were another positive interaction. He suggested that one electron could interact with the atoms of the crystal in a manner that distorted the local atomic potential, creating a potential well into which a second electron could be drawn. He thought the process would work best if the two electrons had opposite spin. The combination of the two electrons, via the atomic interaction, creates the Cooper pair. Because the two electrons have opposite spin, the net spin of the Cooper pair is zero. This makes the Cooper pair into what is called a *composite boson*. Bosons are not required to satisfy the Pauli exclusion principle, so there is no limit to the number of bosons that can exist in any quantum state. Moreover, when we form the Cooper pair, it lies in a lower energy state—this lowering of the energy is the pair formation energy, which we called Δ per electron. The result of all of this process is observable by a gap of 2Δ that opens in the energy spectrum at the Fermi level. A Cooper pair must gain this energy to transition above the gap and become a pair of free electrons. The interesting point is that the two electrons in the Cooper pair need not be close to one another. Instead, it is generally thought that they can be some 100–400 nm apart, a length that is referred to as the *coherence length*. This means that there are a great number of other electrons near to the Cooper pair, or at very low temperature, the large number of Cooper pairs are all intertwined with one another. The energy gap that opens is also temperature dependent, varying as

$$E_G = E_{G0}\left(1 - \frac{T}{T_C}\right)^{1/2},$$

(8.32)

where this gap is the 2Δ mentioned above.

It has been suggested that the Cooper pairs are the basic unit of superconductivity [44, 45]. In the BCS theory, named for these three authors, the critical temperature is found to depend upon the density of states at the Fermi energy and the strength of the interaction between the lattice and the electron that leads to forming the Cooper pairs. Because all the states are full below the energy gap, one would not expect conduction, but the nature of the bosonic Cooper pairs seems to indicate they can move through the lattice without dissipation. That is, they cannot be scattered, so move dissipation free. And this seems to be the case. If one creates a coil of superconducting wire and induces a current through the wire, and then shorts the leads together, this current will continue to flow without dissipation. This current is known as a *persistent current*, and is the heart of superconducting magnets.

8.4.1 Josephson tunneling

The Josephson junction [46] is a tunnel junction in which the materials on either side of the insulator are superconductors. In the Josephson junction, these superconducting leads produce zero loss in dc operation. The formation of the Cooper pairs lowers the energy of the entire electron gas below the Fermi energy, and this leads to a gap opening at the Fermi energy, as pointed out above.

Normally, with Josephson junctions, the tunneling is carried out by these Cooper pairs, but both the Cooper pair tunneling and single-electron tunneling can occur under the appropriate conditions. The Josephson junction itself operates so that, no tunneling occurs until $eV_a > 2\Delta$ at low temperature. As the temperature increases, both unpaired single electrons and Cooper pairs exist below the gap, and the normal (unpaired) can tunnel giving a very small current. As with normal tunneling, if the barrier is relatively thin, say 1–2 nm, then the superconducting wave functions can extend through the barrier, so that they interact with those wave functions on the other side. An unusual effect in the Josephson junction is that this coherent mixing of the wave functions on either side of the junction can produce a current at zero bias! This is termed the dc Josephson current. This current has a fixed magnitude that can be modulated by a magnetic field as

$$I_J(B) = I_{J0}\cos\left(\frac{eBA}{h}\right), \tag{8.33}$$

where $BA = \Phi$ is the flux flowing through the area of the junction. As before, h/e is the quantum unit of flux, so that the flux can be quantized in this quantum system. The interesting aspect is that the current peak I_{J0} is proportional to the single-particle tunneling coefficient through the junction [47]. We note that if a bias is applied, nothing much happens to the normal tunneling curve, except a few unpaired electrons may tunnel through. But, this voltage has a significant effect on the coherent flux of the Josephson current (8.33). This follows from the relationship

$$\hbar\frac{d\Phi}{dt} = 2eV_a, \tag{8.34}$$

where the factor of 2 arises from the two electrons in the Cooper pair. Now, what this means is the associated flux is changed by the voltage and this produces an ac signal, known as the ac Josephson effect. This leads to the important relation

$$\hbar\omega = 2eV_a. \tag{8.35}$$

Hence, the Josephson junction can be a microwave source. It has also been noticed that there is a peak in the spectrum at $\omega = 4\Delta/\hbar$, which is called a Riedel peak in the response of the junction [48].

8.4.2 SQUIDs

The Superconducting QUantum Interference Device (SQUID) is basically a variation of the circuit in figure 8.4, but with current bias instead of voltage bias, as shown in figure 8.23 [49]. The two capacitors are Josephson tunnel junctions, and the quantum dot indicated may not be a microscopic dot at all (but we return to this below). This is known as a dc SQUID and works with the dc Josephson current (8.33). Shown is a bias current I_a that flows in from the left and out to the right. This current splits between the two arms of the interferometer, so that $I_a/2$ flows through each capacitor. An important point is that all of the leads are made from superconducting material. Initially, there is no flux through the superconducting loop that contains the two tunnel junctions. If we induce a flux in the loop that is less than $\Phi_0/2$, this creates a circulating current, which we designate as I_c, that flows in the loop. The flux can be created by coupling the loop to an external loop that provides the flux. Thus, for example, the currents through the two capacitors become

$$I_1 = \frac{1}{2}I_a + I_c$$

$$I_2 = \frac{1}{2}I_a - I_c.$$

(8.36)

When the flux is increased to $> \Phi_0/2$, the induced circulating current has to change sign according to equation (8.33). And, when the flux reaches a full flux quantum, the cycle repeats. As a result, the SQUID is very sensitive to magnetic fields and can measure fields as small as a few times 10^{-18} T, so are used in a great many applications [50]. As the currents oscillate with the flux, the induced voltage, that arises from the junction resistance and the loop inductance, becomes

$$\Delta V = \frac{R}{L}\Delta\Phi.$$

(8.37)

There is also an RF-SQUID which utilizes only a single Josephson tunneling junction and the superconducting loop [51]. However, most applications of the SQUID for quantum computing applications use the two junction (or more junctions) version of the dc SQUID. However, in the drive to make the qubits small, one actually introduces the quantum dot, as shown in figure 8.23. That is,

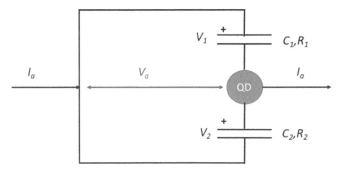

Figure 8.23. A DC SQUID showing the bias current and the measured voltage that arises from a variation in the flux coupled into the loop.

there is a long part of the loop and a short part of the loop, the latter of which forms the quantum dot, along which the capacitors also show single-electron(pair) tunneling. Hence, there is a competition in various energies in the system. There is the single-electron(pair) energy

$$E_c = e/2C. \tag{8.38}$$

(we may assume the two junctions are equal capacitances), where the two arises for the Cooper pair tunneling. There is also the Josephson energy arising from the current (8.33) which becomes

$$E_J = \Phi_0 I_{J0}/2\pi. \tag{8.39}$$

The operation of the qubit and the SQUID depends upon the relative size of these two energies.

To explain this further, one needs to distinguish two types of qubits: microscopic and macroscopic [52]. The microscopic qubit is based upon an internal charge or spin, much like the Si qubits of the previous section, or upon two natural levels as in an atom. Macroscopic qubits are based quantum levels representing collective degrees of freedom (often called *macroscopic quantum states*) like the persistent current in a superconducting ring [53] or excess charge on a superconducting island [54, 55]. Hence, one might think that the macroscopic form will be physically larger, but that can be totally misleading. When superconducting components are used in electrical circuits, then these circuits become quantum objects, which can then be used in either type of qubit. The pursuit of superconducting qubits can be a Cooper pair box charge qubit, the persistent current flux qubit, or a hybrid charge-flux qubit [52, 56].

8.4.3 Charge qubits

The charge qubit is composed primarily of the Cooper pair box, where the box is formed by a pair of small Josephson junctions so that single electron tunneling governs the excitation of the box itself, which is isolated by the two junctions. A single Cooper pair box [57] is a unique artificial solid-state system [54]. Although there are quite a few electrons on the island (box), they all form Cooper pairs in the superconducting state and then condense into a single macroscopic quantum state. This state is separated from the normal electrons by the superconducting gap Δ, discussed above. The only low-energy excitations arise from Cooper pair tunneling when the gap is larger than the charging energy $E_C = e^2/2C$. On the other hand, if the charging energy is larger than the gap energy, and the thermal fluctuations are suppressed, the system can be considered to be a two level system in which the lowest two energy states differ by one Cooper pair. The separation of the two levels of interest can be controlled by an additional gate voltage. In figure 8.24, we show such a charge qubit from the work of Lehnert *et al* [58].

When the island is small, the charging energy dominates the qubit operation, but one can write a simple Hamiltonian as [54]

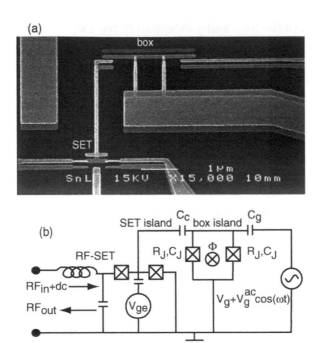

Figure 8.24. (a) An SET electrometer. The device is made from an evaporated aluminum film (light gray regions) on an insulating SiO$_2$ substrate (dark gray regions) by the technique of double angle evaporation, which gives the double image. The aluminum has BCS gap $\Delta/k_B = 2.4$ K. (b) A circuit diagram of the box and RF-SET electrometer. The SET gate voltage V_{ge}, the 500 MHz oscillatory bias, and the dc bias (RF$_{in}$ + dc) determine the electrometer's operating point. The charge on the box is inferred from variation in the amount of applied RF power that is reflected (RF$_{out}$) from the SET electrometer, which is a sensitive function of SET's conductance. The tunnel junctions (crosses in boxes) are characterized by a junction resistance R_J and capacitance C_J, which enter the box's Hamiltonian through $C_T = C_c + 2C_J + C_g$ and $R_T = R_J/2$. (Reprinted with permission from [58]. Copyright 2003 the American Physical Society.)

$$H = 4E_C(N - n_g)^2 - 2E_J\cos(\vartheta)\cos\left(\frac{1}{2}\varphi_e\right), \tag{8.40}$$

where E_J is the Josephson energy, n_g is indicative of the gate voltage (through the capacitor coupling), φ_e is the external flux and θ is the difference in the Josephson phases of the two junctions. With the two junctions, the box can be tuned with the external flux passing by the reservoir, much like a SQUID, which is discussed below.

When the charge qubit is shunted by a capacitor, as a method of reducing current noise, the device has been called a *transmon* [59]. The shunt capacitor also reduces the charging energy and hence increases the size of the Josephson energy relative to the charging energy. Measurement and control of the box qubit is commonly done by means of microwave resonators with the techniques usual in quantum electro-dynamics, as in other superconductor qubits [60]. As a result, it is possible to couple a pair of transmons to photons and produce nonlinear optical effects such as photon blockade [61].

8.4.4 Flux qubit

One of the most interesting aspects of circuits is that a resonant circuit can be a quantum oscillator, much akin to the harmonic oscillator [1]. In the absence of dissipation, the Hamiltonian can be written as

$$H = \frac{q^2}{2C} + \frac{\Phi^2}{2L},\tag{8.41}$$

where the charge q and flux Φ are the conjugate variables, and satisfy a commutator relation with one another. Of course, to reach the quantum limit, we must work at low temperature, which naturally makes a connection with superconducting circuits. In the presence of dissipation, we need to have a good quality factor Q $(= \omega\tau \gg 1)$. Hence, when we couple a Josephson junction to a superconducting ring, we obtain the persistent current flux qubit, sometimes called an RF superconducting quantum interference device (RF-SQUID, discussed above). Here, we need to have $E_J \gg E_C$ [53, 62], so the charging energy cannot be too large. The eigenstates of the ring represent two counter-rotating persistent currents, corresponding to a fixed number of flux quanta (h/e) in the loop. The inductance of the ring gives rise to a parabolic potential, like the harmonic oscillator, and adding the Josephson oscillating potential provides the needed nonlinearity to separate off the lowest states from the linear chain, as discussed earlier. For the two levels of the qubit, the Hamiltonian can be written as [54]

$$H = 4E_C n^2 + E_l(\varphi - \varphi_e + \varphi_{int}\sigma_z)^2 - E_J\cos(\varphi),\tag{8.42}$$

where the interaction term arises from the effect of the external bias flux on the two persistent current states. When $\varphi_e = \Phi_0/2$, the lower part of the potential is a symmetric double well creating nearly degenerate $|L\rangle$ and $|R\rangle$ states which lead to bonding and anti-bonding hybrids of these two states. These latter two states are created by the macroscopic tunneling through the Josephson junction which couples the two persistent current states, and these hybrid states provide the two levels of the qubit. The Wigner function in number-phase representation shows that the state of the system evolves into a quantum superposition of two coherent states which clearly demonstrate interference and negative values for the Wigner function description of the system. Yet, when the system is represented in the number-phase coordinates, the Wigner function evolves in a classical manner [62]. This ring-junction system was studied further to investigate the role of dissipation on the evolution of the Wigner function description of the system [63]. In this latter case, the dissipation is incorporated by coupling the system to a reservoir, and this expanded system is then projected back onto the reduced density matrix for the ring-junction system. They conclude that the two coherent states survive even in the presence of dissipation, at least for weak dissipation.

The flux qubit can also be prepared with a three junction ring, much like the hybrid qubit of the next section. This qubit can then be coupled to a transmission line resonator to produce cavity QED interactions. The capacitance between the qubit and the resonator can be controlled by varying the width of the capacitance

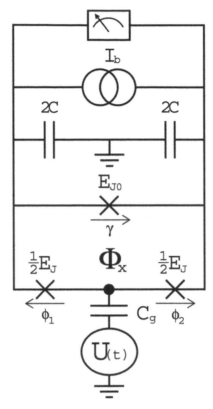

Figure 8.25. Circuit diagram of the quantronium qubit. (Reprinted with permission from [67]. Copyright 2006 the American Physical Society.)

line, so that the coupling depends upon the number of qubits placed in the overall circuit [64, 65]. We note that this can also be done with single junction qubits [66].

8.4.5 The hybrid charge-flux qubit

If we add a second Josephson junction in the ring of the flux qubit (as in the charge qubit), we can have both the charge box between the two junctions and flux coupled through the ring. If the Coulomb energy E_C dominates the Josephson energy, we have the charge qubit limit. If the Josephson energy dominates the Coulomb energy, we reach the flux qubit limit, but if we have $E_J \geqslant E_C$, we are in the charge-phase, or hybrid charge-flux qubit regime [52]. In many situations, the system acts as a two-level atom, which is called a quantronium [55]. A quite similar version of the quantronium is shown in figure 8.25 [67]. The two lower Josephson junctions create the charge box part of the circuit, but the ring from these to the larger third junction creates the supercurrent, flux reservoir. This loop is coupled (not shown) to an external flux circuit that is used to vary the flux enclosed in this ring. The box charge is varied through the coupling capacitor and gate voltage. A pair of capacitors is used to reduce the ac noise, while the current source provides the current drive to the qubit itself. The Hamiltonian for the circuit is given as [67]

$$H = E_C(N - n_g)^2 + E_J\left[1 - \cos\left(\frac{\varphi + \gamma}{2}\right)\cos(\varphi)\right]$$
$$+ \frac{Q^2}{2C} + E_{J0}(1 - \cos\gamma),$$

(8.43)

where the different parameters are defined in the figure in terms of the fluxes through the various junctions with $\varphi = (\varphi_1 + \varphi_2)/2$ and n_g has its former meaning. The fluxes are normalized to the flux quantum. The lowest and first excited states of the quantronium are taken to be the $|0\rangle$ and $|1\rangle$ states. In this configuration, the circuit is operated so that the applied flux Φ_x is set to zero and the normalized gate charge is set to 1/2 so that the qubit levels are separated by the Josephson junction coupling energy. Then the Hamiltonian can be written as

$$H' = E_J\left[1 - \cos\left(\frac{\gamma}{2}\right)\right]\sigma_z + E_{C0}N^2 + E_{J0}(1 - \cos\gamma),$$

(8.44)

where $E_{C_0} = 2e^2/C$ and N is the charge operator for the large Josephson junction. This Hamiltonian describes a two-level system, with the levels separated by an energy given by the first term on the right of (8.44). The phase in this term provides the coupling between the qubit and the readout junction.

As discussed above, the charge-flux qubit is operated normally at this charge degeneracy point and is control and switched via microwave pulses in order to demonstrate the presence of Rabi oscillations between the two qubit states. It often can have a fairly long coherence time, and thus can perform hundreds of well-controlled single qubit gate operations. In simulations of the circuit, it is found that during the readout pulse for the circuit of figure 8.25, the coherence time is of the order of 0.2 ns after the bias current has been applied (the role of the bias current is to diphase the qubit). The simulation results have been compared with the experimental data obtained by Vion et al [55].

8.4.6 Novel qubits

It is possible to also make Josephson junctions by using a semiconductor as an insulator. If the semiconductor can undergo induced superconductivity from a nearby superconductor, then there is the possibility of using gate control on the junction. This can become more interesting if there is an interplay between induced superconductivity, spin–orbit coupling, and topological edge states [68, 69]. In this quest, quasi-two-dimensional semiconductors have been of intense interest for the creation of gate controlled junctions, that may have the possibility to create topological states of interest. The interest in topological quantum computing lies in the fact that the topological states are protected and such qubits should be free from random noise. A novel three junction Josephson device, which utilizes a semiconductor electrode and is gate switchable, has recently been proposed in a quest for such topological states [70]. The device is depicted in figure 8.26. The device uses an InAs quantum well and a 10 nm aluminum superconducting layer to generate the proximity superconductivity in the InAs. The structure is complex, as the lattice constants of the InP substrate and the InAs quantum well are dramatically different. The first layer grown (starting from the substrate) is a graded $In_xAl_{1-x}As$ layer, where the grading begins at $x = 0.52$ (lattice matched to the substrate) up to $x = 0.81$. Then, a 25 nm $In_{0.81}Ga_{0.19}As/In_{0.81}Al_{0.19}As$ superlattice layer is grown, followed by 100 nm of $In_{0.81}Al_{0.19}As$ with Si δ-doping at 2×10^{12} cm^{-2}. Then, a 6 nm bottom barrier of $In_{0.75}Ga_{0.25}As$ is followed by the 7 nm InAs quantum well and a 10 nm $In_{0.75}Ga_{0.25}As$ top barrier. The sample has a measured carrier concentration of about 10^{12} cm^{-2} in the quantum well and a mobility of about 30 000 cm^2 V^{-1} s^{-1}. Electron-beam lithography and wet etching are used to finish the device fabrication.

The authors performed dc current-biased measurements in a dilution refrigerator with a base temperature of $T \sim 14$ mK. The current applied between terminals 1 and 0 (see figure 8.26) is referred to as I_1, the current applied between 2 and 0 as I_2, and the current between 1 and 2 as I_{12}. By making the top-gate voltage more negative, the electron density in the interstitial 2DEG is gradually depleted, which tunes the

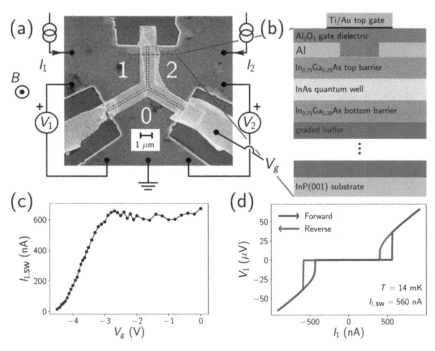

Figure 8.26. (a) False-color scanning electron microscope image of gated three-terminal junction with measurement schematic. Blue areas are aluminum and grey areas are etched to the insulating buffer layers to create the device mesas. The Ti/Au top gate (yellow) overlays the 200-nm-wide Y-shaped junction formed by selectively etching the Al layer only (dotted lines). (b) Cross-sectional view along the purple horizontal line in (a) (not to scale). (c) Gate dependence of the switching current when biasing between terminals 1 and 2 showing pinch-off occurring at $V_g \sim -4.5$ V. (d) Two-terminal I_1 versus V_1 curve with $I_2 = 0$ showing hysteresis. (Reprinted with permission from [70]. Copyright 2020 the American Physical Society.)

switching current from ~560 nA to 0 [panel (c)]. As shown in panel (d) the junctions exhibit hysteresis with respect to the current sweep direction. Hysteretic Josephson *I–V* curves can occur either due to the presence of a shunt capacitance (as described by the resistively and capacitively shunted junction model), or due to Joule self-heating which sets on as the junction becomes dissipative. Although 2DEG-based lateral Josephson junctions can in principle have large shunt capacitances due to conducting underlayers in the heterostructures or capacitive coupling of each superconducting terminal to the top gate, we estimate that the capacitance in our device is small [70].

The authors then mapped out the device performance by varying the different excitations at the three terminals. The ability to find novel distinguishing features, that are due to mesoscopic superconducting transport in such a multiterminal Josephson junction is hoped to prove useful in future studies aiming at topological effects [70]. These effects could provide a route forward in the development of qubits with greater defense against decoherence.

Problems

1. Consider a trapezoidal potential well that is V_0 deep on the left-hand side and V_1 deep on the right-hand side. The well is located in the region $0 < x < a$. Using the Wentzel–Kramers–Brillouin (WKB) approximation, determine the bound states within the well. If $V_0 = 0.3$ V and $V_1 = 0.4$ V, with $m^* = 0.067m_0$ and $a = 5$ nm, what are the bound state energies?

2. When a voltage (V_{sd}) is applied across a QPC it is typical to assume that the quasi-Fermi level on one side of the barrier is raised by αeV_{sd} while that on the other drops by $(1 - \alpha)eV_{sd}$, where α is a phenomenological parameter that, in a truly symmetrical structure, should be equal to 1/2. If we consider a device in which only the lowest subband contributes to transport, then the current flow through the QPC may be written as:

$$I_{sd} = \frac{2e}{h}\left[\int_L T(E)dE - \int_R T(E)dE\right],$$

where $T(E)$ is the energy-dependent transmission coefficient of the lowest subband and 'L' and 'R' denote the left and right reservoirs, respectively. If we assume low temperatures, we can treat the transmission as a step function, $T(E) = u_0(E - E_1)$, where E_1 is the threshold energy for the lowest subband. (a) Write this integral with limits appropriate to determine the current. (b) Use this information to obtain an expression for the current flowing through the QPC when the source drain voltage is such that both reservoirs populate the lowest subband, and when it populates the subband from just the higher-energy reservoir.

3. Consider an AlGaAs–GaAs–AlGaAs RTD with barrier widths of 5 nm and a well width of 5 nm. Assume the WKB approach used in the previous problem. (a) Assuming that the barrier height is 300 meV, estimate Γ_L and Γ_R, the effective tunneling rates through the two different barriers, under

conditions corresponding to the first tunneling resonance. (b) Calculate the fraction of the current that is coherent if the phase relaxation time is 1 ps.

4. Consider a SET whose individual tunnel barriers have capacitances $C_1 = C$ and $C_2 = 2C$ and whose gate capacitance $C_g = C/3$. You may assume that any background charge can be taken to be zero in this problem. (a) Write the set of four equations that can be used to define the charge-stability diagram for this device. (b) Plot the charge-stability diagram for this device and indicate the regions where current flow is Coulomb blockaded. (c) An important parameter for any transistor is its voltage gain, i.e., the change in source–drain voltage arising from a change in gain voltage. For the SET considered here, what can you say about its voltage gain?

5. Consider a parabolic quantum dot implemented in a GaAs 2DEG with $\hbar\omega_0 = 2$ meV, where ω_0 is characterized by the harmonic oscillator itself (See appendix G). Now consider the situation in which 12 electrons are present in this dot and assume also that spin splitting of any electron states can be neglected. Plot the energy levels of the dot as a function of magnetic field (for $0 < B < 4$ T) and indicate the variation of the highest filled electron state as a function of magnetic field in the situation where the electrons occupy the ground state of the dot.

Appendix F Klein tunneling

In classical semiconductors, a potential barrier which is higher than the energy of the incident particle generally blocks transmission of the particle through the barrier. However, if the barrier has a finite thickness, then it is possible for the particle to tunnel through the barrier [1]. This is because, with the wave interpretation, the part of the wave that penetrates the potential barrier does not fully decay before the back edge of the barrier is reached. Nevertheless, the transmission probability decays exponentially both with the height of the potential barrier and with its thickness. In the relativistic world, however, the Klein paradox leads to a situation in which an incoming electron can penetrate a potential barrier if the height exceeds the rest energy mc^2 [71]. When this happens, the transparency of the barrier depends only weakly on the barrier height and actually increases as the barrier height increases. The physics of the process is that the penetrating electron can couple to positrons under the barrier to affect the transmission, and matching between the two sets of wave functions leads to the high transparency [72].

Graphene has the Dirac bands and the zero energy gap leads to zero rest energy. Consequently, any barrier height would lead to the same behavior as the Klein paradox, a result that has been worked out for the chiral particles in graphene [73]. Thus, under a wide range of conditions, a potential barrier poses no obstacle to an electron or hole in graphene, and the concept of a Schottky barrier just does not work well, as discussed earlier in the chapter. Here, we follow the treatment of [74] to illustrate the conditions under which Klein tunneling appears. The basic premise is, of course, that the electrons and holes in graphene accurately follow the properties of the electrons and positrons that Klein analyzed. Let us consider a barrier of height

V_0, that exists in the region $0 < x < L$. An electron at the Fermi energy approaches the barrier with wave vector k_F and with an angle ϕ defined by the longitudinal and transverse components of the Fermi wave vector through

$$k_x = k_F\cos(\phi), \quad k_y = k_F\sin(\phi). \tag{F.1}$$

As in the classical tunneling problem, it is assumed that the barrier has infinitely sharp edges so that no disorder is introduced by these edges. The valley degeneracy gives the equivalent Dirac spinor of two wave functions corresponding to the pseudo-spin of the composite wave function. These two wave functions are written as [70]

$$\psi_1 = \begin{cases} (e^{ik_x x} + re^{-ik_x x})e^{ik_y y}, & x < 0, \\ (ae^{iq_x x} + be^{-iq_x x})e^{ik_y y}, & 0 < x < L, \\ te^{-ik_x x}e^{ik_y y}, & x > L, \end{cases}$$

$$\psi_2 = \begin{cases} s(e^{ik_x x+i\phi} + re^{-ik_x x-i\phi})e^{ik_y y}, & x < 0, \\ s'(ae^{iq_x x+i\vartheta} + be^{-iq_x x-i\vartheta})e^{ik_y y}, & 0 < x < L, \\ ste^{-ik_x x+i\phi}e^{ik_y y}, & x > L. \end{cases} \tag{F.2}$$

Here, we use

$$q_x = \sqrt{\frac{(E - V_0)^2}{\hbar^2 v_F^2} - k_y^2},$$

$$\vartheta = \arctan\left(\frac{k_y}{q_x}\right), \tag{F.3}$$

$$s = \text{sign}(E), \quad s' = \text{sign}(E - V_0).$$

The angle ϑ is the refraction angle of the wave. Matching the various coefficients lead to the reflection coefficient

$$r = 2ie^{i\phi}\sin(q_x L)\frac{\sin(\phi) - ss'\sin(\vartheta)}{ss'[e^{-iq_x L}\cos(\phi + \vartheta) + e^{iq_x L}\cos(\phi - \vartheta)] - 2i\sin(q_x L)}. \tag{F.4}$$

The transmission through the barrier is given by $T = |t|^2 = 1 - |r|^2$.

Appendix G The Darwin–Fock spectrum

In appendix B, we introduced the operator solution to the harmonic oscillator. Here, we treat a cylindrically symmetric quantum dot in two dimensions, which actually makes use of the harmonic oscillator solutions. In principle, we write the Schrödinger equation in the (x,y)-plane as [1]

$$H = -\frac{\hbar^2}{2m^*}\left(\frac{\partial^2}{\partial x^2} + \frac{\partial^2}{\partial y^2}\right) + \frac{m^*\omega^2}{2}(x^2 + y^2). \tag{G.1}$$

This, of course, just doubles the number of variables in the overall harmonic oscillator, and one can immediately write the energy as

$$E = \hbar\omega\left(a_x^\dagger a_x + \frac{1}{2}\right) + \hbar\omega\left(a_y^\dagger a_y + \frac{1}{2}\right)$$
$$= \hbar\omega(n_x + n_y + 1).$$

$$\text{(G.2)}$$

But, this is not the most efficient way to proceed, especially when we add a magnetic field. The cylindrical quantum dot in a magnetic field was solved early on in quantum mechanics by Darwin [75] and Fock [76]. To proceed, it is better to use the symmetric gauge (see appendix C), in which the vector potential is written as

$$\mathbf{A} = \frac{1}{2}(-By\mathbf{a}_x + Bx\mathbf{a}_y).$$

$$\text{(G.3)}$$

When this is added via the Peierls' substitution, two additional terms in the Hamiltonian arise as

$$\delta H = -ie\hbar B\left(x\frac{\partial}{\partial y} - y\frac{\partial}{\partial x}\right) + \frac{e^2 B^2}{4}(x^2 + y^2).$$

$$\text{(G.4)}$$

The second term here just adds to the last term in equation (G.1) to give a hybrid frequency for the combined oscillator as

$$\Omega = \sqrt{\omega^2 + \left(\frac{\omega_c}{2}\right)^2}, \qquad \omega_c = \frac{eB}{m^*}.$$

$$\text{(G.5)}$$

The first term in (G.4) is a new term that breaks the symmetry of our two-dimensional harmonic oscillator. If we recognize that $\mathbf{L} = \mathbf{r} \times \mathbf{p}$, then this first term is just eBL_z. It is well known in atomic physics that the magnetic field splits the angular momentum states, which are usually degenerate. To generate the same thing in this problem, we need to go away from our x and y oscillators and rewrite everything in cylindrical coordinates. To achieve this, we introduce a new set of operators as

$$a = \frac{1}{\sqrt{2}}(a_x - ia_y) \quad a^\dagger = \frac{1}{\sqrt{2}}\left(a_x^\dagger + ia_y^\dagger\right)$$
$$b = \frac{1}{\sqrt{2}}(a_x + ia_y) \quad b^\dagger = \frac{1}{\sqrt{2}}\left(a_x^\dagger - ia_y^\dagger\right).$$

$$\text{(G.6)}$$

To understand these a little better, we can look at the expansions of these operators

$$a^\dagger a = n_a = \frac{1}{2}\left(a_x^\dagger a_x + a_y^\dagger a_y - ia_y^\dagger a_x + ia_x^\dagger a_y\right)$$
$$b^\dagger b = n_b = \frac{1}{2}\left(a_x^\dagger a_x + a_y^\dagger a_y + ia_y^\dagger a_x - ia_x^\dagger a_y\right).$$

$$\text{(G.7)}$$

Hence, we still achieve the same energy, with the total index

$$n = n_x + n_y = n_a + n_b. \tag{G.8}$$

If we rewrite the coordinate operators in terms of their actual coordinates, then we can write

$$L_z = (a^\dagger a - b^\dagger b)\hbar = (n_a - n_b)\hbar. \tag{G.9}$$

We now have two sets of operators which work individually on the positive angular momentum states and the negative angular momentum states. These four operators work on both the energy and the angular momentum in the following way:

$$
\begin{array}{ccc}
\text{Operator} & \text{energy} & L_z \\
a^\dagger & \uparrow & \uparrow \\
a & \downarrow & \downarrow \\
b^\dagger & \uparrow & \downarrow \\
b & \downarrow & \uparrow
\end{array} \tag{G.10}
$$

where the arrows indicate a raising or lowering of the state. A given energy level n has both an energy and a degeneracy of $n + 1$, where $n = 0, 1, 2, \ldots$. The angular momentum here runs from $-n$ to $+n$ in steps of two units, and the magnetic field splits the degeneracy. If we write $m_z = n_a - n_b$, we can write the total energy as

$$
\begin{aligned}
E &= \hbar\Omega(n_a + n_b + 1) + \frac{1}{2}\omega_c L_z \\
&= \hbar\Omega(n + 1) + \frac{1}{2}\hbar\omega_c m_z.
\end{aligned} \tag{G.11}
$$

The heart of the spectrum of states lies in the hybrid frequency Ω. For small values of the magnetic field, we can expand this term as

$$\Omega \sim \omega\left[1 + \frac{1}{2}\left(\frac{\omega_c}{2\omega}\right)^2\right] \sim \omega + \frac{\omega_c^2}{8\omega}, \tag{G.12}$$

and the energy is

$$E \sim \hbar\omega(n + 1) + \hbar\omega_c\left[m_z + (n + 1)\frac{\omega_c}{8\omega}\right]. \tag{G.13}$$

So, while the magnetic field does split the angular momentum states, it also gives them an upward curvature. In the opposite limit, where the magnetic field is large, we can approximate the hybrid frequency as

$$\Omega \sim \frac{\omega_c}{2}\left[1 + \frac{1}{2}\left(\frac{2\omega}{\omega_c}\right)^2\right] \sim \frac{\omega_c}{2} + \frac{\omega^2}{\omega_c^2}, \tag{G.14}$$

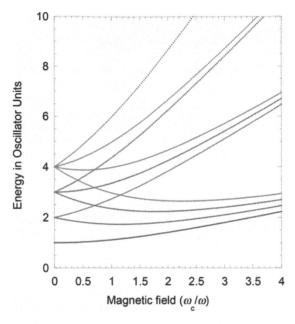

Figure G1. The various angular momentum levels for the lowest four dot energies. One can see the splitting in the magnetic field at low values and the tendency to form Landau levels at high values of the magnetic field.

and the energy is given as

$$E \sim \hbar\omega_c\left(n_a + \frac{1}{2}\right) + \hbar\frac{\omega^2}{\omega_c}(n + 1). \tag{G.15}$$

This last form is interesting, as it indicates that the major feature of the spectrum is the formation of Landau levels, which all have the similar positive angular momentum, but the levels are split by the interaction of the confining potential. We see these two limits in the structure shown in figure G1.

References

[1] Ferry D K 2001 *Quantum Mechanics* 2nd edn (Bristol: IOP Publishing)

[2] Gorter C 1951 *Physica* **17** 777

[3] Darmois E 1956 *J. Phys. Radium* **17** 210

[4] Neugebauer C A and Webb M B 1962 *J. Appl. Phys.* **33** 74

[5] Averin D V and Likharev K K 1986 *J. Low Temp. Phys.* **62** 345

[6] Iansiti K, Johnson A T, Lobb C J and Tinkham M 1988 *Phys. Rev. Lett.* **60** 2414

[7] Fultan T A and Dolan G J 1987 *Phys. Rev. Lett.* **59** 109

[8] van Bentum P J M, van Kempen H, van der Leemput L E C and Teunissen P A A 1988 *Phys. Rev. Lett.* **60** 369

[9] Ohshiro T and Umezawa Y 2006 *Proc. Natl. Acad. Sci. USA* **103** 10

[10] Lindsay S, He J, Sankey O, Hapala P, Jelinek P, Zhang P, Chang S and Huang S 2010 *Nanotechnology* **21** 262001

[11] Tsutsui M, Taniguchi M, Tokota K and Kawaii T 2010 *Nat. Nanotechnol.* **5** 286

[12] Huang S, He J, Chang S, Zhang P, Liang F, Li S, Tuchband M, Fuhrmann A, Ros R and Lindsay S 2010 *Nat. Nanotechnol.* **5** 868
[13] Zhao Y *et al* 2014 *Nat. Nanotechnol.* **9** 466
[14] Pang P *et al* 2014 *ACS Nano* **8** 11994
[15] Tarucha S, Austing D G, Honda T, van der Hage R J and Kouwenhoven L P 1999 *Phys. Rev. Lett.* **77** 3613
[16] Kouwenhoven L P, Austing D G and Tarucha S 2001 *Rep. Prog. Phys.* **64** 701
[17] Kang S, Kim D-H, Park I-H, Kim J-H, Lee J-E, Lee J D and Park B-G 2008 *Japan. J. Appl. Phys.* **47** 3118
[18] Fuechsle M, Miwa J A, Mahapatra S, Ryu H, Lee S, Warkschow O, Hollenberg L C L, Klimeck G and Simmons M Y 2012 *Nat. Nanotechnol.* **7** 242
[19] Kane B E 1998 *Nature* **393**
[20] Lyding J W, Shen T C, Hubacek J S, Tucker J R and Abeln G C 1994 *Appl. Phys. Lett.* **64** 2010
[21] van der Wiel W G, De Franceschi S, Elzerman J M, Fujisawa T, Tarucha S and Kouwenhoven L P 2002 *Rev. Mod. Phys.* **75** 1
[22] Pothier H, Lafarge P, Urbina C, Esteve D and Devoret M H 1992 *Europhys. Lett.* **17** 249
[23] Ruzin I M, Chandrasekhar V, Levin E I and Glazman L I 1992 *Phys. Rev.* B **45** 13469
[24] Dixon D C, Kouwenhoven L P, McEuen P L, Nagamune Y, Motohisa J and Sakaki H 1996 *Phys. Rev.* B **53** 12625
[25] Moriyama S, Tsuya D, Watanabe E, Uji S, Shimizu M, Mori T, Yamaguchi T and Ishibashi K 2009 *Nano Lett.* **9** 2891
[26] Nielsen M A and Chuang I L 2000 *Quantum Computation and Quantum Information* (Cambridge: Cambridge University Press)
[27] Marinescu D C and Marinescu G M 2005 *Approaching Quantum Computing* (Upper Saddle River, NJ: Pearson)
[28] Akis R, Vasileska D and Ferry D K 2002 *Proc. 13th Workshop on Physical Simulation of Semiconductor Devices (Ilkey, UK, March 25–26 2002)*
[29] Loss D and DiVincenzo D P 1998 *Phys. Rev.* B **57** 120
[30] Brunner R, Shin Y-S, Obata T, Pioro-Ladrière M, Kubo T, Yoshida K, Taniyama T, Tokura Y and Tarucha S 2011 *Phys. Rev. Lett.* **107** 146801
[31] Hollenberg L C L, Dzurak A S, Wellard C, Hamilton A R, Reilly D J, Milburn G J and Clark R G 2004 *Phys. Rev.* B **69** 113301
[32] Pia J J, Tan K Y, Dehollain J P, Lim W H, Morton J J L, Jamieson D N, Dzurak A S and Morello A 2012 *Nature* **489** 541
[33] Kim D *et al* 2014 *Nature* **511** 70
[34] Thorgrimsson B *et al* 2017 *npj Quantum Inform.* **3** 32
[35] Nichol J M, Orona L A, Harvey S P, Fallahi S, Gardner G C, Manfra M J and Yacoby A 2017 *npj Quantum Inform.* **3** 3
[36] Ferdous R *et al* 2018 *npj Quantum Inform.* **4** 26
[37] Ferdous R, Chan K W, Veldhorst M, Hwang J C C, Yang C H, Sahasrabudhe H, Klimeck G, Morello A, Dzurak A S and Rahman R 2018 *Phys. Rev.* B **97** 241401
[38] Abadillo-Uriel J C *et al* 2018 *Phys. Rev.* B **98** 165438
[39] Wolfe M A, Calderon-Vargas F A and Kestner J P 2017 *Phys. Rev.* B **96** 201307
[40] Veldhorst M, Ruskov R, Yang C H, Hwang J C C, Hudson F E, Flatté M E, Tahan C, Itoh K M, Morello A and Dzurak A S 2015 *Phys. Rev.* B **92** 201401
[41] Gamble J K *et al* 2016 *Appl. Phys. Lett.* **109** 253101
[42] Neyens S F *et al* 2018 *Appl. Phys. Lett.* **112** 243107

[43] Cooper L N 1956 *Phys. Rev.* **104** 1189

[44] Bardeen J, Cooper L N and Schrieffer J R 1957 *Phys. Rev.* **106** 162

[45] Bardeen J, Cooper L N and Schrieffer J R 1957 *Phys. Rev.* **108** 1175

[46] Josephson B D 1962 *Phys. Lett.* **1** 251

[47] Kuper C G 1968 *An Introduction to the Theory of Superconductivity* (Oxford: Clarendon) section 8.2

[48] Riedel E 1964 *Z. Naturforsch.* **19a** 1634

[49] Jaklevic R C, Lambe J, Silver A H and Mercereau J E 1964 *Phys. Rev. Lett.* **12** 159

[50] Drung D, Aßmann C, Beyer J, Kirste A, Peters M, Ruede F and Schurig Th 2007 *IEEE Trans. Appl. Supercond.* **17** 699

[51] Jaklevic R C, Lambe J, Mercereau J E and Silver A H 1965 *Phys. Rev.* **140** A1628

[52] Wendin G 2003 *Phil. Trans. Roy. Soc. London* A **361** 1323

[53] Mooij J E, Orlando T P, Levitov L, Tian L, van der Wal C H and Lloyd S 1999 *Science* **285** 1036

[54] Nakamura Y, Pashkin Y A and Tsai J S 1999 *Nature* **398** 786

[55] Vion D, Aassime A, Cottet A, Joyez P, Pothier H, Urbina C, Esteve D and Devoret M H 2002 *Science* **296** 886

[56] Steffen M, DiVincenzo D P, Chow J M, Theis T N and Ketchen M B 2011 *IBM J. Res. Develop.* **55** 13

[57] Bouchiat V, Vion D, Joyez P, Esteve D and Devoret M H 1998 *Phys. Scripta* T **76** 165

[58] Lehnert K W, Bladh K, Spietz L F, Gunnarsson D, Schuster D I, Delsin P and Schoelkopf R J 2003 *Phys. Rev. Lett.* **90** 027002

[59] Koch J, Yu T M, Gambetta J, Houck A A, Schuster D I, Majer J, Blais A, Devoret M H, Girvin S M and Schoelkopf R J 2007 *Phys. Rev.* A **76** 042319

[60] Zhu M-Z and Ye L 2016 *Ann. Phys.* **373** 512

[61] Wang Y, Zhang G-Q and You W-L 2018 *Laser Phys. Lett.* **15** 105201

[62] Joshi A 2000 *Phys. Lett.* A **270** 249

[63] Zou J, Shao B and Su W-Y 2001 *Phys. Lett.* A **285** 401

[64] Kim M D 2015 *Quantum Inf. Processes* **14** 3677

[65] Orgiazzi J-L, Deng C, Layden D, Marchildon R, Kitapli F, Shen F, Bal M, Ong R and Lupascu A 2016 *Phys. Rev.* B **93** 104518

[66] Hofheinz M *et al* 2009 *Nature* **459** 546

[67] Hutchinson G D, Holmes C A, Stace T, Spiller T P, Milburn G J, Barrett S D, Hasko D G and Williams D A 2006 *Phys. Rev.* A **74** 062302

[68] Doh Y-J, van Dam J A, Roest A L, Bakkers E P A M, Kouwenhoven L P and De Franceschi S 2005 *Science* **309** 272

[69] Lutchyn R A, Bakkers E P, Kouwenhoven L P, Krogstrup P, Marcus C M and Oreg Y 2018 *Nat. Rev. Mater.* **3** 52

[70] Graziano G V, Lee J S, Pendharkar M, Palmstrøm C J and Pribiag V S 2020 *Phys. Rev.* B **101** 054510

[71] Klein O 1929 *Z. Phys.* **53** 157

[72] Krekora P, Su Q and Grobe R 2004 *Phys. Rev. Lett.* **92** 040406

[73] Katsnelson M I, Novoselov K S and Geim A K 2006 *Nat. Phys.* **2** 620

[74] Charlier J-C, Blasé X and Roche S 2007 *Rev. Mod. Phys.* **79** 677

[75] Darwin C G 1931 *Proc. Cambridge Phil. Soc.* **27** 86

[76] Fock V 1928 *Z. Phys.* **47** 446

IOP Publishing

Transport in Semiconductor Mesoscopic Devices
(Second Edition)

David K Ferry

Chapter 9

Open quantum dots

The description of a quantum dot can be applied to a great many types of structure, some of which are defined by physical characteristics, such as self-assembled dots [1], and others of which are defined through lithography. This can either be an etched structure [2], or achieved by the imposition of a self-consistent potential which is applied through a set of confining gates [3], much like those which form the quantum point contact (QPC) in chapter 2. Over the last two decades, quantum dots have proven to be a natural test bed in which to probe the understanding of the transition between quantum mechanics and classical mechanics. In an open quantum dot, the interior dot region is coupled to the reservoirs by means of waveguide leads, usually formed by a pair of QPCs. While it might be expected that the charge fluctuations which wash out the Coulomb blockade should also obscure the signatures of this quantum structure, this is not the case. Many states are washed out by interaction with the environment, but it is the remaining states that provide new insight into the fundamental quantum physics. To be sure, open quantum systems, and the ability to make measurements upon them, have been a point of discussion since the early days of quantum mechanics [4–6]. When open quantum dots were first studied, it was found that a series of conductance fluctuations could be observed as current passed through the device [7, 8], much like those of section 5.3. Early on, it was assumed that these conductance fluctuations were the same as the earlier ones, which arise from disordered material [9]. Since the material involved in studies of these quantum dots was not disordered, but of relatively high quality, it was then thought that the conductance fluctuations arose from chaotic behavior within the quantum dots. However, there was a significant difference in that the fluctuations observed in these open quantum dots were nearly periodic in the magnetic field, instead of the aperiodic behavior expected from chaotic effects, and that this period was very close to that found for regular classical trajectories in classically confined structures of the same dimensions [10]. In this regard, it was established that the conductance fluctuations were really connected with regular orbits within the open dots [11], even in typically chaotic structures such as stadiums [12]. Later, it was clearly demonstrated that these regular quantum trajectories were very closely related to equivalent classical trajectories originating from Kolmogorov–Arnold–Moser (KAM) islands and rings of attractors (in arrays) that gave rise to classical orbits [13, 14].

The results of measurements have always been assumed to be classical, as the results appear in the laboratory itself. Yet, when the quantum nature persisted in the dots, it was found that the wave function was heavily scarred. Such 'scars' are imprints of a classical orbit which leads to fringes and a larger magnitude of the wave function along these orbits. The connection between the quantum scarred wave functions arising from the trajectories and the classical orbits turned out to be intrinsic to Zurek's decoherence theory in quantum mechanics [15]. There are two crucial parts to this theory that make the connection. First, when the quantum system is opened, a great many of the normal eigenstates become 'decohered' by interacting strongly with the environment (outside the quantum dot in this case). However, there remained a large number of eigenstates which remained strongly coherent as they did not mix with the environmental states. These states were termed the 'pointer states' [16], and these would leave an imprint upon the environment in such a manner that they could be measured. Second, these pointer states would evolve into the classical regular orbits [15], and should show a classical distribution of states, as a function of level spacing, rather than the quantum Gaussian distributions connected with quantum chaos [17].

In an important early study using Wigner functions, Berry [18] showed that the quantum wave function would be concentrated on the energy surface for which the classical orbit provides an oscillatory correction. That is, the scar is an enhancement of the quantum wave function around a periodic orbit, which itself is a property of the underlying classical system [19]. The existence of this scar is a connection between the quantum system and the equivalent classical system. It has normally been felt that such scars are unstable. However, it was found that these scars were quite stable [20], a result of a process known as quantum Darwinism [16, 21].

9.1 Conductance fluctuations in open quantum dots

As mentioned above, quantum dots have been extensively investigated for a few decades. Our interest is in the case in which the dots are open and relatively strongly coupled to the environment. In most of the earlier studies, considerable effort was expended to try to establish that the basic behavior of these ballistic quantum dots was governed by universal properties that were generic in nature and independent of the specific properties of the individual dots. In fact, averaging over, e.g. the gate voltage, was used to remove the quasi-periodic fluctuations in order to reveal what was believed to be a chaotic background. In fact, this process removed the significant signatures of what we now believe to be the pointer states, which are the surviving quantum states in the dot. In principle, the underlying physics of the ballistic quantum dots is described by characteristics that are dependent upon the individual dot under study, so that there is a universal behavior that is characteristic of such dots, but it is not the generic behavior described by the pseudo-chaotic mathematics of these previous studies. Instead, the underlying properties, and the characteristic transport through the dots, is governed by the basic regular nature of the semi-classical orbits in the dot [22]. This regular nature is observed as reproducible fluctuations.

The transport is dominated in the typical quantum dot at low temperature by reproducible fluctuations which are observed when a magnetic field, or the gate voltage applied to the dots, is varied. These fluctuations exhibit quasi-periodic oscillations, whose nearly single-frequency character is easily discernible in the correlation functions and in the Fourier transforms, and which are endemic to the semiclassical regular orbits in the dots. Crucial to this result is the excitation of the ballistic dots by open QPCs, as the latter provide a collimated excitation of the particles within the dot. This collimation provides a specific excitation of quantum structure related to the semiclassical orbits.

In figure 9.1, we show a variety of open quantum dots to illustrate that there are various ways in which to fabricate these structures [23]. In all cases, the dot size is much smaller than the elastic mean free path, so that the underlying motion is ballistic in nature. These structures are all fabricated on a GaAs/AlGaAs hetero-structure, and include: (1) a device with surface Schottky barrier gates to define the dots [24, 25], (2) a structure in which trenches have been etched into the material so that other areas of the structure can be used as gates [26], and (3) a structure which has been etched to define the dot [27]. We have also studied these dots in an InAs/AlInAs heterostructure and an InGaAs/GaAs heterostructure. The basic observations are similar in all of these cases, and independent of the material system and gate technology. For the moment, we will concentrate on dots of the first sort. For these dots, the gate regions are defined by liftoff of metallic Schottky barrier metals. The dots are basically square in nature, and sizes ranged from 1.0 to 0.4 μm (the electrical sizes are somewhat smaller due to edge depletion around the gates). In nearly all the cases studied, the carrier density was $3–5 \times 10^{11}$ cm^{-2} and mobilities

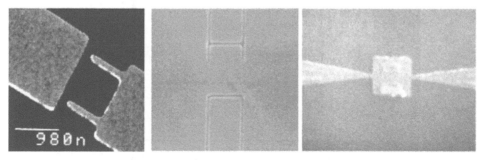

Figure 9.1. Three different methods for forming a quantum dot. Left: we use surface Schottky gates. Middle: we etched trenches so that other regions of the heterostructure can be used as gates. Right: a dot in which other regions of the heterostructure have been removed by etching. Left and middle: reprinted with permission from [30], Copyright 2005 IOP Publishing. Right: reprinted with permission from [27]. Copyright 1997 American Physical Society.

are typically 40–200 m^2 V^{-1}s^{-1}. The gate design allows electrons to be trapped in the central square cavity, but with the open QPCs, this density is set by the two-dimensional electron density outside the dot. This also means that the Fermi energy in the system is the same and set by these regions outside the dot. Measurements are typically carried out at 10–30 mK, the base temperature in the dilution refrigerator, and source–drain voltage is kept well below the thermal voltage with lock-in techniques utilized.

9.1.1 Magnetotransport

In figure 9.2, the conductance fluctuation that results from a sweep of the magnetic field for a typical sample is shown. This particular sweep came from a dot with Schottky gates used to define it. These reproducible fluctuations persist across a wide range of magnetic fields, and for a variety of gate voltages (applied to the Schottky gates). At higher magnetic fields, a transition into edge-state behavior and the quantum Hall effect occur. At still higher magnetic fields, AB oscillations from edge-state interference within the dot are observed [28], and this allows us to unambiguously determine the electrical size of the ballistic cavity. It is found that the basic nature of the reproducible fluctuations is independent of the details of the sample material and gate technology. Instead, this behavior seems to be a basic property of the dot geometry and size, a result of the intrinsic dot properties and quite distinct from the generic results of chaos theory [22]. Here, the basic oscillatory properties are universal, although the specific frequency content is very dot-dependent.

Although not shown in the above figure, the low-field magnetoresistance often exhibits a peak at zero magnetic field, which is unrelated to weak localization but instead is an intrinsic property of the states within the dot [29]. The actual line shape of this peak is found to vary with contact opening, and similar behavior has been observed in simple QPC structures which cannot support chaotic behavior. Thus, it is improper to assume *a priori* that such peaks are connected with either chaotic

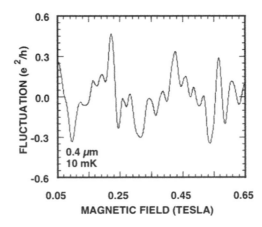

Figure 9.2. A sweep of the magnetic field produces the conductance fluctuations shown here (a background conductance has been removed). These are obtained for a 0.4 μm GaAs dot defined by Schottky gates. (Reprinted with permission from [23]. Copyright 2005 IOP Publishing.)

behavior or weak localization. Indeed, phase-space filling can easily be generated by the integrable motion within these dots due to the magnetic-field-induced precession of the orbits. Thus, it is not surprising that the quasi-periodic fluctuations observed in the magnetoconductance are quite different from those observed in chaotic systems.

We have carried out calculations to simulate the behavior of these dots, using a stable variant of the transfer matrix approach, which was discussed in section 2.5. For example, conductance fluctuations as a function of magnetic field are studied for a 0.3 µm square dot (this size reflects the electrical size, which is less than the actual structure due to gate depletion) with 0.04 µm port openings. This allows two modes to enter and exit the dot. Instead of a random aperiodic variation with magnetic field, a series of nearly periodic oscillations is evident, as may be seen in figure 9.3. Also apparent are several resonance features. In particular, a set of resonances occurs at $B \sim 0.069, 0.173, 0.283$ and 0.397 T, in which the wave function is heavily 'scarred', in that the quantum mechanical amplitude appears to follow a single underlying classical orbit, as shown in figure 9.4. Given that the period for the reappearance of the diamond is $\Delta B \sim 0.11$ T, and using the criterion familiar from the AB effect, that $\Delta\varphi/\varphi_0 = 2\pi$ for the difference in magnetic flux, one obtains $A \sim 0.04$ µm^2 for the enclosed area, which corresponds well with the enclosed area of the diamond-shaped wave function, depicted in figure 9.4.

Periodic orbits have played a huge theoretical role in the computation of semiclassical quantization of bound states for many years, dating back to the Einstein–Brillouin–Keller view of quantization. More recent studies indicate that the 'imprints' of these orbits persist up through thousands of states [18], a result that suggests the closed orbits are quite stable in regular systems and only become unstable as one passes to the ergodic regime, which seems to be replicated as one passes from the semiclassical to the quantum regime. The QPC imposes a boundary condition on the particles, in which the entry angle is determined by the wave

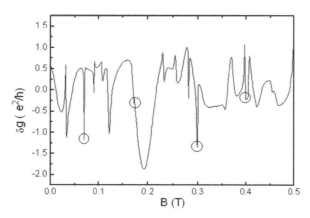

Figure 9.3. The computed conductance fluctuations at $T = 0$ K for a 0.3 µm dot in GaAs. The circles indicate the periodic resonances which display the heavily scarred wave function of figure 9.4 below. (Reprinted with permission from [23]. Copyright 2005 IOP Publishing.)

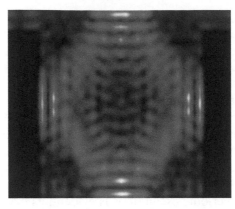

Figure 9.4. The squared magnitude of the wave function within the dot whose magnetic field sweep is shown in figure 9.3. This wave function recurs at several magnetic fields as a resonance. (Reprinted with permission from [30]. Copyright 2011 IOP Publishing.)

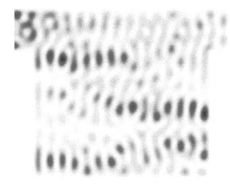

Figure 9.5. This movie illustrates the sharp resonance at which the scar of figure 9.4 recurs in the magnetic field. The scar appears four times in the sweep of the magnetic field. Available at https://iopscience.iop.org/book/978-0-7503-3139-5.

mechanical nature of propagation through the QPC. It is this collimation that is important for exciting a particular set of regular orbits of the particles. From the study of classical orbits in these dots, we can compute the transmission, and hence the conductance, as well as the fluctuations. When we say that the transmission, as denoted in figure 9.3, exhibits resonances, we mean that a particular scarred orbit, that shown in figure 9.4, recurs on a regular basis [30]. It is this recurrence that gives rise to the regular and periodic orbits corresponding to a dominant wave function. The fact that it recurs at a regular set of eigen-energies points strongly to the idea of quantum Darwinism [21]. To illustrate this recurrence, we show in figure 9.5 a video which depicts the wave function in the dot as the magnetic field is varied over the range illustrated in figure 9.3.

9.1.2 Gate-induced fluctuations

It is also possible to sweep the Fermi energy by varying the gate voltage. In this situation, making the gate voltage (on the confining gates) more negative. As the gate voltage is made more negative, the dot is reduced in size and the various energy levels are pushed up through the Fermi energy. The results are quite similar in behavior to that seen as the magnetic field is varied. This leads to a series of conductance oscillations, which ride on top of a monotonic background, and which disappear at a few degrees Kelvin. The oscillations often are observed over the entire range of gate voltage and persist to conductance values as high as $15e^2/h$, which represents a very open dot. Again, the experimental results and quantum simulations (discussed below) yield the same dominant frequency in the dots. Here, we will focus upon a smaller dot, whose gate defined dimension varied from 0.2–0.3 µm, depending upon the value of the applied gate voltage, a picture of which is shown in the inset to figure 9.6. The conductance through the quantum dot as the gate voltage is varied is shown as the main panel in figure 9.6 as the blue curve exhibiting the fluctuations. It may be seen that the fluctuations ride on a uniformly increasing (for increasing gate voltage) conductance background. Rather than try to smooth the curve, the temperature is raised above 2 K, a point at which the fluctuations are largely damped out, as shown in the red dotted curve in the figure. This latter curve is then subtracted from the low temperature curve to isolate the fluctuations themselves, which are plotted as the upper (blue) trace in figure 9.7. These oscillations are very nearly periodic with a dominant period of about 15 V^{-1}.

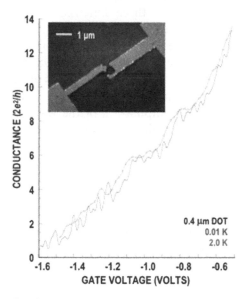

Figure 9.6. Conductance as a function of the gate voltage for a dot with staggered QPCs (shown in the inset, the two gates are kept at the same potential). The solid (blue) curve is the conductance at 10 mK, while the dashed (red) curve is at 2 K, where the fluctuations have been temperature damped. (Reprinted with permission from [30]. Copyright 2011 IOP Publishing.)

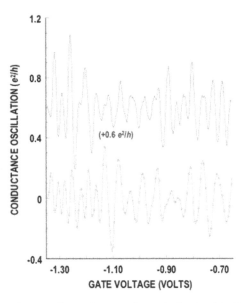

Figure 9.7. The measured conductance fluctuations are shown in the top (blue) trace, while those calculated from the recursive scattering matrix are shown in the lower (red) trace. (Reprinted with permission from [30]. Copyright 2011 IOP Publishing.)

To account for the behavior seen in the experiment, we have used our recursive scattering matrix simulation to study these fluctuations. These begin by first computing the exact self-consistent potential for the dot at more than 300 gate voltages. This is accomplished with a three-dimensional Poisson solver, which uses only experimental values for the various parameters of the heterostructure and gate layout. Once the potential profile is known, we simulate the quantum transport through the device. The calculations are performed for a fixed Fermi level which is defined by its value in the reservoirs to which the dot is attached through the QPC. The influence of these contacts is included via the exact potential profile of the open dots. This approach also allows computation of the wave function within the dot. This latter can be decomposed into the states of the closed dot for later use by an eigenfunction decomposition. The computed conductance fluctuation is also shown in figure 9.7 as the lower (red) trace. While there is a variation in amplitude, the frequency and general periodicity agree quite well between the experiment and theory. This may be confirmed by comparing the Fourier transforms of these traces, and this is shown in figure 9.8.

The same self-consistent potential has also been used to study the classical dynamics of electrons injected into these dots [10]. As with most systems of this nature, the dynamics is non-hyperbolic in that there are regions of chaotic scattering which coexist with non-escaping KAM islands surrounding stable orbits in phase space. Hence, these dots have a mixed phase space [31]. As mentioned above, when the quasi-periodic fluctuations are removed by averaging over gate voltage, only the chaotic background remains. It may be presumed that this chaotic sea provides the background conductance through the dot, such as that given by the 2.0 K curve

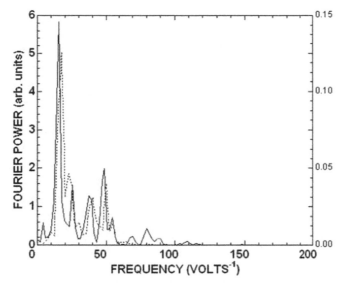

Figure 9.8. Fourier transform of the fluctuations seen in figure 9.7. Those from the experiment are shown as the solid curve. The simulated fluctuations are given by the dotted curve. (Reprinted with permission from [23]. Copyright 2005 IOP Publishing.)

shown in figure 9.6. It is thought that this chaotic background provides the regions of chaotic scattering seen in the phase-space portraits of the classical dynamics within these dots [10]. These phase-space portraits will be discussed below. Hence, the periodic orbits which are enclosed within the KAM island must correspond to the quasi-periodic fluctuations that we have been describing here. This was checked by varying the gate voltage, and observing the change in size of these periodic orbits. The results indicate that this periodicity agrees exceedingly well with that found for both the experiment and the quantum simulation. The conclusion is that it is those scarred quantum wave functions, which give rise to the periodicity, that relax into the classical periodic orbits on the KAM island. This is important, as it is generally felt that the pointer states are essentially the classical remains of the quantum states.

9.1.3 Phase-breaking processes

It is already clear from the plots in figure 9.6 that the fluctuations decay relatively rapidly with temperature. While there may be some semblance left at 2.0 K, the large amplitude of the fluctuations is basically gone. So the thermal spread of the discrete energy levels of the selected eigenstates that have not decayed due to interaction with the environment is still relatively narrow. However, it is relatively hard to estimate this quantity, as concepts such as the mean level separation (Fermi energy in the dot divided by the number of electrons) have no meaning when a portion of the spectrum has been decohered via the interaction with the environment. So, it is not clear at once whether the thermal effect leads to overlap of the pointer state energy levels, or leads to a hybridization of these states with the decohered states. However, we must remember that the states in the closed dot are all a set of orthonormal states; that is, there is a well-determined set of energy levels, which are not highly degenerate. When we open the dot, a large fraction of these states do hybridize with the environment and are no longer eigenstates of the dot. However, one can study this effect within the quantum transport simulation that has been mentioned several times in this chapter. When this is done, it is found that the pointer states are very stable and persist within the dot [32]. Moreover, these can move right through the decohered spectrum without any hybridization with these latter states [33], which just reinforces the fact that these are eigenstates. So, the thermal effects do not arise from hybridizing the pointer states, and must arise from the broadening of these states so that they interfere with one another. This implies that there is a fairly sizable number of these pointer states within the dot. The effective thermal spread at 2.0 K is a fraction of a meV, so one expects there to be sufficiently many pointer states that their individual level spacing is smaller than this.

Nevertheless, the dots and their pointer states undergo phase-breaking processes just as those discussed in chapter 5. But, as mentioned there, the dominant phase-breaking processes may arise from electron–electron interactions. But, these cannot be interactions where both carriers are within the dot, as these would exist even in the closed dot, where the eigenstates are stable, time-independent states. Just opening the dots to the outside world would not change this, but certainly introduces an interaction between the carriers in the pointer states and those lying outside the dot. The persistence of the pointer states means that this interaction is weak, and at most is a small perturbation. Nevertheless, the effective phase-breaking time can be determined precisely by the same techniques as discussed in chapter 5.

In figure 9.9, the dependence of the phase-breaking time for two dots, whose lithographic dimensions are 0.6 and 1.0 µm, is shown. The shape of the dot is that of the left panel of figure 9.1. The phase-breaking time was determined by measuring the magnetic field dependence of the correlation magnetic field for the fluctuations in the magnetoconductance, as described in section 5.5. As mentioned, the

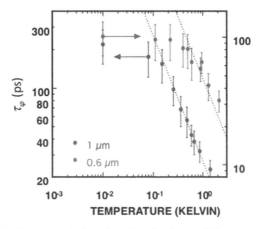

Figure 9.9. The variation in the measured phase-breaking time in two different quantum dots. The dotted line is a guide to the eye indicating a slope of $1/T$. (Reprinted with permission from [23]. Copyright 2005 IOP Publishing.)

measurements indicate a saturation in the phase-breaking time below a transition temperature, which is sample-size-dependent, and a decay at higher temperatures. This saturation at low temperatures is not due to a saturation of the sample temperature, as the amplitude of the conductance oscillations continue to increase as the temperature is lowered. This saturation also occurs in other dimensions as well, although the temperature decay portion has a different exponent for different dimensionalities. In the figure, the temperature decay is as T^{-1}, which is indicative of a two-dimensional system. From figure 9.1 it is clear that the dot is connected through two QPCs to regions which behave as Q2DEGs, and here the temperature decay appears to correspond to the dimensionality of those regions.

Let us now consider a quantum dot coupled to quantum wires. The gated dot is defined by electron-beam lithography and then etching away the extra material as shown in the third panel of figure 9.1. The dot has a lithographic dimension of 0.8 μm² with ~150 nm QPC leads positioned at either side. These leads open to quantum wires which gradually increase in width to 2 μm, over a distance of 10 μm, before reaching a 2DEG patterned into a Hall bar structure. The wires provide adiabatic coupling to the dot. The saturated phase-breaking time figure 9.10, at low temperature, is found to depend on the nature of the coupling between the dot and the wire, as this is varied by changing the bias applied to a top gate on the structure. From measurements at high magnetic field of the trapped AB orbits, it is determined that the electrical dot size depends upon this applied voltage from the top gate. At low temperatures, the phase-breaking time saturates at a value which depends upon the actual dot size. At higher temperatures, the phase-breaking time is found to exhibit an apparent $T^{-2/3}$ behavior [27], which is consistent with the coupling of the dot to a one-dimensional quantum wire [34], as shown in the figure. However, some caution is advised as only a few data points are present. These suggest that the behavior has this shape, but much more work is necessary to confirm this. One

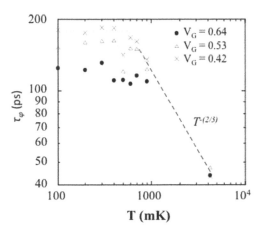

Figure 9.10. The phase-breaking time in a single etched dot, which is attached to quantum wires as shown in the third panel of figure 9.1. Here the coupling is varied as the gate voltage is made more negative, which increases the phase-breaking time in the saturation regime. (Reprinted with permission from [23]. Copyright 2005 IOP Publishing.)

expects to have the $T^{-2/3}$ behavior for phase-breaking processes in a one-dimensional system. Once again, we seem to see that the decoherence of the pointer states is occurring in the environmental system to which the quantum dot is coupled.

Surprisingly, the above behavior persists even when multiple dots are coupled together as an array of open quantum dots. Such a three dot array is shown in the inset to figure 9.11. We have carried out measurements of the phase-breaking time in three or four identical dots in series, separated by evenly spaced QPC. The lithographic dimensions of the dots were 1.4×1.0 μm^2 for the three dot array and 1.0×0.6 μm^2 for the four dot array [35]. The structures were designed using the split-gate technique and all gates were tied together to provide uniform formation of the dots. Measurements were performed in magnetic fields up to 4 T. In general, a very low sample current was used. The phase-breaking time as a function of the lattice temperature is shown in the main panel of figure 9.11 for the two dot arrays. Once again, the temperature decay seems to vary as T^{-1}, in keeping with the fact that the arrays are embedded in a Q2DEG. The pointer states apparently do not significantly decohere within one dot or the dot array, but this decoherence occurs within the environment as represented by this Q2D contact system. This is a very interesting result, and is in agreement with studies which have shown that the fluctuations themselves can be composed of orbits which are coherent across the dots of the array [36]. Indeed, there is some indication in this figure that the phase-breaking time is longer in the four dot array than in the three dot array. As these two

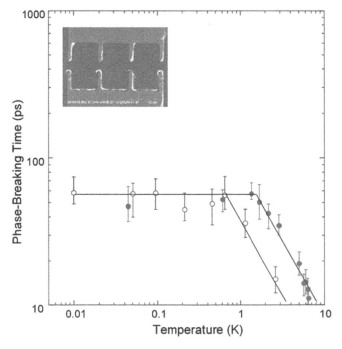

Figure 9.11. The phase-breaking time for a three dot (open circles) and four dot (closed circles) array. The inset is a micrograph of the three dot array. (Reprinted with permission from [23]. Copyright 2005 IOP Publishing.)

devices were made on the same sample, at the same time, this observation may be valid, although much more work in this area needs to be undertaken before such a conclusion can be drawn.

9.2 Einselection and the environment

Decoherence is thought to be an important part of the measurement process, especially in selecting the classical results, that is, in passing from the quantum states to the measured classical states of a system [15]. However, the description (and interpretation) of the decoherence process has varied widely, but key is the interaction of the system with the environment, as well as the interaction of the environment with the system. When the quantum dot is closed to the outside world, a complete set of eigenstates exist within the dot. All of these eigenstates are orthogonal to one another. Consider the dot shown in figure 9.4, in which the squared magnitude of one of the pointer states is illustrated. In this dot, the leads are located at the top of the two sides of the dot. What is clear is that this eigenstate has no weight near the QPCs. If we divide the entire set of states of the closed dot into two groups, we can label these groups by their weights near the QPC. Those which have significant wave function amplitude near the QPC have a chance to hybridize with the environment states when the QPC is open. The other set of states do not recognize whether or not the QPC is open, as they have no wave function amplitude near the QPC. It is this latter group of states which become the pointer states as they survive the process of the opening of the QPCs.

As the first group of states hybridizes with the environment states, they are no longer eigenstates of the quantum dot. Moreover, as these states provide continuity between the regions outside the dot (the environment) and regions within the dot, these become the current-carrying states. However, the portion of the wave function of these current-carrying states that lies within the dot remains orthogonal to the pointer states. Hence, there is no natural connection between the current-carrying states and the pointer states. Their mutual orthogonality remains. If the position and orientation of the QPCs is altered, the main effect is to modify the details of the electron collimation so that *a different set of pointer states will occur, giving rise to different scarring properties* [12]. As a result, any coupling between the current-carrying states and the pointer states must occur via phase-space tunneling. Phase-space, or dynamical, tunneling through regular phase-space structures, such as KAM islands [17], fundamentally determines the characteristics of the conductance fluctuations in these quantum dots. Theoretical analysis, discussed further below, based on the tunneling mechanism gives quantitative predictions (the average frequency of the fluctuations) in excellent agreement with experimental measurements [10].

In this section, we will examine the classical set of states that can exist, even in a closed dot. Then, we will turn to a mathematical approach to showing how the environment affects the dot when it is opened. This will lead us to examine some new states that arise in the open dot, but do not appear in the closed dot; these are known as hybrid states.

9.2.1 Classical orbits

A special place in the description of the classical motion is given by the specific features of the phase-space description of such motion. The classical phase space is coordinate space versus momentum space. Dynamical systems which follow regular closed orbits trace reproducible paths through this phase space, while chaotic orbits follow random paths through the phase space. While dynamical systems may be purely chaotic or regular, the most ubiquitous in nature are those whose phase space is mixed [31], in which regular and chaotic orbits exist in adjacent regions of phase space. The nature of each orbit may be categorized within the nonlinear transport regime through the KAM theory [17], which suggests that for small, smooth perturbations around a quasi-periodic motion, there will be an invariant torus in phase space. Such an investigation can lead to an understanding of the nature of the motion of the classical representation of the quantum wave function, as we will discuss in this section. But, first we want to examine the classical behavior of the closed dot [37].

When Bohr first provided a quantum picture of discrete energy levels and atomic radii, he gave life to the emerging quantum world. Shortly thereafter, Einstein turned the question around and asked which classical mechanical systems could be subject to such quantum behavior [38]. The result was that when the Hamiltonian was rewritten in terms of action-angle variables, then there existed conserved action integrals. These appear already as in equation (7.26), but he wrote them as

$$\oint_{C_i} p\,dx = 2\pi\hbar n_i, \tag{9.1}$$

where n_i is an integer, and i refers to the particular conserved action. In equation (9.1), p is the generalized momentum and x is the generalized coordinate. The paths C_i refer to *extremal* orbits on the invariant torus of the system, but they can be almost any closed path. In two dimensions, we should have two such integrals. Each of the periodic orbits can have as many integers in its description as there are degrees of freedom in the Hamiltonian. Further understanding of equation (9.1) has been provided by Brillouin [39] and Keller [40], so that this equation is often called the EBK quantization condition. In a more modern scenario, we know in general that the index on the RHS of equation (9.1) is modified, such as by the addition of a factor of ½ in WKB approximations. Hence, it has generally been found that equation (9.1) should be written as

$$\oint_{C_i} p\,dx = 2\pi\hbar\left(n_i + \frac{\beta_i}{4}\right), \tag{9.2}$$

where β_j is the Morse or Maslov index [41]. In the former version, it is related to the number of conjugate points in the trajectory, and this relates it to WKB theory, where this index is the number of turning points. What this means is that closed orbits lead to quantized states in the formal quantum theory. So, already we have a close connection between the classical closed orbits and the quantum eigenstates. We also know, from the discussion in chapter 7, that the EBK quantization rule, and equation (9.2), are closely related to the Berry phase that is encountered in classical periodic orbits [18].

The Maslov index is more related to the topology of the system [42], for integrable systems. The motion takes place on what is called a Lagrangian manifold in phase space, and the Maslov index relates to the topology of this manifold. For non-integrable systems, this can be related to the Morse index for a particular trajectory. The closed trajectories which give rise to equation (9.2) are closely related to the wave functions in the quantum world. Berry and Balazs [43] showed that the classical phase space is convenient for semiclassical quantization, and used the quantum Wigner function [44] for this purpose. Away from turning points, the Wigner function is found to localize on the classical trajectory. At turning points, however, the normal WKB approximation breaks down, and a more complicated approach is required. In these latter regions, the wave function develops complicated interferences, which these authors refer to as *whorls* and *tendrils*. At these regions, the regular motion is affected by non-integrable regions, and the wave function has trouble resolving the details of the phase space structure. Nevertheless, the closed orbits giving rise to equation (9.2) remain related to the regular bound spectrum of the quantum world [45]. The idea that the regular orbits would be imprinted on the quantum wave function was given by Heller as the idea of a *scar* [46]. Berry [18] has extended this to show that any periodic orbit will yield a scar (wave function) which is centered on the orbit, but will have fringes whose characteristics will depend upon whether or not the orbit is on the energy shell.

The above point is important, because it suggests that, as the classical regular spectrum is dominated by the closed orbits, the frequencies related to these orbits can be seen in experiment. For any trajectory, we can form the correlation function in time as

$$C(\tau) = \lim_{T \to \infty} \frac{1}{2\pi T} \int_0^T x(t)x(t + \tau)dt. \tag{9.3}$$

Here, T is the period for a periodic function and in this case the limit can be dropped. The Fourier transform of C has some very interesting characteristics. If the trajectories are regular, then the Fourier transform will consist of only a few frequencies, which are related to the time variation of the angle variables in action-angle coordinates. That is, these frequencies will provide the dominant variations of the trajectories in the phase space. It is clear from figure 9.4, that there are very regular parts of the function. Indeed, this figure may have only a single frequency. It is clear then that the quantum behavior is dominated by a few (one or two) closed orbits, and these will be related to the pointer states discussed above.

When we look at the transport through the open dot, the oscillations that are seen arise from the density of states within the dot. The latter can include regular orbits as well as a broader background arising from the density of chaotic orbits within and passing through the dot. Clearly, the oscillations, as discussed in the preceding sections, are not washed out with the open dot, and we can use this fact to begin to understand the source of the oscillations and fluctuations. The conductance is easily shown to be an integral over the density of states at the Fermi energy, and this density of states is defined by the orbits within the dot. When the dot is opened, it is the closed orbits which are confined in the dot, but which connect to the ports through tunneling to the open dot trajectories. It is important to note that these must be derived from the orbits of the closed dots and make a large, but finite, number of orbits around the trajectory before exiting; that is, these trajectories return to a starting point, as has been conjectured in much semiclassical transport theory [47].

The eigenvalue spectrum of the closed dot is a series of δ-functions located (in energy space) at each of the resonant eigen-energies of the dot cavity. Carrying this further, it is then possible to say that the density of states for the closed dot is

$$\rho(E) = \sum_n \delta(E - E_n). \tag{9.4}$$

where the E_n are the various energy levels for the available states. It is assumed here that the sum runs even over degenerate levels. The connection between the density of states and semiclassical trajectories is easily obtained through semiclassical quantum mechanics. The δ-function is replaced by its Fourier representation (in energy space) via the Poisson summation formula [48]

$$\sum_n f(n) = \sum_{M=-\infty}^{\infty} \int_0^{\infty} f(n)e^{i2\pi nM}dn + \frac{1}{2}f(0). \tag{9.5}$$

The various integrals can be evaluated via the saddle-point method, and the density of states can be expanded into the form [48], for a two-dimensional dot of side a

$$\rho(E) = \frac{m^*a^2}{2\pi\hbar^2} \sum_{M_1, M_2=-\infty}^{\infty} J_0\left(\frac{1}{\hbar}S_{M_1M_2}\right) - \frac{a}{4\pi\hbar} \sum_{M_1=-\infty}^{\infty} \cos(X), \tag{9.6}$$

where

$$X = \frac{2M_1a}{\hbar}\sqrt{2m^*E} \tag{9.7}$$

is the effective phase when only a single M_i is present. The general action appearing in the Bessel function of the first term is given as

$$\frac{1}{\hbar}S_{M_1M_2} = \frac{2a}{\hbar}\sqrt{2m^*E}\sqrt{M_1^2 + M_2^2}. \tag{9.8}$$

Here, M_1 and M_2 are the multiplicities of the primary periodic orbits defined through the quantization condition in equation (9.2). As mentioned following equation (9.7), the second term in equation (9.6) is a boundary correction for those orbits which do not sample all four walls of the dot.

The major point of these equations is that the density of states has oscillatory terms, which are needed to reproduce the delta functions in equation (9.4). As the system is more heavily damped, or the levels broadened, these oscillations are gradually smoothed and in the ultimate case, only the term for $M_1 = M_2 = 1$ survives from the first term of equation (9.6). This is the Thomas–Fermi density of states, given merely by the so-called mean level separation. That is, the mean level separation is the smoothed value appropriate to the Thomas–Fermi approximation of a smooth two-dimensional density of states. If we keep the boundary (second) term in equation (9.6), we obtain the extended Thomas–Fermi approximation

$$\rho(E) = \frac{m_* a^2}{2\pi\hbar^2} - \sqrt{\frac{m_* a^2}{2\pi^2\hbar^2 E}} .$$

(9.9)

We note that both the correction term in equation (9.9) and the argument of the oscillatory terms in equation (9.6) depend upon the *length* of the orbit, a point to which we will return later. The important issue here is that the Thomas–Fermi approximation discards the oscillatory contributions to the density of states, hence *discards the fluctuations that arise from the details of the quantization of the system.* That is, using the mean level separation, or equivalently the Thomas–Fermi approximation, will miss the important oscillatory terms arising from the trapped regular orbits of the pointer states.

The fact that the density of states has oscillatory terms is information that can be probed experimentally. Moreover, the regular and the chaotic states have different statistics [17]. In particular, the regular states, and their quantum pointer state counterparts, have classical Poissonian statistics, while the chaotic states are characterized by one of the Gaussian ensembles (the choice depends upon whether or not time reversal symmetry is broken by a magnetic field). This difference in statistics has been quantified in a quantum simulation of the quantum dot [49]. This further serves as a clear connection between the classical and the quantum behavior.

As mentioned, both equations (9.6) and (9.9) have terms which vary with the area of the orbit as well as with the length of the orbit. What is often missed in such trace formulas is that the action integral must be quantized in the dot, and hence only certain values of the energy (or momentum) are allowed. As a result, only some of the terms contribute at each energy. It is this quantization of the energy that will allow a particular term in equation (9.6) to recur as the energy is varied (or as a magnetic field is varied). It has been suggested that one can introduce the magnetic field variation by multiplying each term in equation (9.6) by a factor given by [50]

$$\cos\left(\frac{n\Phi}{2}\right), \quad \Phi = \frac{ea^2 B}{M_1 M_2 \hbar},$$

(9.10)

but this is not entirely correct. The problem with this formulation may be seen by the plots in figure 9.12. The two closed trajectories, labeled (a) and (b), both have $M_1 = M_2 = 1$, and therefore have the same length around the orbit. Hence, their action integral is the same in the absence of a magnetic field, and they are degenerate at the same energy level. However, these two orbits enclose different areas, so that their behavior in a magnetic field will be different. In essence, the magnetic field raises the degeneracy of the levels, and (for a circular dot) the different areas correspond to different angular momentum in the orbit. In the square, this degeneracy is still raised, but the results are not thought to be pure angular

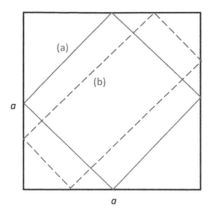

Figure 9.12. Two trajectories embedded in a dot of side a. The trajectories (a) and (b) both have the same length and action integral in the absence of a magnetic field. However, they have different areas and will respond differently in a magnetic field. Reprinted with permission from [37], copyright 2012 by IOP Publishing.

momentum states. Equation (9.10) defines the magnetic dependence only through M_1 and M_2, but clearly another quantum number is required to specify the area or something like the angular momentum.

The multiplicities M_1, M_2 lead to an interesting interpretation of the orbits which circulate around the dot multiple times before closing. Quite generally, the wave function of the closed dot may be expanded in a Fourier series as

$$\psi(x, y) = \sum_{m, n} A_{mn} \sin\left(\frac{n\pi x}{a}\right) \sin\left(\frac{m\pi y}{a}\right), \tag{9.11}$$

and the coefficients are determined by the particular energy eigenstate. The factors in the parentheses are in fact related to a set of reciprocal lattice vectors, in which the first Brillouin zone is defined by

$$-\frac{\pi}{a} < k_x, k_y \leqslant \frac{\pi}{a}. \tag{9.12}$$

Where there is a reciprocal lattice, there must be a real-space lattice, and this is just the multiplicity of the simple square unit cell of side a. But, this is just one cell of the periodic lattice that can be considered to exist. The multiplicities M_1, M_2 can then be said to correspond to orbits which traverse more than the single unit cell of this lattice.

It is easy to understand how this view can arise. Consider, for example, figure 9.13, in which we embed the dot in a periodic lattice. If we continue the orbit through the boundary wall, instead of reflecting it back into the dot, we see that the orbit which starts at point c ends at point c', which is a translation of the original point c by a set of *vectors corresponding to the sides of the dot, and to the lattice* [51]. In a sense then, all closed trajectories create a lattice in which the basic unit cell has the edge length of the original dot. A sum over all trajectories represents a sum over all possible lattice vectors in figure 9.13, and this results in a localized state in momentum space

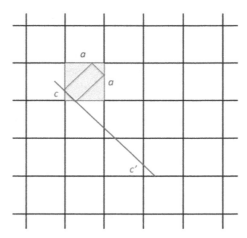

Figure 9.13. Periodic lattice corresponding to extension of the basic square unit cell. The red line is the extension, by reflection, of the enclosed trajectory in the single unit cell. Reprinted with permission from [37]. Copyright 2012 by IOP Publishing.

(due to the principle of *closure of a complete set*), which is the δ-function on the left-hand side of equation (9.4). If the longer trajectories are damped by, for example, phase breaking processes, then the δ-function is broadened. This is just the scattering broadened density of states, which further contributes to the oscillatory (fluctuating) density of states. On the other hand, the quantum point contacts can also affect which trajectories are selected.

The idea that the trajectory is extended by reflection takes on added meaning when the magnetic field is added. With a magnetic field, the trajectories take on curvature, and each section of the trajectory is an arc taken from the cyclotron orbit appropriate for that energy and magnetic field. Now, however, when that segment is reflected through the boundary of the unit cell, *the curvature is reversed*. It takes two reflections at subsequent boundaries before the curvature is returned to the original value. This means that the unit cell in the presence of the magnetic field is a supercell of dimension $2a \times 2a$. To account for the reversed curvature, this means that the magnetic field is reversed in the second and fourth quadrants of this supercell. As a result, we have breaking of time-reversal symmetry (which leads to magnetic ordering) without any net magnetic field in the supercell. Consequently, the electron states retain their usual Bloch state character. Such a situation was considered by Haldane for the 2D graphite (what we now know as graphene) structure [52], which is unique in that the hexagonal lattice supports points in the Brillouin zone where the conduction and valence bands meet with a zero gap [53]. The square lattice does not have such peculiarities. Since the net flux in the supercell vanishes, one can chose to use a periodic vector potential with some flexibility in choice of phase.

A two-dimensional system can be wrapped around a two-torus (a donut), and with a magnetic field present, trajectories will close only for so-called *rational* values of the magnetic field relative to the lattice [54]. The most famous study of these systems was done by Hofstadter [53], and the fractal-looking band structure has been called the 'Hofstadter butterfly.' It has been shown that even for open quantum dots in a real array, the energy structure maintains major elements of the Hofstadter butterfly in that the energy bands become split when the magnetic field takes rational values [55], and the current amplitudes (conductance) also obeys a type of Harper's equation [56]. This seems to hold even into the edge state regime [57]. The conclusion is that transport in the single quantum dot has similarities to that expected in a superlattice of such dots, even in the open dot regime, as we have attempted to show throughout this chapter.

9.2.2 Coupling the dot to the environment

Decoherence is thought to be an important part of the measurement process, especially in selecting the classical results; that is, in passing from the quantum states to the measured classical states of a system [58]. However, the description (and interpretation) of the decoherence process has varied widely, but key is the interaction of the system upon the environment, as well as the interaction of the environment upon the system. Zurek has proposed that the interaction of the system on the environment leads to a preferred, *discrete* set of quantum states, known as *pointer* states, which remain robust, as their superposition with other states, and among themselves, is reduced by the decoherence process [59]. This decoherence-induced selection of the preferred pointer states was termed *einselection* [58]. While this describes the physics of einselection, the mathematics can be shown by the use of projection operators. We give a brief overview here. We consider a system S, interacting with its environment E, so that the combined system plus environment ($S+E$) is either closed, or influenced by external driving fields that are assumed known and unaffected by the feedback from this combined $S+E$. The Hilbert spaces of both the environment and the system are assumed to be finite dimensional, although this is not critical. These two spaces form a tensor-product Hilbert space of the system plus environment. The operators in which we are interested are called *superoperators*, and exist in an expanded space often called the *Liouville* space. When we open the dots, there will be an interaction between the dot and the environment, so that we can write the total Hamiltonian as [60]

$$H = H_S + H_E + H_{\text{int}}, \tag{9.13}$$

where the three terms represent the system, the environment, and the interaction between these two, respectively.

There are two crucial steps in defining a reduced density matrix for just the pointer states. The first is to project out these states via a projection superoperator. The second is to then trace over the environmental states yielding just the reduced set of pointer states. This procedure has been known for a considerable time [61–63]. There are many ways to define, or create, the necessary projection operator, which, as mentioned above, is a commutator-generating superoperator [64]. We want to choose the particular projection operator such that

$$\hat{P}\hat{H}\hat{P} = \hat{H}_{\text{red}}, \tag{9.14}$$

where the last term is the reduced Hamiltonian for the pointer states, and the carets over the operators indicate that these are superoperators. Now, these projection operators are idempotent ($\hat{P}^2 = \hat{P}$), and consequently have eigenvalues of 0 or 1, and it is a central tenet of quantum mechanics that we can build the system through knowledge of the eigenfunctions and their eigen-values. For this subsystem, all the eigenvectors are stationary states in the Heisenberg representation [65]. We have used this in an earlier study of open quantum systems [66], and this approach has been used to create so-called decoherence-free subspaces in quantum information processing [67–69]. With this approach, we recognize that the pointer states really are a set of isolated states within the dot, and are not directly coupled.

We begin by writing the Liouville equation in terms of the composite density matrix ρ, which is defined on a tensor product Hilbert space of the system density matrix and the environment density matrix, as

$$\rho = \rho_E \otimes \rho_S, \tag{9.15}$$

for which the Liouville equation can be written

$$i\hbar \frac{\partial \rho}{\partial t} = \hat{H}\rho. \tag{9.16}$$

In particular, the Hamiltonian is a commutator-generating superoperator. Equation (9.16) is easier to understand when we see that the Hamiltonian is now a 4th rank tensor, which generates the commutator relation normally seen in this equation via

$$(\hat{H}\rho)_{kn} = \sum_{rs} \hat{H}_{kn,rs}\rho_{rs} = \sum_{rs}(H_{kr}\rho_{rn} - \rho_{ks}H_{sn}). \tag{9.17}$$

If the dimension of ρ_S is d_s and the dimension of ρ_E is d_e, then the dimension of the superoperator is $d_s^2 d_e^2$ [66]. To simplify the approach, we Laplace transform (9.16), and then trace over the environment variables to give [64]

$$(i\hbar s - \hat{H}_S)\rho_S = Tr_E\{\hat{H}_{\text{int}}\rho\} + \rho_S(0). \tag{9.18}$$

As discussed above, we now use a projection operator which yields the pointer states as eigenfunctions. This operator has the basic properties

$$\rho_{S,\text{red}} = \hat{P}\rho_S, \quad \hat{P}^2 = \hat{P}, \quad \hat{Q} = \hat{1} - \hat{P}. \tag{9.19}$$

To proceed, we need only to use an identity that is obtained by projecting the Liouville equation with both $\hat{P}$ and $\hat{Q}$, solving for $\hat{Q}\rho_s$ to formally decouple these two equations and recombining the terms [66, 70]. This identity is

$$\frac{1}{i\hbar s - \hat{H}} = (\hat{P} + \hat{Q}\hat{R}\hat{Q}\hat{H}\hat{P})\frac{1}{i\hbar s - \hat{C} - \hat{P}\hat{H}\hat{P}}(\hat{P} + \hat{P}\hat{H}\hat{Q}\hat{R}) \\ + \hat{P}\hat{H}\hat{Q}\hat{R}\hat{Q}\hat{H}\hat{P}, \tag{9.20}$$

where

$$\hat{R} = \frac{1}{i\hbar s - \hat{Q}\hat{H}\hat{Q}},$$

$$\hat{C} = \hat{P}\hat{H}\hat{Q}\hat{R}\hat{Q}\hat{H}\hat{P}.$$

(9.21)

The last term is a 'collision' type term which connects the environment to the device via the off-diagonal elements of the superoperator. If we now define some further reduced parameters as

$$\hat{\Sigma}\rho_{S,\text{red}} = Tr_E\{\hat{C}\rho_{S,\text{red}}\},$$

$$\hat{H}'_{\text{int}} = Tr_E\{\hat{H}_{\text{int}}\rho_{S,\text{red}}\},$$

(9.22)

we can then write the reduced equation as

$$i\hbar\rho_{S,\text{red}} - \rho_S(0) = \hat{H}_{S,\text{red}}\rho_{S,\text{red}} + \hat{\Sigma}\rho_{S,\text{red}} + \hat{H}'_{\text{int}}\rho_{S,\text{red}}$$

$$+ i\hbar Tr_E\{\hat{P}\hat{H}\hat{Q}\hat{R}\hat{Q}\rho_S(0)\}.$$

(9.23)

In general, we are seeking the steady-state, long-time limit, and would ignore the initial conditions. However, the last term in equation (9.23) has been suggested as contributing to the random force [71] that appears in e.g. the Langevin equation as well as to screening [72], so that it may not be proper to totally ignore it. However, the final long-time limit equation becomes, after inverting the Laplace transform,

$$ih\frac{\partial\rho_{S,\text{red}}}{\partial t} = (\hat{H}_{S,\text{red}} + \hat{H}'_{\text{int}})\rho_{S,\text{red}} + \hat{\Sigma}\rho_{S,\text{red}}.$$

(9.24)

The second term on the right hand side is not a scattering, or decoherence term, as that would appear in the last term on the right. Instead, it represents a weak interaction between the pointer states and the environment via *the decohered states*. That is, it represents primarily the environment interaction on the non-pointer states, which can then weakly couple to the pointer states. This term must represent the phase space tunneling by which the pointer states appear in experiment [10]. It is true, however, that this can bring decoherence into the world of the pointer states via electron-electron interactions in the environmental states, a point we have reviewed previously [23]. Normally, the pointer states do not interact with the environment, so that the last term would vanish, but the interaction of the pointer states, through the decohered states, to the environment produces the phase breaking discussed in the previous section. This means that the last term *does not vanish*, but represents this phase breaking process through an imaginary term in this self-description. Since this scattering term is part of the Hamiltonian of the pointer states (the reduced set of states), it is a diagonal term, but has an imaginary part, which makes the Hamiltonian non-Hermitian.

It is important to remark here that this form of Hamiltonian is not what one normally refers to as non-Hermitian. Here, the imaginary term which breaks up the

Hermitian properties lies on the diagonal of the matrix. In essence, this will also make the resulting reduced density matrix non-norm conserving as it breaks up the unity of the trace. But, this is because the germane interactions are from states within the dot and states of the environment, which can include renormalized decohered states of the dot. This environment sensitive dephasing interaction is seen in experiment [23], as discussed above. The steady-state situation has to then account for lost amplitude in the dot being replaced by electrons injected from the environment itself. These source terms must be incorporated within the first term on the right-hand side of equation (9.24) as in any other transport problem.

Now, there are further problems with equation (9.24). While it appears to be quite simple conceptually, this is not really the case. First, within the partial trace in both terms of equation (9.22), there is an explicit dependence on the choice of the projection operator $\hat{P}$ or, equivalently, on the environment density matrix that induces the projection operator, so one must make a choice of the latter to actually be able to use equation (9.24). At the end of the day, the equation of motion for $\rho_S(t)$ should not depend on it. And yet it does through its affects on the projection operator. The rationale for believing that the equation of motion for $\rho_S(t)$ should not depend on the details of the environment goes back to our statement above that the eigen-states of $\rho_S(t)$ are our pointer states, and these are properties of the specific dot, and exist in the dot for a variety of different environments. We have previously shown [66] that quite generally, the eigen-space of the projection operator, corresponding to those states with eigenvalue 1, must be isomorphic to the states of $\rho_S(t)$, which are our pointer states.

Now, in addition, to giving rise to the phase-space tunneling interaction between the environment states and the pointer states, the interaction term can also give rise to hybrid states that would not exist either in the environment or in the closed system. We will see examples of these in the following sections.

9.2.3 Relating classical and quantum orbits

When we study motion in multiple dimensions, the phase space also has more dimensions. For example, if we study particle flow in two spatial dimensions, then we have four dimensions in the phase space. How are we to plot these? First, we do not simply plot the entire phase space, and the Poincaré map becomes essential to understanding the dynamics of this two-dimensional motion. Typically, a plane is chosen to represent the plane of the Poincaré section, such as the $y = 0$ axis. The next choice is where to choose the critical point. In the following, we will see two-dimensional Poincaré sections and three-dimensional ones. For the two-dimensional plots, we chose the time at which the trajectory passes through the $y = 0$ plane with a positive velocity in the y-direction. At times, we will also plot the value of v_y and this leads to a three-dimensional Poincaré section. These choices are, of course, arbitrary, but have become values that are widely used. With this bit of introductory material, let us now turn to the classical study of the motion within the quantum dot potential.

To study the connection, we first describe a method of carrying out the classical simulation of orbits. With Schottky barrier confinement walls, such as shown in the left panel of figure 9.1, the actual potential within the dot is a relatively soft wall potential (much like a harmonic oscillator potential). In numerically modeling the classical motion, a good fit to the experimental magnetoresistance is only obtained by using such a soft wall potential. Within the dot, the potential is modeled as a two-dimensional harmonic oscillator via the potential

$$V_{dot}(x, y) = \frac{m^*}{2}\left[\omega_x^2(x - x_{0,i})^2 + \omega_y^2 y^2\right]. \tag{9.25}$$

The dot shape is somewhat elliptical, as shown by the different frequencies in the two directions. Here, the x-axis is taken along the major axis of, for example, a seven dot array, in which the center of each dot is indicated by $x_{0,i}$. This array is used to probe the locality and coupling of various states between the dots and with the environment. Between the dots, as well as between the end dots and the 2DEG environment, there is a quantum QPC, which is modeled by the saddle potential

$$V_{QPC}(x, y) = \frac{m^*}{2}\left[-\omega_{x,\,QPC}^2(x - x_{Q,j})^2 + \omega_y^2 y^2\right] + V_{0,QPC}, \tag{9.26}$$

where $V_{0,QPC}$ is the saddle potential height at the center of the QPC and $x_{Q,j}$ is the position of the saddle potential peak in QPC j. The various frequencies in the two potential forms are adjusted so that these potentials match the fully self-consistent potentials obtained in the quantum calculations [20]. An example of such a potential is shown in figure 9.14. Each dot has a length of $2l_d$ and each constriction has a length of $2l_c$. The classical simulations consider a ballistic motion of an electron in this potential and a normal (to the plane of the motion) magnetic field B_z, which produces the cyclotron frequency $\omega_c = eB_z/m^*$. These potentials produce the equations of motion

$$\begin{aligned}
\frac{d^2x}{dt^2} &= -m^*\omega_x^2(x - x_{0,i}) + \omega_C\frac{dy}{dt} \\
\frac{d^2y}{dt^2} &= -\omega_y^2 y - \omega_C\frac{dx}{dt}
\end{aligned} \tag{9.27}$$

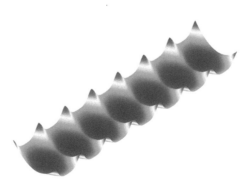

Figure 9.14. An example of a seven dot confining potential. Coupling between the dots and with the environment is via the saddle potentials along the major axis.

within a dot, and

$$\frac{d^2x}{dt^2} = m^*\omega_{x,\,QPC}^2(x - x_{Qj}) + \omega_C\frac{dy}{dt}$$

$$\frac{d^2y}{dt^2} = -\omega_y^2 y - \omega_C\frac{dx}{dt}$$

(9.28)

in the constriction. Just as in a two-dimensional harmonic oscillator, the coupling of the harmonic motion with the cyclotron motion leads to a pair of hybrid frequencies [73]

$$\omega_{\pm}^2 = \beta \pm \sqrt{\beta^2 - \omega_x^2\omega_y^2}$$

$$\beta = \frac{1}{2}\left(\omega_x^2 + \omega_y^2 + \omega_C^2\right)$$

(9.29)

For a constriction region, the motion is somewhat more complicated, with a modification of these hybrid frequencies as

$$\omega_+^2 = -\beta' + \sqrt{\beta'^2 + \omega_{x,\,QPC}^2\omega_y^2},$$

$$\omega_-^2 = -\beta' - \sqrt{\beta'^2 + \omega_{x,\,QPC}^2\omega_y^2},$$

$$\beta' = \frac{1}{2}\left(\omega_{x,\,QPC}^2 - \omega_y^2 - \omega_C^2\right).$$

(9.30)

The motion is followed by plotting each passage of the $y = 0$ plane with a positive velocity in the y-direction. For each passage of this plane, the x-position and velocity are plotted. This yields a Poincaré plot that can be used to study the characteristics of the classical dynamics. As discussed above, a perfectly regular orbit will always pass through precisely the same spot in this plot. Hence, a regular orbit creates a single point in the plot. At the other extreme, a chaotic orbit will never pass through the same point, and so creates a sea of points, which is often referred to as the 'sea of chaos'. Other orbits can pass through the plane many times before coming back to the same point, and these can create closed lines in the Poincaré plot. These latter are quasi-periodic orbits, which together with the single points usually are surrounded by regions where no trajectory passage can be found. These are the KAM islands. In figure 9.15, we show the Poincaré plot for one of the inner dots of the seven dot array [37]. This is calculated for an off-resonance condition where $\omega_+/\omega_- = 2.156$ ($B_z \sim 0.28$ T). Here there are two larger KAM islands at the left and right side of the image, along the $v_x = 0$ axis. These two KAM islands are embedded in a dense sea of chaos, but each is surrounded by a 'sticky' layer [74], that emerges from trajectories that oscillate between a given dot and one of its QPCs for a long time before finally entering the sea of chaos. All the trajectories that begin in the sea of chaos are unstable and escape the dot array, contributing to the current-carrying decohered states (we illustrate these later).

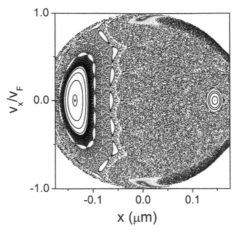

Figure 9.15. The Poincaré section for one of the interior dots in the array. This is for an off-resonance condition, and two major KAM islands can be seen near the left and right edges of the image. (Reprinted with permission from [37]. Copyright 2012 IOP Publishing.)

We can illustrate the nature of the states on the KAM islands a little better by using the full three-dimensional Poincaré plot where we include also the y-velocity, instead of just taking the point where it passes through the $y = 0$ plane. This is shown in figure 9.16(a), for the first two dots of the array. Three KAM islands now lie on the top of the three-dimensional surface [37], as well as near the lower front of the sphere. In the classical phase space, different KAM islands may be degenerate in energy, as is the case here. Quantum mechanics will generally split this degeneracy by the presence of eigenstates with significant amplitude (but different phases) on both KAM islands. This does not occur classically, because there is an energy barrier between the two islands which cannot be surpassed by the classical motion at constant energy. In the present case, the pointer states typically correspond to closed, periodic orbits that are normally classically inaccessible. But, in the open dot, phase-space tunneling [10] can occur between these KAM islands, and this leads to the fluctuations that have been discussed above. That is, there is a very good probability that an incoming electron will pass through the regular regions in phase space via a tunneling process. In figure 9.16(a), six distinct KAM islands are shown. Three of these have been labeled as 1–3. The individual orbits for the states lying at the center of the KAM islands are illustrated in figure 9.16(b). The orbits found in the center of islands 2 and 3 (in the first and second dots, respectively) are essentially the same, although they have reversed velocities, so that they can be expected to be coupled by a quantum wave function. Thus, we expect phase-space tunneling to be possible between these KAM islands, as shown by the arrow in figure 9.16(a).

It is possible to connect the classical motion to that obtained from a full quantum simulation of the same seven dot array. In figure 9.17, we do just that. In panel (a), we plot the magnitude of the quantum mechanical wave function for the condition of point 2 in figure 9.16(b). Here, the wave is colored according to the x-directed velocity, red for positive and blue for negative. The arrows indicate the direction of

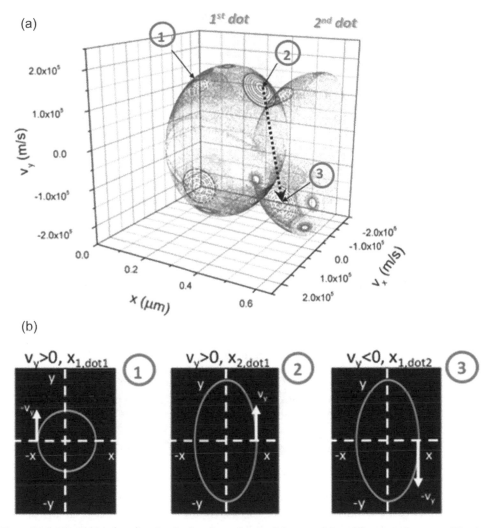

Figure 9.16. The Poincaré surface for the first dot and part of the second dot of the seven dot array. The three KAM islands are identified by the possibility of phase-space tunneling between them. (b) The Poincaré plots for each of the three islands showing the different shapes and velocity vectors (arrows) in the y-direction. (Reprinted with permission from [37]. Copyright 2012 IOP Publishing.)

the total velocity for this eigenstate. In panel (b), we replot the classical motion from figure 9.16(b) for orbit 2. It is clear that there is a strong one-to-one correlation between the classical and the quantum motion at this point. In order to better visualize the motion, we plot the so-called quiver plot for the quantum wave function. The individual arrows give the direction of the velocity and their size gives the relative magnitude of this velocity. The fuzziness that appears around the $y = 0$ axis in panel (a) is clearly associated with the slow velocity as the trajectories cross this axis, as indicated by the shorter arrows and the broader area over which they are spread in panel (c).

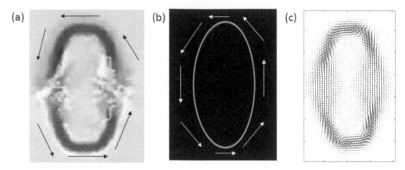

Figure 9.17. The magnitude of the quantum wave function (a) and the classical orbit (b) taken from the orbit labeled 2 in figure 9.15(b). Arrows have been added to these to show the velocity direction. (c) The quiver plot in which the arrows indicate the velocity direction and magnitude for the quantum wave function.

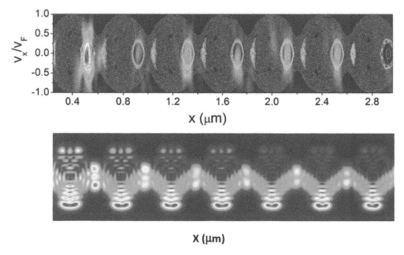

Figure 9.18. Top: the Poincaré plot for the entire seven dot array. As earlier there is a large sea of chaos, with two KAM islands located in each dot. Bottom: the magnitude squared of the wave function for the frequency ratio of 3, showing a skipping orbit that arises from the decohered, current-carrying states. (Reprinted with permission from [13]. Copyright 2007 the American Physical Society.)

As mentioned above, the decohered states are coupled to the environment and serve as current-carrying states for the background current that flows through the dot array. These states arise as one varies the ratio of the two hybrid frequencies and often show up when this ratio is an integer. We can see this for a couple of ratios in figure 9.18. In panel (a), we plot the Poincaré plot for the entire seven dot array for a value of $\omega_+/\omega_- = 1.6269$ ($B_z \sim 0.16$ T). In each dot, there is a large chaotic sea and two different types of KAM islands, as discussed above. However, in panel (b), we show the case for $\omega_+/\omega_- = 3.0$ ($B_z \sim 0.43$ T), where a large skipping orbit passes entirely through the array. In this latter case, the probability obtained from the quantum wave function is plotted rather than the Poincaré plot. The classical trajectory for this type of current-carrying state does not contribute to the sea of chaos; in fact, at this value of magnetic field, the sea of chaos has largely been drained (e.g., reduced to a very small part of the phase space). As a result, the phase-space probability is largely concentrated in the QPCs which connect the dots [13].

For a comparison of the classical phase-space dynamics and the quantum mechanical analog, the Husumi function is a useful concept [75]. The wave function obtained from the quantum simulation is smoothed and transformed via a coherent state, which is a Gaussian wave packet with a finite momentum, such as

$$\xi(x, y) = \left(\frac{1}{\pi\sigma^2}\right)^{1/4} \exp\left(-\frac{x^2 + y^2}{2\sigma^2} + i\mathbf{k} \cdot \mathbf{r}\right). \qquad (9.31)$$

Here, the wave packet is centered at a convenient point, taken to be the center of each dot, and $\mathbf{k}$ and $\mathbf{r}$ are both two-dimensional vectors, and σ is the half-width of the wave packet. The imaginary term provides a uniform velocity for the packet as $\mathbf{v} = \hbar\mathbf{k}/m^*$, the customary form in semiconductors. This coherent wave packet is then used to transform the quantum wave function $\psi(x,y)$ into the Husumi phase-space function via the integral transform

$$H(\mathbf{r}, \mathbf{k}) = |\int d\mathbf{r}'\psi(\mathbf{r}')\xi(\mathbf{r} - \mathbf{r}', \mathbf{k})|^2. \qquad (9.32)$$

The projection of this onto the Poincaré plot is obtained by taking $y = 0$ (to correspond to the equivalent classical plot) and then integrating over k_y. In the top panel of figure 9.18 above, the Husumi function is overlaid on the classical phase space, and appears as the color parcels over the normal blue–white Poincaré plot. The wave function for this case is precisely the one obtained for the same set of hybrid frequencies, potential, and magnetic field as the classical case. Hence, it may

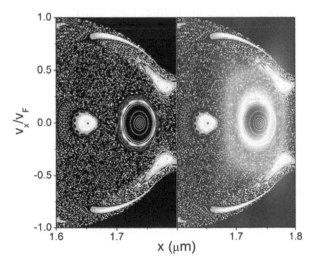

Figure 9.19. A blowup of the first dot from figure 9.17, with the Husumi function overlaid for comparison. The left side shows the normal Poincaré plot, while the right side shows the plot with the Husumi function overlaid on it.

be seen that the maximum of the Husumi function (bright red) lies precisely in the KAM island where the phase-space point of the classical orbit resides. We show a blowup of this region in figure 9.19. In the left-hand side of the figure is the blowup of the classical Poincaré plot. In the right-hand side, the black–white image is converted to blue–white and the colored Husumi plot is overlaid, so that one can clearly see the convergence of the classical and quantum pictures. It is clear that the quantum–classical correspondence is very strong and this strengthens the argument that the pointer states are the connection to the classical orbits in these open quantum dots. One further connection arises in the next section.

9.2.4 Pointer state statistics

In the above discussion, we have dealt with an important issue in quantum measurement theory, namely how the quantum states evolve into classical states, in this case through the connection between the quantum pointer states and the classical trajectories in the same potential system. It is important to recall that it has been suggested that these pointer states are indeed the basis of the transition to classical behavior, and even possess classical properties [15]. The relationship between the classical dynamics of a system and the spectral statistics of its quantum analog has been a primary concern in the study of quantum chaos [17, 41]. Systems with integrable dynamics are expected to have uncorrelated energy levels that yield Poisson statistics [41], while completely chaotic dynamics is associated with the Wigner statistics [76] of one of the random matrix ensembles, the Gaussian orthogonal ensemble (GOE) when time-reversal symmetry is preserved and the Gaussian unitary ensemble when it is broken (for example, by a magnetic field). However, we have shown that when a stadium dot, which is fully chaotic in the closed state, is opened to the environment and only the pointer states with amplitude localized to in the interior remain, the pointer state distribution becomes Poissonian, indicating that the pointer states are intimately associated with the regular orbits [49]. Here, we will show that the same is true for the nearly square quantum dots discussed in this chapter.

For the analysis of the spectral statistics of a quantum dot system, a set of states is chosen which lies in a given energy that covers that expected to be probed in the experimental situation. Since typically only a limited number of eigenstates occur over this range, one also varies the perpendicular magnetic field to generate many different ensembles of eigenstates, thus greatly increasing the number of energy levels available for the analysis. It is important that the number of QPC modes does not vary over the range of magnetic field. Moreover, the values used for B should be sufficiently small that no substantial change in the nature of the electron dynamics is expected [33]. Once the set of relevant energy levels is determined, the raw energy levels are mapped onto a dimensionless, unfolded set of eigenvalues which have a local density of unity [77]. This is achieved through the conversion

$$e_i(B) = \frac{E_i(B)}{\langle \Delta E \rangle}, \tag{9.33}$$

where $\langle \Delta E \rangle$ is the average level spacing for the set of energy levels obtained at a given magnetic field. This now yields a dimensionless sequence of nearest-neighbor energy level spacings

$$s_i = e_{i+1}(B) - e_i(B). \tag{9.34}$$

This is computed for each magnetic field. The statistics of the number of dimensionless energy levels that have a given spacing s_i can be determined and the resulting plot yields the nature of these statistics. In figure 9.20, we plot the nearest-neighbor distribution function $P(s)$ for the pointer states [78]. It may be seen that $P(s)$ for pointer states very clearly follows the Poisson distribution (the GOE is also shown for comparison). If the total set of energy levels is considered, then the GOE distribution is recovered, as there are many more decohered eigenstates than pointer states.

It is clear from figure 9.20 that pointer states in an open quantum dot yield Poisson spectral statistics associated with classically regular behavior. On the other hand, if we use all the eigenstates of the closed dot, the GOE distribution associated with quantum chaos is recovered. While there have been previous theoretical spectral studies that did not observe such a change in behavior, a key factor that enabled us to see a dramatic shift to a regular distribution is that we studied very open quantum dots strongly coupled to the external environment. This means that the width of the QPCs is such that the latter pass several modes, and thus have a very

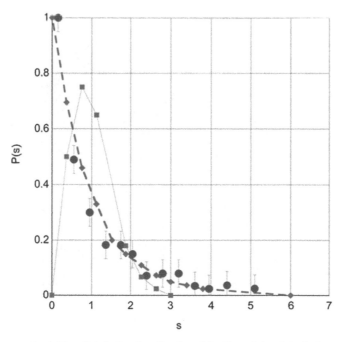

Figure 9.20. The nearest-neighbor distribution function found for the pointer states is shown by the blue dots. The red dashed line is a Poissonian distribution, while 'the green curve represents the GOE distribution. (Reprinted with permission from [78]. Copyright 2015 the Swedish Academy of Sciences and IOP Publishing.)

important effect on the nature of its confinement. In this limit, the pointer states that survive are localized in the interior of the dot, and are typically scarred by relatively simple periodic orbits.

9.2.5 Hybrid states

In the previous sections, we have talked about states which evolve from the eigenstates of the closed dot. However, opening the dot to the environment, in this case the Q2DEG that exists in the remainder of the heterostructure, can lead to other states, where a particle enters the dot through one QPC and bounces around to exit at the same QPC, or, in multiple dots, just sit in the connecting QPCs. This requires a magnetic field in order to make the trajectory curve sufficiently well to create this unique back-scattered trajectory. Of course, there will be a quantum equivalent of this trajectory. It is important to note that this is very different from the weak localization effect, introduced in section 3.7. In weak localization, one looks at the conductance or resistance at zero magnetic field and the effect arises from the interference of two time-reversed paths, which are scattered by multiple impurities until they return to the source. In this case, the magnetic field breaks the time-reversed symmetry so that the two paths do not trace each other any longer, and the effect goes away. Here, the magnetic field is critical to the enhanced resistance that arises from the back-scattering, and there are no time-reversed paths due to the presence of the magnetic field. This effect was apparently first discovered by Ochiai *et al* [79], in an array of quantum dots. Of course, these trajectories can bounce around and leave through the other end of the dot or array of dots, as

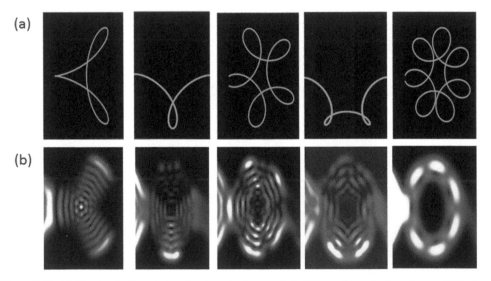

Figure 9.21. (a) The classically calculated trajectories for certain initial conditions in the entrance QPC which lead to either back-scattered or transmitted trajectories. The conditions of frequencies are $\omega_+/\omega_- = 2, 3, 4, 5, 6$ for the plots from left to right, respectively. Other parameters are given in the text. (b) The quantum mechanical density probabilities for the same conditions. (Reprinted with permission from [13]. Copyright 2007 the American Physical Society.)

already shown in figure 9.18(b). The choice of whether the entering trajectory leaves at the entrance of the QPC or at the far end of the array depends upon the potentials and the exact magnetic field. We illustrate these trajectories for various conditions in figure 9.21. According to the details of the self-consistent potential, the parameters used in equations (9.25) and (9.26) are $\omega_x = 1.06 \times 10^{12}\,\mathrm{s}^{-1}$, $\omega_y = 0.85 \times 10^{12}\,\mathrm{s}^{-1}$, and $\omega_{x,QPC} = 2.16 \times 10^{12}\,\mathrm{s}^{-1}$. The dots have an extension in the x-direction of 0.32 μm in the dot and 0.076 μm in the QPC, which corresponds to the Fermi energy. This is less than the lithographic dimension of 0.4 μm for each dot. The magnetic field for the first, third, and fifth images are approximately 0.2, 0.5, and 0.7 T. For the trajectories in the figure, these three images show back-scattered trajectories, which make successively larger numbers of bounces before returning to the entrance QPC. On the other hand, the second and fourth images illustrate transmitted trajectories which exit at the last QPC, and correspond, for example, to the plot shown in figure 9.18(b). The top row of images are the classical trajectories, while the lower row of images are the quantum mechanical density probability plots for the same conditions.

In a dot array, the possibility for back-scattering trajectories can be more complicated as the reflection may come from a dot other than the one closest to

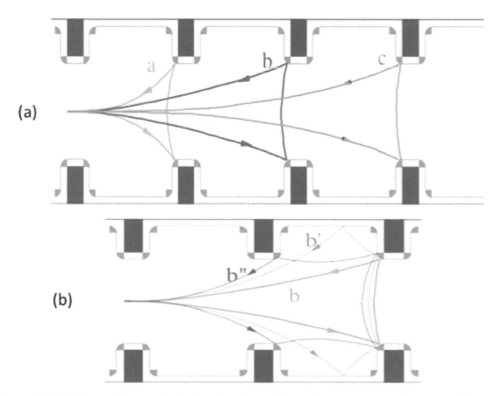

Figure 9.22. (a) Three different back reflecting trajectories that can arise in a dot array. Of course, each occurs at a different magnetic field. (b) Expansion of a set of possible trajectories for the 'b' described in panel (a). (Adapted with permission from [79]. Copyright 1997 the American Physical Society.)

the entrance QPC. We illustrate how multiple back reflections can occur in a dot array in figure 9.22. In panel (a) of this figure, we illustrate back reflections from the sequence of dots that can occur. Of course, each back-reflected orbit occurs for a slightly different magnetic field, as we will show below. In panel (b) of the figure, we show back-reflected orbits from the second dot, which are basically parallel with those shown in figure 9.22(a). As may be expected, the details of these various orbits will be sample-dependent and depend upon the details of both the potentials and the magnetic field.

By carefully determining the size of the dot array and determining the depletion inside the metal edges from exact self-consistent potentials, one can model the various trajectories and compute the conductance and resistance of the dot array quantum mechanically [20]. This allows a good comparison with the available experimental data. In figure 9.23, we illustrate this comparison using the data from Ochiai *et al* [79]. The theory curve shows more fine structure, presumably due to the fact that this curve has not been thermal averaged.

In these back-scattered orbits, one might wonder why they have to occur from the entrance QPC. In fact, if you look at the trajectory in the first panel of figure 9.22(a), and consider how this might look in an interior QPC, you can reach a striking conclusion. If one rotates the array around the center point of an interior QPC, you realize that rotating the trajectory of this latter figure would lead to a trapped trajectory centered upon the interior QPC. This is not a state that arises from one of the closed dot states, but is a state that can appear only in the extended dot array. Nevertheless, it is an eigenstate of the dot array which does not couple to the environment. The actual trajectory that is found to be stable in these dot arrays is

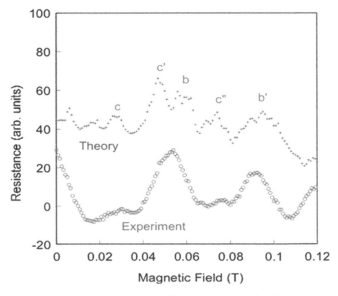

Figure 9.23. Comparison of a quantum simulation of the back-scattered trajectories with the experimental data from [79]. The labels correspond to various trajectories shown in figure 9.21. (Adapted with permission from [79]. Copyright 1997 the American Physical Society.)

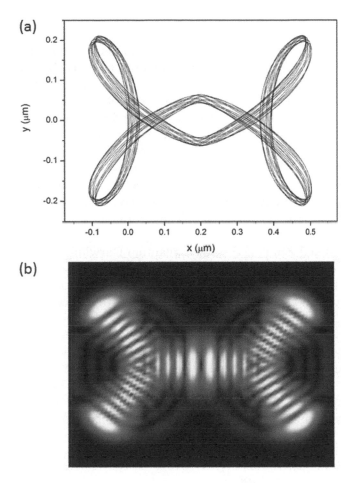

Figure 9.24. (a) Classical trajectories for the bipartite state that exists at a QPC interior to the dot array. Multiple trajectories correspond to the states in the basin of the attractor. (b) The quantum wave function corresponding to this bipartite state.

slightly more complex, but not much. The classical and quantum versions of these internal reflecting states are shown in figure 9.24. These new states are distributed over two interior dots and have been termed 'bipartite states' [14]. These states are found to be quite robust, and exist over a range of variation in the QPC parameters, and are also found to correspond to a classically stable trajectory. However, instead of corresponding to a KAM island state, it is found that the bipartite states correspond to an island of stable attractors. That is, every attractor has its own basin, where all the trajectories within this region of phase space become attracted. In some sense, the basin of attraction corresponds to the size of the bipartite state in real space. This suggests that the bipartite states also arise from the same einselection process that is important for the regular pointer states, but are hybrid states that can only arise in a dot array.

9.2.6 Quantum Darwinism

The promotion of certain information in a quantum system due to a natural selection process is known as quantum Darwinism [16, 21]. It is inspired by the Darwinian concept [80] for the rules of reproduction, heredity, and variation. The pointer states are characterized not only by their robustness, despite the existing environment, but also by their ability to create 'offspring' of the states, which means that they advertise information about themselves. This ability makes it possible for different observers to measure the same information. The resulting objectivity arises from the classical states. That is, in order to measure a quantum system objectively, one has to design a system where the transition between the classical and quantum world is observable. The open quantum dots we have been studying in this chapter are ideal for this purpose. In particular, the hybrid states are the easiest to study in this regard.

Blume-Kohout and Zurek [21] studied quantum Darwinism in zero-temperature quantum Brownian motion, where they partitioned the environment as a whole into smaller subspaces and then observed the *imprint* of the pointer states on these individual subspaces. This observation demonstrates the objective existence of the pointer states. Similarly, we can study the transmission amplitudes of individual propagating modes through the QOCs. These modes can be seen as being analogous to the individual subspaces in [21]. Our calculations basically show that all of the transmitting modes become resonant at the same time. That is, by looking at the *imprint* of the bipartite pointer state, shown in figure 9.24, in the individual propagating modes, we gain confirmation that *offspring* of the robust state exist (other modes yield similar results). This is illustrated in figure 9.25 for a two dot

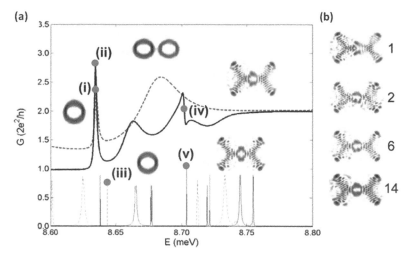

Figure 9.25. (a) Conductance for a single-dot (red, thick dashed line) and a two-dot array (black, thick line). Insets: probability density of the quantum wave function (at the solid blue circles) for regular pointer states (i)–(iii), and for the bipartite states (iv)–(v). (b) probability density of the quantum wave function for the 1st, 2nd, 6th, and 14th modes at (iv). A magnetic field of 0.2 T is present in all cases. (Adapted from [14]. Copyright 2008 by the American Physical Society.)

array, in which the bipartite state of figure 9.24 can be studied in the region between the two dots. For every mode, we find a display of the bipartite state. Therefore, the information is objective [16], and every 'observer' is able to see the same result. Consequently, quantum Darwinism is in action in the open dot array. Now, by observer, we point out that in transport, most of the conductance occurs at the Fermi energy. In these open dots, the Fermi energy is set by the quasi-two-dimensional electron gas that defines the environment. By varying the opening of the QPC, we are adjusting where the Fermi energy sits relative to the harmonic oscillator states of the transverse potential of the QPC. Thus, as we sweep from a negative to a more positive QPC potential, we are moving the Fermi energy from the lowest harmonic oscillator state to higher lying states. These states are the modes that we probe with the conductance. Thus, the bipartite state is present in each and every transverse mode of the QPC. We regard the first mode as the 'mother' state, and the replicas seen in each higher order mode the daughter, or replica, of that mother mode. Each of these modes represents a different 'subspace' of the system, which is examined as the Fermi energy is swept through the modes. It is thus felt that this system represents a valid theoretical and experimental test bed for the ideas of quantum Darwinism.

9.3 Imaging the pointer state scar

In chapter 2, we first discussed the use of SGM, whereby a rastered AFM tip is modified by metallizing the tip so that a voltage can be applied to the tip [81, 82]. This voltage is used to perturb the local surface potential and carrier density, while monitoring the conductance of the sample as a function of tip position. This generates a map of conductance change, which should be proportional to the local density near the probe tip. The use of this approach with the quantum dot is illustrated in figure 9.26, where we simulate the conductance change as the biased tip is rastered over the dot. In this case, the negative bias tends to break up the scar and push carriers into the current-carrying states, thus raising the conductance. This development has allowed one to probe the interior of quantum structures with the intent of mapping the local density around the high regions of probability density given by the magnitude of the wave function. This is based upon the idea that this property of the wave function should correlate with peaks in the local density. In particular, this has allowed the investigation of quantum behaviors within quantum dots in heterostructures such as are used here [83, 84]. These early attempts did not yield clear images, perhaps due to the larger effective mass that the electrons have in GaAs, or the large dot size used. We subsequently achieved a better image of what was believed to be a real scar of one of the pointer states in a quantum dot in InAs [85]. InAs has a larger g value than GaAs and will therefore have a more pronounced response to the magnetic field. To achieve this result, however, the fact that the scar recurs periodically in the magnetic field was used to make the measurement. We recall that the pointer states survive opening the dot to the environment primarily because they do not couple to the environment. With the SGM, we are bringing the environment into the dot itself. There is certainly

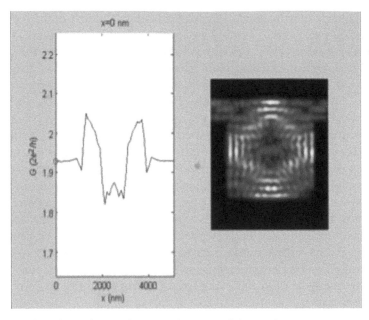

Figure 9.26. Simulation of scanning conductance microscopy of the scar in an open quantum dot. Here the biased tip breaks up the scar, pushing carriers into the current-carrying states and raising the conductance. Animation available at https://iopscience.iop.org/book/978-0-7503-3139-5.

the danger that this will destroy the pointer states during the scan, but this did not seem to be the case. Another problem is that the classical trajectory correlating with the pointer state will deform in the presence of the tip voltage. The voltage creates a local minimum in the potential which deflects the trajectories. Hence, the scar was imaged by making a large series of SGM scans with a small change in magnetic field for each scan. This image was then Fourier transformed in magnetic field, and these Fourier transforms searched for periodic parts. Thus, the transform could be filtered around discrete frequencies, and the filtered transforms then used to generate a real space image corresponding to that magnetic frequency. It was these images which showed the presence of the scars appropriate to the dots.

What we would like to have seen, however, was the scar represented by figure 9.4. This is clearly one of the dominant pointer states. But, the InAs dot used above had the QPCs centered along the central line of the dot, and this would couple the scar of figure 9.4 to the environment. This should destroy that particular pointer state, although the resultant image obtained had significant similarity to that of figure 9.4. Nevertheless, a search was continued to find the state for the dot with the QPCs along the top, as indicated in this previous figure. For this approach, a 12 nm InAs quantum well was imbedded with an $Al_{0.7}Ga_{0.3}Sb$ strain relaxed bottom cladding layer and a thin $In_{0.2}Al_{0.8}Sb$ top cladding layer [86]. The sample was then etched to produce a ribbon of InAs with etched trenches to isolate the dot region within the ribbon. From Shubnikov–de Haas measurements it was determined that the density and mobility were 1.5×10^{12} cm^{-2} and 65 000 cm^2 V^{-1}s^{-1} at 280 mK. The dot size

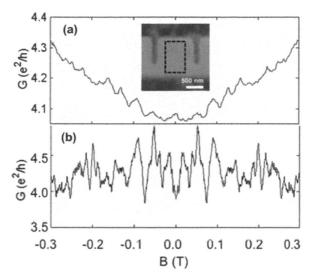

Figure 9.27. (a) The measured magnetoconductance through the dot as a function of magnetic field. The inset is an AFM scan of the dot itself. The dashed rectangle is the SGM scan area. (b) The conductance fluctuations from a quantum simulation of the dot. (Reprinted with permission from [87]. Copyright 2010 the American Physical Society.)

after patterning was nominally 1.1×1.1 μm^2, while the QPCs had 0.45 μm openings and a length of 0.1 μm. The measured magnetoconductance is shown in figure 9.27, where an AFM image of the dot is shown as an inset [87]. We also performed a full quantum simulation of the quantum dot, using the techniques discussed previously, and the conductance fluctuations that were computed are shown in figure 9.27(b).

The SGM was a metallized AFM tip as mentioned above. In this case, we used a piezoresistive AFM tip which was coated with 15 nm of Cr. The tip was rastered across the quantum dot at a lift height of 100 nm and with a bias of −0.7 V. This provides a small perturbation to the local density, which results in a conductance change of less than $0.2G_0$ (where G_0 is the ubiquitous $2e^2/h$ associated with quantum conductance studies). The dashed line in the inset image of figure 9.27 indicates the scan area over the sample. A series of SGM images was taken every 10 mT from 0 to 350 mT. Each such image consists of 120 raster scans across the dot area. Of course, features within the scan area are masked by the conductance background. To remove this, the magnetic Fourier transform approach described previously was then used. Because the conductance is symmetrical in a magnetic field, a cosine transform was used. A recurrent feature was found at a magnetic frequency of 13.5 T^{-1}, and the filtered transform at this frequency was inverted to yield a real space image. This state is shown in figure 9.28(a), and strongly resembles the scarred state of figure 9.4, although there are significant differences. This leads to the question of why the SGM images are not better representations of the scarred state of figure 9.4. The answer lies in the perturbation that is affected to the quantum wave function by the rastered tip potential. To probe this effect, we have carried out

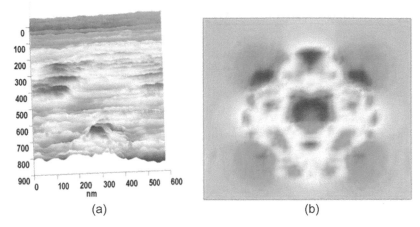

(a)   (b)

Figure 9.28. (a) Resulting SGM image determined from the Fourier analysis. (b) Simulation of the SGM modulated conductance through the dot. ((a) reprinted with permission from [87]. Copyright 2010 the American Physical Society. (b) reprinted with permission from [30]. Copyright 2011 IOP Publishing.)

a simulation with the dot and the rastering potential tip. In this case the conductance is determined at each point of the tip during its raster scan, and the resultant simulation for the SGM scan is computed. This is shown in figure 9.28(b), where it can be compared with the experimental image. Note that the amplitude of the wave function is larger at the four corners, which are points of reflection as there is a turning point in the classical trajectory. This larger amplitude is also seen in the experimental image. In areas where the scar is weaker, there is a corresponding weaker response to the SGM sweep. Another point to note is that there is no signal around the edges of the dot, where the SGM tip potential cannot be distinguished from the confining potential. But the strong correlation between the experimental image and the simulated image is strong support for the type of pointer state shown in figure 9.4.

Problems

1. In figure 9.20, it is apparent that particles (and the wave function) can correspond to an entrance into the dot with a direction that is not along the axis of the QPC. Using the image of figure 2.4, explain how this occurs for QPC modes other than the lowest.

2. Using the model of equations (9.1) and (9.2), construct a potential for a single quantum dot with two QPCs aligned along a central axis. Assume that the Fermi energy within the dot is 15 meV above the bottom of the potential, while it lies 7 meV above the saddle potential maxima in the QPCs. The size of the dot at the Fermi energy is to be 0.3 μm in the transverse direction. Make a three-dimensional plot of this potential. How many modes traverse the QPCs? What is the width of the QPC at the Fermi energy?

3. Consider a physical system, which is of course subject to a differential equation between position and time. By dividing the time into discrete increments, this equation can be written as

$$x_{n+1} = 1 - ax_n^2 + bx_{n-1},$$

where n corresponds to the time increment. Starting from say $x = 1$, make a plot of x_{n+1} as a function of x_n for $a = 1.4$ and $b = 0.3$.

4. Consider a physical system, which is of course subject to a differential equation between position and time. By dividing the time into discrete increments, this equation can be written as

$$x_{n+1} = rx_n(1 - x_n),$$

where n corresponds to the time increment. Plot each value x_{n+1} as a function of r, as r is varied from 2.5 to 4 in steps of 0.01. Plot at least 100 values for each x_{n+1} at each value of r.

5. Consider the potential surface created in problem 2. At what value of the magnetic field does a 'particle' make a complete cyclotron orbit at the Fermi surface within the dot? Assume the effective mass is that for GaAs.

6. For the results of problem 5, estimate the mobility that is required for the 'particle' to complete the cyclotron orbit at the Fermi surface.

References

[1] Bimberg D, Grundmann M and Ledentsov N N 1990 *Quantum Dot Heterostructures* (Chicester: Wiley)

[2] Reed M A, Randall J N, Luscombe J H, Matyi R J, Moore T M and Wetsel A E 1988 *Phys. Rev. Lett.* **60** 535

[3] McEuen P L, Foxman E B, Meirav U, Kastner M A, Meir Y, Wingreen N S and Wind S J 1991 *Phys. Rev. Lett.* **66** 1926

[4] Planck M 1900 *Verhand. Deutsches Physikal. Gesell* **2** 237

[5] Einstein A 1905 *Ann. Phys.* A **17** 132

[6] Wheeler J A and Zurek W H (ed) 1983 *Quantum Theory and Measurement* (Princeton NJ: Princeton University Press)

[7] Marcus C M, Rimberg A J, Westervelt R M, Hopkins P F and Gossard A C 1992 *Phys. Rev. Lett.* **69** 506

[8] Chang A M, Baranger H U, Pfeiffer L N and West K W 1994 *Phys. Rev. Lett.* **73** 2111

[9] Lee P A, Stone A D and Fukuyama H 1987 *Phys. Rev.* B **35** 1039

[10] de Moura A P S, Lai Y-C, Akis R, Bird J P and Ferry D K 2002 *Phys. Rev. Lett.* **88** 236804

[11] Akis R, Ferry D K and Bird J P 1996 *Phys. Rev.* B **54** 17705

[12] Akis R, Ferry D K and Bird J P 1997 *Phys. Rev. Lett.* **79** 123

[13] Brunner R, Meisels R, Kuchar F, Akis R, Ferry D K and Bird J P 2007 *Phys. Rev. Lett.* **98** 204101

[14] Brunner R, Akis R, Ferry D K, Meisels R and Kuchar F 2008 *Phys. Rev. Lett.* **101** 024102

[15] Zurek W H 2003 *Rev. Mod. Phys.* **75** 715

[16] Ollivier H, Poulin D and Zurek W H 2004 *Phys. Rev. Lett.* **93** 220401

[17] Gutzwiller M C 1990 *Chaos in Classical and Quantum Mechanics* (New York: Springer)

[18] Berry M V 1989 *Proc. R. Soc.* **423** 219

[19] Fishman S, Georgeot B and Prange R E 1996 *J. Phys. A: Math. Gen.* **29** 919

[20] Akis R *et al* 2003 *Electron Transport in Open Quantum Dots* ed J P Bird (Boston, MA: Kluwer) pp 209–76

[21] Blume-Kohout R and Zurek W H 2006 *Phys. Rev.* A **76** 062310

[22] Bird J P, Akis R, Ferry D K, De Moura A P S, Lai Y-C and Indlekofer K M 2003 *Rep. Prog. Phys.* **66** 583

[23] Ferry D K, Akis R and Bird J P 2005 *J. Phys.: Condens. Matt.* **17** S1017

[24] Bird J P, Ferry D K, Akis R, Ochiai Y, Ishibashi K, Aoyagi Y and Sugano T 1996 *Europhys. Lett.* **35** 529

[25] Ferry D K *et al* 1996 *Proceedings of the International Conference in High Magnetic Fields in Semiconductors* (Singapore: World Scientific) p 299

[26] Connolly K, Pivin D P Jr, Ferry D K and Wieder H H 1996 *Superlattices Microstruct.* **20** 307

[27] Pivin D P Jr, Andreson A, Bird J P and Ferry D K 1999 *Phys. Rev. Lett.* **82** 4687

[28] Bird J P, Ishibashi K, Stopa M, Aoyagi Y and Sugano T 1994 *Phys. Rev.* B **50** 14983

[29] Akis R, Vasileska D, Ferry D K and Bird J P 1999 *Japan. J. Appl. Phys.* **38** 328

[30] Ferry D K, Burke A M, Akis R, Brunner R, Day T E, Meisels R, Kuchar F, Bird J P and Bennett B R 2011 *Semicond. Sci. Technol.* **26** 043001

[31] Ketzmerick R 1996 *Phys. Rev.* B **54** 10841

[32] Akis R, Bird J P and Ferry D K 2002 *Appl. Phys. Lett.* **81** 129

[33] Ferry D K, Akis R and Bird J P 2004 *Phys. Rev. Lett.* **93** 026803

[34] Altshuler B L, Aronov A G and Khmelnitsky D E 1982 *J. Phys. C: Sol. State Phys.* **15** 7367

[35] Prasad C, Ferry D K, Shailos A, Elhassen M, Bird J P, Lin L-H, Aoki N, Ochiai K, Ishibashi K and Aoyagi Y 2000 *Phys. Rev.* B **62** 15356

[36] Elhassan E 2001 *Phys. Rev.* B **64** 085325

[37] Brunner R, Ferry D K, Akis R, Meisels R, Kuchar F, Burke A M and Bird J P 2012 *J. Phys. Cond. Matter* **24** 343202

[38] Einstein A 1917 *Ver. Deutsche Phys. Ges.* **19** 82

[39] Brillouin L 1926 *J. Phys. Radium* **7** 353

[40] Keller J B 1958 *Ann. Phys. (New York)* **4** 180

[41] Stockmann H-J 1999 *Quantum Chaos: An Introduction* (Cambridge: Cambridge University Press)

[42] Reichl L E 2004 *The Transition to Chaos* (New York: Springer)

[43] Berry M V and Balazs N L 1979 *J. Phys. A: Math. Gen.* **12** 625

[44] Wigner E P 1932 *Phys. Rev.* **40** 749

[45] Berry M V and Tabor M 1976 *Proc. R. Soc. (London)* A **349** 101

[46] Heller E W 1984 *Phys. Rev. Lett.* **53** 1515

[47] Brack M and Bhaduri R K 1997 *Semiclassical Physics* (Reading, MA: Addison-Wesley) section 2.7

[48] Bohm A 1993 *Quantum Mechanics* 3rd edn (New York: Springer) ch 22

[49] Akis R and Ferry D K 2006 *Physica* E **34** 460

[50] Ullmo D, Richert K and Jalabert R A 1995 *Phys. Rev. Lett.* **74** 383

[51] Berry M V and Mount K E 1972 *Rept. Prog. Phys.* **35** 315

[52] Haldane F D M 1988 *Phys. Rev. Lett.* **61** 2015

[53] Hofstadter D 1976 *Phys. Rev.* B **14** 2239

[54] Rauh A, Wannier G H and Obermair G 1974 *Phys. Stat. Sol.* B **68** 215

[55] Akis R, Barnes C and Kirczenow G 1994 *Phys. Rev.* A **50** 4930

[56] Harper P G 1955 *Proc. Phys. Soc. London* A **68** 874

[57] Akis R, Barnes C and Kirczenow G 1995 *Can. J. Phys.* **73** 147

[58] Zurek W H 1982 *Phys. Rev.* D **26** 1862

[59] Zurek W H 1981 *Phys. Rev.* D **24** 1516

[60] Ferry D K, Akis R, Burke A M, Knezevic I, Brunner R, Bird J P, Meisels R and Kuchar F 2013 *Fortschr. Phys.* **61** 291

[61] Nakajima S 1958 *Prog. Theor. Phys.* **20** 948

[62] Zwanzig R 1960 *J. Chem. Phys.* **33** 1338

[63] Mori H 1965 *Prog. Theor. Phys.* **33** 423

[64] Ferry D K 1991 *Semiconductors* (New York: Macmillan) section 15.3

[65] Dirac P A M 1958 *The Principles of Quantum Mechanics* 4th edn (Oxford: Oxford University Press)

[66] Knezevic I and Ferry D K 2002 *Phys. Rev.* E **66** 016131

[67] Duan L M and Guo G C 1998 *Phys. Rev.* A **57** 737

[68] Zanardi P and Rasetti M 1997 *Phys. Rev. Lett.* **79** 3306

[69] Lidar D A, Chuang I L and Whaley K B 1998 *Phys. Rev. Lett.* **81** 2594

[70] Barker J R 1978 *Sol.-State Electron.* **21** 197

[71] Pottier N 1983 *Physica* A **A117** 243

[72] Barker J R 1979 *Physics of Nonlinear Transport in Semiconductors* ed D K Ferry, J R Barker and C Jacoboni (New York: Plenum Press)

[73] Brunner R, Meisels R, Kuchar F, Akis R, Ferry D K and Bird J P 2007 *J. Comput. Electron.* **6** 93

[74] Zaslavskii G M, Sagdeev R Z, Usikov D A, Chernikov A A and Sagdeeva A R 1991 *Weak Chaos and Quasi-Regular Patterns* (Cambridge: Cambridge University Press)

[75] Husumi K 1940 *Proc. Phys. Math. Soc. Japan* **22** 246

[76] Wigner E P 1955 *Ann. Math.* **62** 548

[77] Brody T A, Flores J, French J B, Mello P A, Pandey A and Wong S S M 1981 *Rev. Mod. Phys.* **53** 385

[78] Ferry D K, Akis R and Brunner R 2015 *Phys. Scr.* A **T165** 014010

[79] Ochiai Y, Widjaja A W, Sasaki N, Yamamoto K, Akis R, Ferry D K, Bird J P, Ishibashi K, Aoyagi Y and Sugano T 1997 *Phys. Rev.* B **56** 1073

[80] Darwin C 1859 *The Origins of the Species* (London: John Murray)

[81] Topinka M A, LeRoy B J, Heller E J, Westervelt R M, Maranowski K D and Gossard A C 2000 *Science* **289** 2323

[82] Crook R, Smith C G, Barnes D H W, Simmons M Y and Ritchie D A 2000 *J. Phys.: Condens. Matter* **12** L167

[83] Crook R, Smith C G, Graham A C, Farrer I, Beere H E and Ritchie D A 2003 *Phys. Rev. Lett.* **91** 246803

[84] Burke A M, Aoki N, Akis R, Ochiai Y and Ferry D K 2008 *J. Vac. Sci. Technol.* B **26** 1488

[85] Burke A M, Akis R, Day T, Speyer G, Ferry D K and Bennett B R 2009 *J. Phys.: Condens. Matter* **21** 212201

[86] Bennett B R, Magno R, Roos J B, Krupa W and Ancona M 2005 *Sol.-State Electron.* **49** 1875

[87] Burke A M, Akis R, Day T, Speyer G, Ferry D K and Bennett B R 2010 *Phys. Rev. Lett.* **104** 176801

IOP Publishing

Transport in Semiconductor Mesoscopic Devices
(Second Edition)

David K Ferry

Chapter 10

Hot carriers in mesoscopic devices

In the preceding chapters, we have often discussed the role of decoherence, which is a process where the quantum-mechanical interference properties of a mesoscopic system decay, and the system begins to behave classically. The characterization of this phenomenon is based upon measurement of the phase-breaking time τ_φ of the system, which determines the temperature (and length) scales over which quantum interference is observed. Generally, the phase-breaking time arises from the inelastic scattering processes in the device. But, these inelastic processes give rise to another characteristic time when the system is driven out of equilibrium, for example by passing a large current through it. In this latter case, the system returns to equilibrium by a process called energy relaxation. The energy-relaxation time τ_e is another parameter of importance as it describes the characteristic time for the system to return to equilibrium. There have been many experimental and theoretical studies directed toward understanding the process of dephasing and its variations with lattice temperature, and these have been discussed in the preceding chapters. There are few studies, however, in which both the phase breaking and energy relaxation have been studied in the same system [1–3]. Generally, such measurements provide us with the experimental means to determine the equivalent carrier temperature that arises from the heating by the current or voltage, and to determine the effective relaxation times, as well as studying how these two times differ.

The study of carrier heating in mesoscopic devices is relatively old, but not as old as the study of carrier heating in bulk materials. The latter were first studied to try to ascertain the breakdown properties of dielectrics [4], as well as silicon [5]. This came to mesoscopics through studies of the detailed properties of Si MOSFETs at low temperatures, where electron heating [6] and velocity saturation [7] were observed, just as in bulk materials, but with different characteristics.

Studies of the phase-breaking time generally involve varying the lattice temperature T and studying how the phase breaking process depends upon this temperature. The phase breaking process is then obtained from various effects, such as weak localization or the change in conductance fluctuations as a magnetic field is varied.

Studies of energy relaxation involve heating experiments where Joule heating of, for example, a 2DEG is used to extract energy-loss rates and the energy-relaxation time. In order to analyze the heating, and to determine things like the average power input per electron and the energy-relaxation time, one needs to have a 'thermometer', a mechanism by which the actual electron temperature can be determined from the measurements. One of the earliest methods was to measure the dependence of the mobility on both the electric field and the lattice temperature (at low electric field) [8]. At low temperatures, particularly in bulk material, the mobility is dominated by scattering from the ionized impurities. In this interaction, only a single temperature—that of the carriers—is involved, so that variation of this temperature by either of the two excitations leads to a scale that can be used to convert the mobility at a given electric field to that at a given lattice temperature. Hence, the carrier temperature is determined. As high mobility material became available, the effect of the impurities was dramatically reduced by the use of modulation doping and the resulting increase in effective screening. Thus, the mobility ceased to be an useful method in mesoscopic studies. Alternatively, it was shown that using the Shubnikov–de Haas oscillations provided an effective thermometer. Here, varying the temperature and applied fields could provide a good measure of obtaining the carrier temperature from the amplitude of the oscillations [8]. This has remained the method of choice for estimating the carrier temperature. If these are studied at various temperatures and currents, these two sets of data may be used to extract the temperature dependence of the energy-loss rate $P(T)$, or equivalently the power being dissipated within the mesoscopic system. This appears in the literature in two forms, the first being the loss per electron, and the second is just the total loss in the device. Obviously, these differ by the area or volume of the device and the carrier density at which the measurements are made. This energy-loss rate is normally a function of a power law of the hot carrier temperature, and the exponent often considered to be an indicator of the type of electron–phonon scattering that is dominating the inelastic process in the device. In spite of years of work in both normal semiconductor systems, as well as mesoscopic devices, there is not yet a clear understanding of the entire process.

In this chapter, we will present the approach that is most often used in the study of hot carriers in these systems, along with some of the available experimental data to illustrate the approach, and that of the general field of hot carriers.

10.1 Energy-loss rates

As mentioned, one traditional approach to the study of energy-loss rates involves measuring the Shubnikov–de Haas curves of the sample under study, both as a function of the lattice temperature and the excitation current (or voltage) of the sample. We illustrate this for a heterostructure grown on a lattice matched InP substrate. This consists of a 25 nm wide InGaAs quantum-well sandwiched between two InAlAs barrier layers. The top barrier layer consists of a 30 nm *n*-doped layer and a 10 nm undoped spacer layer (closest to the InGaAs quantum well). The bottom barrier layer is a 250 nm undoped layer. The doping in the top barrier provides sufficient carriers in the quantum well to populate two subbands, but this does not affect the experiment. The total electron concentration was 1.3×10^{12} cm^{-2} with a mobility of about 140 000 cm^2 V^{-1}s^{-1}. Measurements were made at a base temperature of 4.2 K. A micrograph of the sample is shown in the inset to figure 10.1. As one can see, it is configured as a Hall bar, with current passed through the end leads and the voltages measured on the side arms. Data are recorded for various lattice temperatures in the range 4.2–30 K with an excitation current of 500 nA. Then sample heating experiments are conducted with currents ranging from 0.5–250 µA. The measurements for varying temperature are shown in the main panel of figure 10.1. These measurements clearly show a dual frequency (in $1/B$) behavior, by which the density in each of the two subbands may be determined. The interest is in the major peaks that appear in the plot and how these change with temperature, as indicated by the arrow in the figure. Three of these peaks, for 1.6, 2.0, and 3.2 T, were carefully measured for their temperature dependence. At each magnetic field, the neighboring peak and trough are both measured, and this difference is used as the appropriate measure. Then, the measurements were made at base temperature with varying current. There is a small magnetic field dependence on the carrier

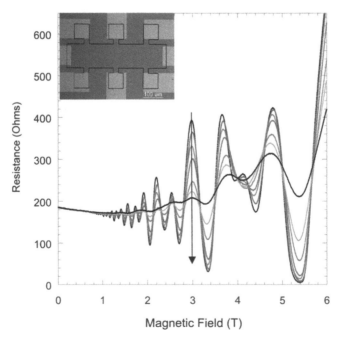

Figure 10.1. Magnetoresistance measurements in the 2DEG show Shubnikov–de Haas oscillations. As the temperature rises, the amplitude decays as indicated by the arrow. The curves are for 4.2, 6, 8, 10, 15.5, 20 and 30 K. The inset is a micrograph of the sample. (Adapted with permission from [3]. Copyright 2003 the American Vacuum Society.)

heating, so using these three values of magnetic field also in the current-dependent measurements allows one to both gain more data points and to compensate for the small field dependence. The resulting change in the amplitude of the oscillations is used to correlate a particular temperature with a particular current. This comparison is shown in figure 10.2. Here, one can see that at the lowest current level, there is no apparent heating of the sample. The straight line drawn in this figure suggests that the temperature varies approximately as the square root of the current.

When measurements such as these are made at low temperatures, one cannot generally assert that the phonons are within the equipartition limit, where the full details of the Bose–Einstein distribution can be ignored. This leads to an important point, by which we may define the Bloch–Gruneisen temperature. In general, the largest q vector for a phonon that can be generated in an electron–phonon scattering event is the one that spans the Fermi surface from one side to the other, basically a value of $2k_F$. With this wavelength, the maximum acoustic phonon energy is then $2\hbar k_F s$ where s is the sound velocity. The issue is whether or not the exponential in the denominator of the Bose–Einstein statistic can be expanded to simplify the expression. The argument of the exponential is the ratio of the phonon energy in the collision to the thermal energy. Thus, the critical temperature is said to be the Bloch–Gruneisen temperature

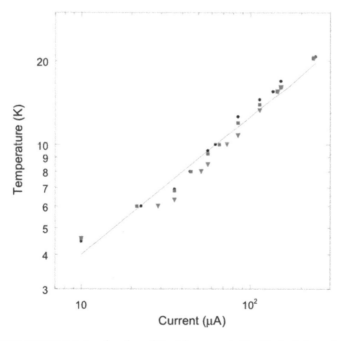

Figure 10.2. Electron temperature as a function of the drive current. The black circles, red squares, and blue triangles denote data centered at 1.6, 2.0, and 3.2 T. (Adapted with permission from [3]. Copyright 2003 the American Vacuum Society.)

$$T_{BG} = \frac{2\hbar k_F s}{k_B},$$ (10.1)

For temperatures below this, the full Bose–Einstein distribution has to be kept, and this complicates the determination of the loss to phonons. So, for a density of 10^{12} cm^{-2}, one finds that in a 2DEG for a typical semiconductor, the Fermi wave number is 2.5×10^6 cm^{-1}. The sound velocity in InGaAs is about 4.7×10^5 cm s^{-1}, so that the Bloch–Gruneisen temperature is about 10 K. Now, this straddles the temperature range involved in the experiments, so that it is unsafe to do anything other than keeping the full Bose–Einstein distribution. When this is done, it has been shown that the general formula for the energy-loss rate can be written as [9]

$$P(T_e) = F(T_e) - F(T) = A(T_e^p - T^p),$$ (10.2)

where F is a general function, and A is a combination of material parameters and fundamental constants. The exponent p (and the prefactor A) depend upon the specific scattering mechanism that governs energy relaxation. For example in heterostructures for the III–V compound semiconductors, one expects $p = 7$ for acoustic phonon interactions, and $p = 5$ for the screened piezoelectric interaction, and $p = 3$ for the unscreened piezoelectric interaction [10].

Using the above approach, the data for the devices discussed above are plotted as a function of the inferred electron temperature in figure 10.3. The data come from

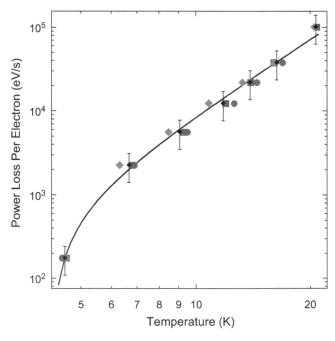

Figure 10.3. The power dissipated per electron for the devices of figures 10.1 and 10.2. The circles are for a magnetic field of 1.6 T, the squares for 2.0 T, and the triangles for 3.2 T. (Adapted with permission from [3]. Copyright 2003 the American Vacuum Society.)

the three values of magnetic field that were used in figure 10.2. It may be seen that the input power per electron all fits on a single curve, although it is difficult to see that this curve fits to $p = 3$ (or any other value). Nevertheless, the slope of the curve at the higher values of electron temperature do correspond to this value of the exponent. To see this better, we re-plot the data in figure 10.4 as a function of the temperature expression in the parentheses of equation (10.2) with $p = 3$. Here, it may be seen that the fit to the power law is very good, although there is a little more scatter in the data. However, it may be seen that slope agrees with each of the three sets of data for different magnetic fields. Thus, it may be inferred that the major source of the energy relaxation of the excited carriers is due to phonons via the unscreened piezoelectric interaction.

In figure 10.5, we show the energy loss per electron for a bilayer of graphene [11]. Here, the samples were prepared by exfoliating Kish graphite onto a doped Si substrate upon which 300 nm of SiO_2 were grown. Monolayer and bilayer devices were formed in different samples by this process, with the number of layers being confirmed by optical microscopy and Raman spectroscopy. The samples were contacted with Cr/Au electrodes in a six probe configuration which allowed the electric field to be determined in a manner free from the effect of current contact resistances (as is done in the Hall effect, discussed in chapter 6). The samples typically exhibited mobilities in the range 10^3–10^4 cm^2 V^{-1}s^{-1} at low temperature. The measurements in the figure are on a bilayer device with a base temperature

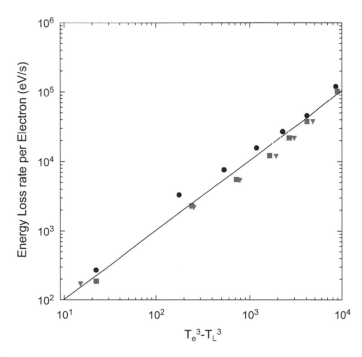

Figure 10.4. The power loss per electron from figure 10.3 re-plotted so that it corresponds to the form of equation (10.2), with the parameter p equal to 3.

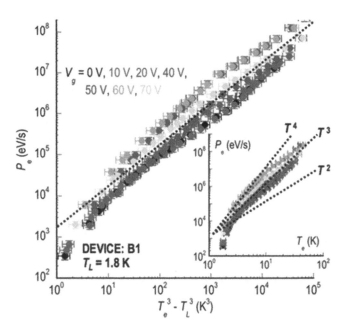

Figure 10.5. The carrier power loss (per electron) as a function of temperature according to the form of equation (10.2) at different gate voltages (color coded as indicated). The inset compares the data for slopes of different p values, showing that 3 is the most likely. This is indicated in the main panel as the dotted line. (Reprinted with permission from [11]. Copyright 2013 American Chemical Society.)

of 1.8 K. In this case, the samples exhibited conductance fluctuations, as discussed in chapter 5. As these fluctuations varied with the base temperature and the current passed through the sample, the amplitude of the fluctuations was used as the thermometer to ascertain the electron temperature as a function of the drive current through the device. In the main panel of the figure, the input power per electron is plotted as a function of the temperature relation of equation (10.2) with $p = 3$, the data sets corresponding to different gate voltages (bias on the Si substrate), whose values are shown by the color coding indicated in the figure. The inset shows different values of p for the same data, with each value shown as a dotted line. The conclusion is that the best fit is to a value of $p = 3$. While this is the same exponent found in the InGaAs heterostructure discussed above, the interpretation of this is much different. First, graphene is not piezoelectric, so the loss process found above does not exist in this material. More importantly, the unique Dirac-like band structure of graphene changes the exponents found for the various scattering mechanisms. For loss via the acoustic phonons, one expects an exponent of 4 in graphene [12–14]. In earlier work, the Oxford group had examined the energy-loss process in chemical-vapor deposition grown graphene on the Si-terminated face of SiC [15]. In this latter work, the authors claimed a dependence with $p = 4$, although some of the data suggest that it was closer to $p = 3$. Similar results were recently found in bilayer graphene as well [16]. The data in figure 10.5, however, clearly do not fit with $p = 4$, and have a much better fit to $p = 3$. Very similar behavior was found for monolayer graphene in several samples. It has been suggested that a second-order impurity mediated phonon scattering could lead to an exponent of 3 [17]. However, it was found that the strength of this interaction was much too weak to explain the data in the figure. More recently, it has been found that plasmon-mediated energy relaxation can explain the energy-loss rate seen in this latter figure [18]. Plasmons are excitations of the electron gas itself, so the plasmon loss process actually only redistributes the energy among the electrons. But, in the presence of current, the Fermi sea moves to account for the current flow and is no longer centered about the zero of momentum. At high currents, this becomes highly accentuated, with many electrons above the equilibrium Fermi energy. Hence, the plasmon process accounts for scattering of these excited electrons back to states mostly vacated by the shift in the Fermi sea. This process reduces the current and the energy input to the electron gas, but ultimate relaxation of the energy must occur by some other process, such as the acoustic phonons. The derivation of the plasmon energy loss follows equation (10.2), with the matrix element [18, 19]

$$W_{\pm} = \frac{2\pi N}{\hbar} Im\left\{ \frac{V(q)}{\epsilon(q, \omega)} \right\} \delta\left(\omega - \frac{E_k - E_{k\pm q}}{\hbar} \right), \tag{10.3}$$

where N is either the Bose–Einstein distribution, in the absorption term, or 1 plus this distribution, in the emission term. The scattering function arises from the imaginary part of the screened Coulomb potential $V(q) = e/\varepsilon_s q$, where e is the electron charge and ε_s is the permittivity of graphene [20, 21]. Here, the screening is done by the full frequency- and momentum-dependent dielectric function. As may be expected, the dielectric function is dominated by the plasmon pole, but in using the inverse of the dielectric function in equation (10.3), we incorporate the collision broadening of this resonance. The broadening arises from the total scattering rates, which yield a scattering time τ. Carrying out various integrations leads to the relaxation rate

$$P(E) = \left\langle -\frac{dE}{dt} \right\rangle = \frac{2.4 k_B^3}{2\pi\hbar^2 v_{FT}^2}(T_e^3 - T^3). \tag{10.4}$$

Since the mobility is easily measured, the relaxation time can be determined for the sample [18, 19], so that equation (10.4) contains no adjustable constants. The electron temperature can be determined either from thermometer type measurements discussed above, or from using the definition of the energy relaxation time, discussed in the next section. If we know the latter quantity from experimental measurements, then the temperature is given by

$$P(E) = \frac{k_B(T_e - T)}{\tau_E}. \tag{10.5}$$

In figure 10.6, we compare the variation of $\tau_E(T_e)$, obtained in the experiment of [11], with that predicted by equation (10.4) for the base temperature of 1.8 K, as used in the experiments. Theoretical values of τ_E are determined by taking the

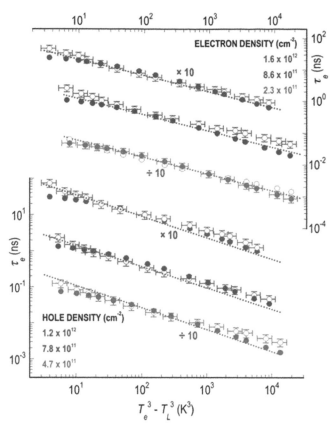

Figure 10.6. The energy relaxation time for several values of the electron and hole density. Experimental values from figure 10.5 [11] are plotted as open symbols while theoretical values obtained from equations (10.4) and (10.5) are the closed symbols. Error bars are shown for the experimental data only. Dotted lines through the theoretical data are guides to the eye that indicate a linear variation of power input with T^3. In each case, the red and green curves have been offset by a factor of 10 for clarity as indicated. Reprinted with permission from [18], copyright 2015 by AIP Publishing.

experimentally-determined energy-loss rate for the given density and T_e and introducing this into equations (10.4) and (10.5) to compute τ_E. From this figure, we see that the plasmon-based model reproduces not only the low-temperature magnitude of τ_E, for both electrons and holes, but that it also captures its quantitative dependence on temperature.

The plasmon-loss mechanism rearranges the energy distribution function, but this energy remains in the electron gas as a whole. Under the excitation current (or electric field), the distribution function is shifted in momentum space to reflect the current, and it spreads as the electron temperature rises. The shift can also distort the distribution to produce an elongated extension in the direction of the field or current. Between the shift in momentum space, and the elongation, the distribution function has more carriers in the forward direction than the backward direction. The plasmon loss takes carriers from the streaming forward direction and moves them to the backward direction, in the symmetric terms of the distribution function, on average. This scattering process relaxes the streaming distribution, but the ultimate result still remains that another process must pass the energy to the lattice. This can be due to acoustic-phonon scattering, or even the so-called supercollisions of [17]. However, it is the plasmon interaction that dominates what is seen experimentally in transport, as it is this process which moves the current carrying particles out of the forward extension of the distribution. On the other hand, optical measurements will see only the symmetric part of the distribution, so may well give other variations with temperature.

10.2 The energy-relaxation time

The energy input per electron is just one measure of the manner in which the energy relaxes to the lattice in a timely manner. The strength of this measure is the exponent that appears when matching to equation (10.2), as this exponent is an indicator of the dominant relaxation process. However, there is another measure that is important and that is the energy-relaxation time itself. The idea of an energy-relaxation time is linked to the study of hot electrons in semiconductors in which the dynamics of the process is developed by a set of moment equations [22]. That is, the Boltzmann transport equation is multiplied by powers of the momentum and integrated to yield a set of equations for the density, the momentum, the energy, and so on. Each of these is characterized by an appropriate relaxation time, which can be determined by taking the corresponding momentum of the scattering-induced changes of the distribution with time. For a simple approach, we can use these results to write an expression relating the input power per electron to the energy-relaxation time which is just equation (10.5). This expression now defines the energy-relaxation time from the measured power input per electron and the inferred electron temperature achieved at that power input level.

In figure 10.7, we plot the energy-relaxation time that was found for the three different values of magnetic field in the InGaAs 2DEG discussed above [3]. It may be seen from this figure that a difference now appears between the lowest magnetic field and the other two values, which presumably arises from the tendency to form edge states (associated with the quantum Hall effect) more effectively at the higher magnetic fields. The edge states are quite similar to one-dimensional quantum wires in which the motion is only allowed to be in one direction, which hinders back-scattering processes (discussed in chapter 6). The lines are guides to the eye, with different slopes, but are indicators of the relaxation process. At higher temperatures, the fit appears to be close to the T^{-3} curve, and this behavior has been seen in clean systems where the energy is thought to be relaxed by three-dimensional electron–phonon scattering processes [2, 23]. At lower temperature, a transition is seen to an alternate T^{-1} behavior at all values of the magnetic field. The cause for this behavior is not fully understood at present and is not expected from the $p = 3$ behavior seen in the input power curves, but it could be the quantum wire behavior of the edge states.

In figure 10.8, the energy-relaxation time is shown for three nanowires fabricated in the InGaAs heterostructure discussed above [24]. In this case, the wires were

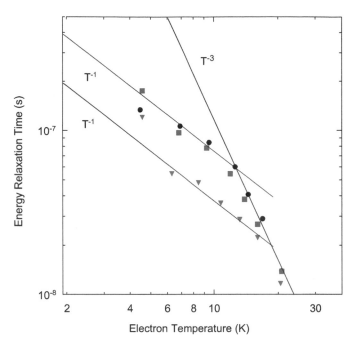

Figure 10.7. The energy-relaxation time determined for the InGaAs device discussed in the previous section. The blue diamonds, red squares, and green circles are for a magnetic field of 1.5 T, 2.0 T, and 3.2 T, respectively. (Adapted with permission from [3]. Copyright 2003 the American Vacuum Society.)

formed by an isotropic etch to remove the InGaAs and upper layers to yield the wires. Each end of the wire is connected to large two-dimensional reservoirs. However, only two-terminal measurements can be made on the wire, so that the results are not guaranteed to be free of contact resistances that form at the reservoir–wire transition. The measurements in the figure are for three different wires, of 215 nm, 545 nm, and 750 nm widths. The electrical widths were inferred from the magnetic field at which the onset of Shubnikov–de Haas oscillations were observed in the narrowest wire. This value was then extrapolated to the other wires. The wider wires show relaxation behavior that is more two-dimensional than one-dimensional, but the narrower wire clearly shows a difference, both in the magnitude and the slope of the relaxation time. The solid curve in the figure shows the T^{-3} behavior observed previously, but it is not at all clear whether any of the wires demonstrate the T^{-1} behavior (dashed curve). In fact, the narrower wire may be approaching a constant value at the lowest temperatures. Interestingly enough, the narrow wire shows a power per electron that appears to also match $p = 3$, much as the larger samples discussed above.

The energy-relaxation time found in graphene samples [11] is displayed in figure 10.9 as a function of the electron density, rather than the temperature. Here, there are curves for both electrons and holes separately, and the data are plotted with the input current level as a parameter. The shaded area around the Dirac point represents a region where the density cannot be accurately ascertained

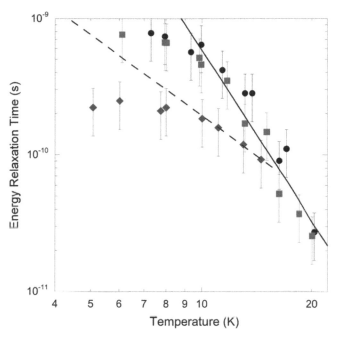

Figure 10.8. The energy-relaxation time found in etched InGaAs nanowires for widths of 215 nm (blue diamonds), 545 nm (black circles), and 750 nm (red squares). The two wider wires are thought to be narrow two-dimensional devices rather than true one-dimensional wires. (Adapted with permission from [24]. Copyright 2004 the American Vacuum Society.)

from Hall measurements and is inferred from the capacitance and gate voltages. Outside this region, the two measurements give comparable values. The data are shown for a bilayer graphene device, but the variations shown are quite generic and seen in all the monolayer and bilayer devices studied. The remarkable behavior indicates that the relaxation time actually decreases rapidly as one approaches the Dirac point from either the conduction or valence band. While unexpected, this behavior can be explained by the plasmon-mediated energy-loss mechanism discussed above, since the total power input is independent of the electron density, due to the unique nature of the Dirac band structure and the resultant dynamic effective mass of the carriers. Hence, the power input per electron increases at lower density, and this leads to a reduced energy-relaxation time from equation (10.5), just as seen in the figure. It may also be due to the fact that puddles of electrons and holes form in graphene near the Dirac point, which likely will complicate the relaxation process [25]. In figure 10.10, the energy-relaxation time is plotted for two electron densities as a function of the electron temperature, and compared with the theoretical prediction for the plasmon-mediated process of energy-relaxation. The open symbols are data taken from that used in figures 10.5 and 10.9, while the closed symbols are the theoretical values obtained. It is clear that the relaxation time decreases with increasing electron temperature, but this decay is not a simple integer value. Rather, it appears the decay in the theory is roughly $T^{-1.75}$. This is not far

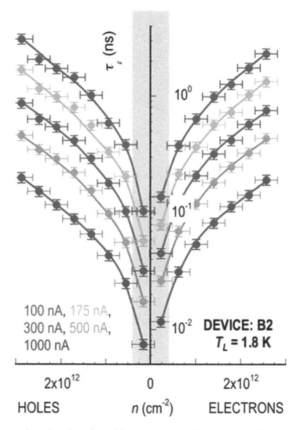

Figure 10.9. The energy-relaxation time for a bilayer graphene device at a lattice temperature of 1.8 K. The various curves are for different values of the excitation current, as indicated at the lower left of the figure. (Reprinted with permission from [11]. Copyright 2013 American Chemical Society.)

from the $T^{-1.5}$ reported in [11]. It is also clear that the relaxation time decreases rapidly as a function of the density, just as indicated in figure 10.10.

10.3 Nonlinear transport

When high electric fields are applied to a semiconductor, the resulting current becomes nonlinear, and is accompanied by a reduction in the differential mobility. In this situation, the system has moved into the far-from-equilibrium condition. The distribution function, characterizing the carriers, is quite far from its low-field Boltzmann form. In this nonlinear transport regime, the ergodic approximation fails. In particular, it fails because the distribution function itself is not a stationary process. It varies in time and its form arises from a delicate balance between the driving forces, the electric field usually, and the dissipative forces, the scattering. As such, it really has no connection to the equilibrium form, and linear response approaches generally do not work.

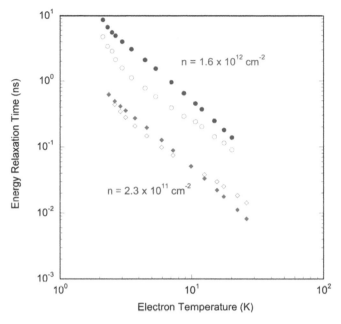

Figure 10.10. The energy-relaxation in a monolayer of graphene as a function of the electron temperature. Here, the data are shown as the open symbols for two densities, and are taken from [11]. The closed symbols are from the plasmon-mediated loss theory [18].

Early studies of high-electric field transport in solids focused mainly on the breakdown studies of dielectrics [4], although the electrical properties of silicon had been studied much earlier [26]. Studies of the variation of the velocity with high electric fields began with Ryder [5], who discovered that the velocity in Si and Ge saturated at high electric fields. The saturated (or nearly saturated) velocity is an important parameter for electron device considerations. For example, with a modern nanoscale MOSFET of say 20 nm gate length, an applied voltage of 1 V produces an average electric field of 0.5 MV cm^{-1} in the channel, which is an incredible electric field, especially when one considers that the saturation velocity sets in at about 20 kV cm^{-1} in Si at room temperature. Then, in many so-called figures of merit for high speed, high frequency, or high power devices, the saturated velocity is an important parameter as it affects the frequency response and power handling properties of the device. The saturated velocity is also important as a mirror into the electron–phonon interactions in the material. Consequently, it certainly affects the distribution function that is found at each electric field. It is natural that these effects are important in mesoscopic devices, as the MOSFET mentioned above can certainly be described as a mesoscopic device, or even a microscopic device. In this section, we want to describe some of the effects that can be studied in the nonlinear regime.

10.3.1 Velocity saturation

(Parts of this section are adapted from *Semiconductor Transport (1st Edition)*, David Ferry, 2000)

In nonlinear transport, one must deal with a multitude of characteristic times for the semiconductor system. The importance of the various time scales is obvious due to the necessity of evaluating the time dependence of the various measurable properties. The non-equilibrium distribution function evolves(and is therefore non-stationary) the time scale for variations of the measurable quantity. This implies that the system is non-ergodic (by which is meant simply that time averages do not equate to ensemble, or distribution, averages, the latter of which are the important averages) over this time scale [27]. In fact, it is important to evaluate carefully the various collection of time scales that are important. In a semiconductor, numerous collisions occur and it is these collisions that provide the mechanism of exchange of energy and momentum and relax these quantities toward their equilibrium values. There are collisions between the carriers which randomize the energy and momentum *within* an ensemble but do not relax either of these quantities for the ensemble as a whole. There are also elastic collisions between the carriers and impurities or acoustic phonons which relax the momentum but do little to relax the energy. Finally, there are inelastic collisions between the carriers and lattice vibrations which relax both the energy and the momentum. In general, four generic time scales can be identified [28, 29]:

$$\tau_c < \tau < \tau_R < \tau_H. \tag{10.6}$$

Here the average *duration* of a collision is denoted by τ_c. Generally, this time scale is quite short and not of importance to most considerations. However, on the scale of fast femtosecond laser experiments, this may no longer be true. The collision duration is the time required to establish the energy-conserving delta function (in the Fermi golden rule for scattering). In distinction, the average time *between* collisions, the mean free time, is denoted by the simple τ. For time scales such that $t \leqslant \tau$, the evolution of the system depends strongly upon the details of the initial state. Generally, $\tau \gg \tau_c$, but this is not always the case at high electric fields, and the breakdown of this inequality can lead to new transport effects, which must be treated in a quantum mechanical manner.

The establishment of equilibrium, or a non-equilibrium steady state, can be achieved within a few or a few tens of τ, and the characteristic time associated with this process is the *relaxation time τ_R*. Typical quantities characterized by a relaxation time are the momentum relaxation process, and in high fields the energy relaxation process. If configuration-space gradients exist, the situation becomes more complex. Relaxation in momentum space proceeds on the scale of τ_R and establishes a 'local' equilibrium over regions smaller than a macroscopic scale, perhaps only a few mean free paths in extent. The achievement of a uniform equilibrium or non-equilibrium steady-state requires a longer time, the hydrodynamic time $\tau_H > \tau_R$. Only for times large compared to this hydrodynamic time can the ensemble truly be said to be

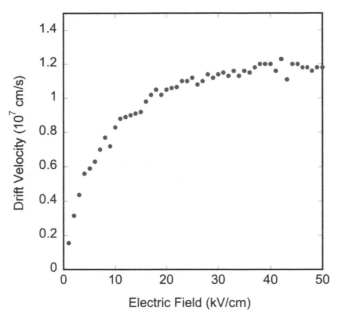

Figure 10.11. Ensemble Monte Carlo simulation of the velocity of electrons in silicon at 300 K. The actual velocity saturation is slightly higher than that seen experimentally.

stationary, and only for times on this scale are the processes even beginning to become ergodic. Examples of the hydrodynamic time scale are diffusion times, arising from recombination processes for excess carriers as well as from local non-homogeneous carrier distributions and sometimes energy relaxation times.

The main experimental observable of the hot carriers, at least in homogeneous semiconductors, is the observation of their velocity saturation. When the carriers are heated by the field, the temperature rises and the distribution spreads with more carriers at higher energy. Since the scattering rate generally increases with energy, there is more carrier scattering (to accommodate the relaxation of the energy input from the field), the mobility is reduced and the velocity does not increase as rapidly. In fact, in Si, the velocity appears to actually saturate at a value near to 10^7 cm s^{-1} at a lattice temperature of 300 K. We show this in figure 10.11, where the results of a Monte Carlo simulation are plotted. In this figure, the velocity is plotted for a range of electric field.

When we study high field effects, many of the important processes set in at 100s of kV cm^{-1}, and it becomes difficult to make measurements on normal sized samples. Hence, considerable effort is necessary to make structures in which the needed high electric fields can be established at reasonable voltages. Such a structure is shown in figure 10.12, used to study the velocity behavior in GaN [30]. In panel (a), the dark areas are the GaN, which has been etched to provide the small wire like structure at the center, and shown expanded in panel (b). In figure 10.13, we show

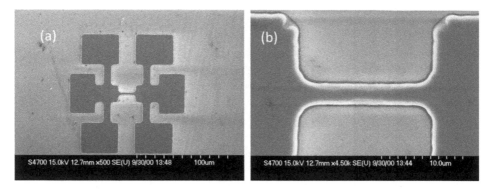

Figure 10.12. (a) Fabrication of a sample to prepare a high field region (the narrow 'wire'). The GaN is the dark region after etching the shape. (b) A blow up of the constriction over which the high electric field will exist. Courtesy of J M Barker.

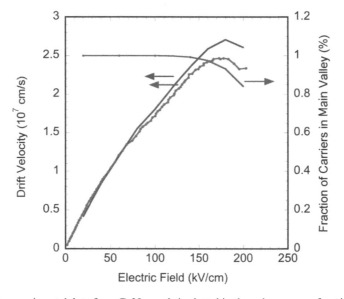

Figure 10.13. The experimental data for a GaN sample is plotted in the red curve as a function of the electric field. The blue curve is the EMC calculation of the velocity, while the green curve is the fraction of carriers that remain in the main conduction band valley (the Γ valley).

the velocity measured in these GaN samples as a function of the applied electric field, for a sample with a constriction of size 15×3 square microns. The measured velocities are compared to those computed via a three valley ensemble Monte Carlo procedure [31]. Also shown is the fraction of carriers that remain in the main conduction band valley as a function of the electric field. It may be seen that the velocity reaches a peak at about 180 kV cm^{-1}, and then begins to decrease. As may be seen from the green curve, this is explained as arising when the carriers leave the main Γ valley and move to satellite valleys in the conduction band, where the mass is heavier and the velocity lower. We return to this point below.

10.3.2 Intervalley transfer

Transfer between different valleys of the full energy band is one of the oldest and most interesting aspects of high-electric-field transport in semiconductors. There are two possible methods by which transfer between nonequivalent sets of valleys can occur. In one, the symmetry-breaking properties of the high electric field are used to break the symmetry between valleys that are normally equivalent. This occurs, for example, in Si at low temperatures when the electric field is not oriented at the same angle with all six mimima of the conduction band. For this case, the carriers in each valley are heated differently and a repopulation appears between the various valleys.

The second method by which intervalley transfer can play a major role is when only a single conduction band minimum, with a small effective mass, is normally occupied, such as in the case of GaN (the principal minimum is at the Γ point), discussed in the previous section. With the application of a high electric field, the carriers are heated to relatively high energies. Then, a fraction of the carriers will actually be at energies above the minima of a secondary set of valleys [which lie along the (111) directions at the set of M-L points of the Brillouin zone in wurtzite GaN, and are some 1.3 eV above the Γ minimum] of the conduction band. Since, the equivalent M and L minima (the difference in notation lies in the basal plane set or the z-axis set of wurtzite) have relatively high values of the effective mass, their density of states is much larger than that of the central minimum, and intervalley scattering will cause electrons to move to the satellite valleys of the conduction band. Consider, for example, the conduction band of InSb shown in figure 10.14. There is a

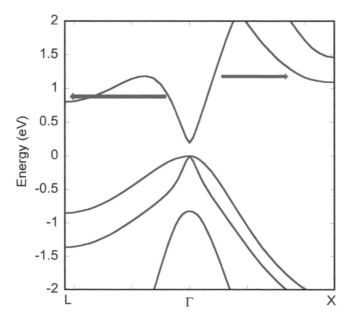

Figure 10.14. The band structure for InSb determined using a nonlocal empirical pseudopotential method that includes the spin-orbit interactions. The arrows show the possible intervalley transfers for hot electrons in the Γ valley.

central valley characterized by a small effective mass of $0.013m_0$. In addition, there are subsidiary minima located at both the L point, lying some 0.55 eV higher, and at the X point, lying some 1.0 eV higher. These valleys have a considerably greater effective mass and density of states, as the former is Ge-like while the latter is similar to Si. Under normal circumstances, the central valley is the only one occupied. However, for an applied field of some 600 V cm^{-1} at 77 K, electrons begin to transfer to the L valleys, a process first observed by Gunn in GaAs [32]. It was identified clearly as intervalley transfer by Ridley and Watkins [33, 34]. In figure 10.15, we show the average drift velocity and kinetic energy as a function of the electric field in InSb at a lattice temperature of 77 K.

Intervalley transfer is seen in a great many semiconductors, particularly those with a direct bandgap, so that the lowest lying conduction band is at the center of the Brillouin zone. Recent examples include observations of this effect in atomically thin transition-metal dichalcogenides [35, 36], and in the absorber material on AlInAs hot-carrier solar cells [37]. As may be seen in figures 10.13 and 10.15, once transfer begins, the velocity-field curve begins to show negative differential conductance (NDC). The resulting NDC that occurs when the carriers are transferred from low-mass, high-velocity states to high-mass, low-velocity states is often referred to as the Gunn effect and is used in transferred electron devices (TEDs, which are used for microwave sources). Transfer occurs as the carriers are heated in an external electric field. Once some of the carriers reach energies near that of the satellite valleys (the L and X valleys), inter-valley scattering can occur. Since the density of states in the satellite valleys is much higher than in the central valley, inter-valley scattering from the Γ valley to the satellite X and L valleys is a dominant process. The transfer in this

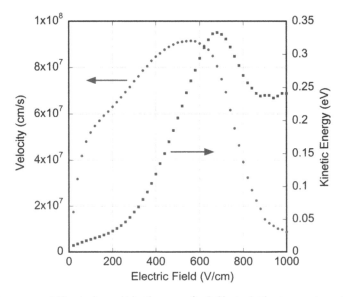

Figure 10.15. The average drift velocity and kinetic energy for InSb at a lattice temperature of 77 K are plotted as a function of the electric field. The drop in velocity and energy above 700 V cm^{-1} is due to intervalley transfer. The energy drops as kinetic energy is converted to the potential energy of the satellite valleys.

particular direction is much more pronounced than the reverse process (recall that the scattering rate is essentially a direct measure of the density of final states). Because of the higher mass and density of states in the satellite valleys, the mobility and velocity are much lower and a negative differential conductivity will occur.

10.3.3 NDC and NDR

Both negative differential conductance (NDC) and negative differential resistance (NDR) are both extreme nonlinear properties of various materials. The difference between the two is mainly one of semantics and how one identifies the principle variable (voltage or current) and the response to this. The non-equilibrium thermodynamics of these two processes were described quite some time ago by Ridley [38]. We illustrate this difference in figure 10.16. In NDC, the electric field is the excitation, while the current density is the response. A given current may appear at multiple values of the electric field, and this gives a characteristic '*N*' shape to the curve. As the curve is the current as a function of the electric field, it naturally displays the conductance, and gives rise to a region of NDC. On the other hand, with NDR, the current is the excitation, and the electric field is the response, since a given electric field may exist at multiple values of the current density. This gives the curve a characteristic '*S*' shape. As the electric field is given as a function of the current, it naturally displays the resistance, and gives rise to a region of NDR.

If we excite a semiconductor with NDC with an increasing voltage, the electric field is usually quite homogeneous (we neglect contact effects) across the length of the device until the peak of the current is reached. The NDC region is an unstable region in which fluctuations in the electric field will lead to an inhomogeneous distribution of the electric field. For example, if we try to hold the electric field in the NDC region (where the current is decreasing with an increasing electric field), the thermodynamics will try to have one electric field on the low-field linear rising part of the curve, and a second much higher electric field on the far-right side of

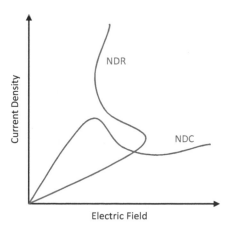

Figure 10.16. Characteristic strong nonlinearities in the current density versus electric field for the cases of NDC and NDR.

figure 10.16. Often there is no point on the curve, and the high electric field 'domain' is limited by the applied voltage V through

$$V = -\int_0^L E(x)dx, \tag{10.7}$$

where the cathode is taken as $x = 0$ and the anode is taken as $x = L$. In many cases, the high field domain is not stationary, but moves with the charge dipoles that creates the high electric field. Then, it can cycle through the device with a period

$$T = \frac{L}{v_{\text{eff}}}, \tag{10.8}$$

Where v_{eff} is an effective velocity lying somewhere between the peak velocity and the valley velocity (lowest value in the curve of figure 10.16). As most compound semiconductors are piezoelectric, the very high electric field can lead to strain in the crystal, and this can ultimately cause the sample to be fractured, as happens in GaN; that's why the experimental data ends where it does in figure 10.13. In other cases, the high electric field can lead to impact ionization and avalanche breakdown; a situation usually seen in InSb [39]. The onset of impact ionization can transition the behavior to the NDR situation.

If we excite a semiconductor with NDR with an increasing current, the current density is usually quite homogeneous (we neglect contact effects) across the cross-section of the device until the peak of the electric field is reached. The NDR region is an unstable region in which fluctuations in the current density will lead to an inhomogeneous distribution of the current density across the cross-section of the device. For example, if we try to hold the current density in the NDR region (where the electric field is decreasing with an increasing current density), the thermody-namics will try to form a filamentary current in the device. These filaments are natural formations in these situations and are called micro-plasmas when observed in the breakdown of p-n junctions [40, 41]. In materials like InSb, the density of the electron-hole plasma that is generated by avalanche breakdown can even lead to a plasma 'pinch' due to the induced magnetic field from the high current in the filament [42]. This leads to an even high current density which may lead to intense local heating and even melting that in turns leads to device destruction from the trapped gasses, whether in a bulk material such as InSb or in a semiconductor device during breakdown [43].

An important point about both NDC and NDR is that, if the appropriate voltage or current is cycled slowly, then the device will exhibit hysteresis. For example, with NDC, the operating point with rising voltage moves up the linear curve to the peak of the current density, and then jumps to the curve on the right. In lowering the voltage, the operating point does not retrace this path. Rather, if follows the right positive conductance path down to the valley current density and them jumps to the linear curve. Thus, the operating point exhibits a clockwise hysteresis in the plane of figure 10.16. On the other hand, for NDR, the current leads the electric field to rise to the peak in the linear region, then jump to higher current density as the filament forms. In lowering the current, the operating point follows the curve down to the 'valley' electric field and then jumps to the linear curve, thus exhibiting and anti-clockwise hysteresis. In particular, this hysteresis in NDR materials has led to interesting memory devices [44, 45].

10.3.4 Velocity overshoot

In high electric fields, it is quite likely that the momentum relaxation time and the energy relaxation time will differ, with the latter being the slower process. The momentum relaxation time τ_m describes the decay of the velocity (and the velocity fluctuations about a local near-equilibrium state). In essence it is the shift in momentum space of the entire distribution that is required to give an average velocity. However, the nonlinear transport in high electric fields arises primarily from the change in the distribution function in the presence of the high electric field, which leads to an increase in the average energy of the carriers. The response of the distribution function, which results in this increase in average energy, is characterized by its own relaxation time, which is referred to as the energy relaxation time τ_E, since the evolution of the distribution function represents the evolution of the average energy of the carrier ensemble.

If the energy relaxation process is slower than the momentum relaxation process, the velocity can overshoot its ultimate steady-state value (the saturation velocity) in high fields. This occurs because the distribution function first shifts (equivalent to the shift studied in previous chapters) in momentum space as the velocity rises to a value characterized mainly by its low-field mobility. As the distribution function then diffuses in energy (or momentum) space to its non-equilibrium form, the mobility decreases to its ultimate high-field value, with a consequent decrease in the velocity. It can readily be shown that this 'overshoot' behavior requires a more complicated behavior than the simple behavior characterized by a Langevin equation

$$\frac{dv}{dt} = \frac{eE}{m^*} - \frac{v}{\tau_m}, \tag{10.9}$$

where we have omitted the random force as it will average to zero. When overshoot occurs, an equation such as equation (10.9) must have at least two zeros—one at a time corresponding to the steady state and one at a time corresponding to the peak velocity. However, the relaxation rate is an increasing function of energy (or velocity), so that the right-hand side has only a single zero, at the steady-state product of the field and the mobility. Thus a second time scale must be involved, which is the characteristic time of the energy relaxation. One approach to this is to let the motion of the particles be governed by a *retarded* Langevin equation, written as [46]

$$\frac{dv}{dt} = \frac{eE}{m^*} - \int_0^t \gamma(t - u)v(u)du, \tag{10.10}$$

where again the random force term has been omitted. The function $\gamma(t)$ is a 'memory function' for the non-equilibrium system, and has a time response related to the energy relaxation time. In fact, this function can be written in the simplest form as

$$\gamma(t) = \frac{1}{\tau_m \tau_E} e^{-t/\tau_E}. \qquad (10.11)$$

These results demonstrate a number of important aspects of hot-carrier behavior. First, the dynamics become retarded with a memory effect because of the extra time behavior corresponding to the evolution of the distribution function, the energy relaxation time. This, in turn, opens the door for velocity overshoot to occur. Moreover, this process, when coupled with the velocity saturation effect, clearly indicates the far-from-equilibrium nature of this nonlinear transport. In the field range where velocity saturation and velocity overshoot can occur, the distribution function is determined by carrier-lattice interactions and by boundary conditions in the form, for example, of applied fields, and is not simply related to the equilibrium form. There is one caveat, however, and that is that the transient velocity (which may be calculated in an ensemble Monte Carlo process) will be a result that depends upon all the scattering processes.

In figure 10.17, we illustrate the transient response of electrons in graphene to a high electric field at room temperature. Here, the graphene is deposited on a SiO_2 (grown on silicon) substrate and has been patterned in a way to support a high electric field, as discussed above, although these are theoretical data taken from an EMC simulation of the transport. Because of the relatively high scattering rates of the surface optical phonons in SiO_2 [47] which strongly affect the graphene

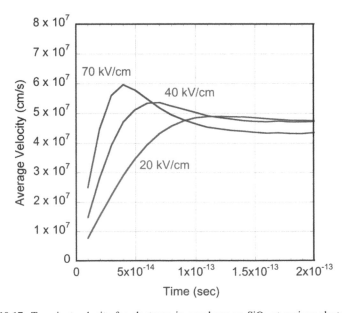

Figure 10.17. Transient velocity for electrons in graphene on SiO_2 at various electric fields.

transport [48], the overshoot lasts for less than a single picosecond. The graphene shows a peak velocity at about 15 kV cm^{-1} and the shows the onset of a weak NDC beyond this field. At the three electric fields in the figure, the 126 meV surface optical phonon is the major scattering process in graphene, being about 5–10 times more effective than the low energy 55 meV mode. Although the threshold energy for the onset of emission of the high-energy phonon is larger than that for the low-energy phonon, it becomes a much stronger scattering process at the high energy the carriers reach in high electric field. The saturated velocity of graphene remains at about 4×10^7 cm s^{-1} across a wide range of electron densities. In fact, for fields at, or above 40 kV cm^{-1}, this dominant scattering rate is above 10^{13} s^{-1} at room temperature.

Problems

1. A particular semiconductor has a zero-field mobility composed of ionized impurity scattering of 3500 cm^2 V^{-1}s^{-1} and of acoustic scattering of 4500 cm^2 V^{-1}s^{-1} at 4.2 K. Impurity scattering varies as $T^{3/2}$ while acoustic scattering varies as $T^{-1/2}$. We can assume that the average temperature of the carriers is given by

$$\frac{3}{3}k_B(T_E - T) = e\mu_e E \tau_E,$$

where E is the electric field and τ_E is the energy-relaxation time and has a value of 1 ps. Plot the electron temperature and the mobility as a function of the electric field E.

2. In a particular semiconductor, it is found that the electron temperature varies linearly with the electric field as

$$T_e = T + \alpha E,$$

where E is the electric field and α has a value of about 10^{-3} cm V^{-1}. In addition, one may represent the mobility as a function of the electric field as

$$\mu_e = \frac{\mu_0}{1 + \dfrac{\mu_0 E}{v_{sat}}},$$

where μ_0 is 10^3 cm^2 V^{-1} s^{-1} and v_{sat} is 10^7 cm s^{-1}. Using the equation from the previous problem, plot the electron temperature and the energy-relaxation time as a function of the electric field. Then, plot the energy-relaxation time as a function of the electron temperature.

References

[1] Linke H, Bird J P, Cooper J, Omling P, Aoyagi Y and Sugano T 1997 *Phys. Rev.* B **56** 14937
[2] Prasad C, Ferry D K, Shailos A, Elhassen M, Bird J P, Lin L-H, Aoki N, Ochiai Y, Ishibashi K and Aoyagi Y 2000 *Phys. Rev.* B **62** 15356
[3] Prasad C, Ferry D K, Vasileska D and Wieder H H 2003 *J. Vac. Sci. Technol.* B **21** 1936

[4] Fröhlich H and Seitz F 1950 *Phys. Rev.* **79** 526

[5] Ryder E J 1953 *Phys. Rev.* **90** 766

[6] Fowler A B, Fang F F, Howard W E and Stiles P J 1966 *Phys. Rev. Lett.* **16** 901

[7] Fang F F and Fowler A B 1970 *Phys. Rev.* **169** 619

[8] Bauer G 1974 *Determination of Electron Temperatures and of Hot Electron Distribution Functions in Semiconductors* (Springer Tracts in Modern Physics vol 74) (Berlin: Springer)

[9] Price P J 1982 *J. Appl. Phys.* **53** 6863

[10] Ma Y, Fletcher R, Zaremba E, D'Iorio M, Foxon C T and Harris J J 1991 *Phys. Rev.* B **43** 9033

[11] Somphonsane R, Ramamoorphy H, Bohra G, He G, Ferry D K, Ochiai Y, Aoki N and Bird J P 2013 *Nano Lett.* **13** 4305

[12] Bistritzer R and MacDonald A H 2009 *Phys. Rev. Lett.* **102** 206410

[13] Kubakaddi S 2009 *Phys. Rev.* B **87** 405414

[14] Tse W K and das Sarma S 2009 *Phys. Rev.* B **79** 205404

[15] Baker A M R *et al* 2013 *Phys. Rev.* B **87** 045414

[16] Huang J, Alexander-Webber J A and Janssen T J B M *et al* 2015 *J. Phys.: Cond. Matt.* **27** 164202

[17] Song J C W, Reizer M and Levitov L S 2012 *Phys. Rev. Lett.* **109** 106602

[18] Ferry D K, Somphonsane R, Ramamoorthy H and Bird J P 2015 *Appl. Phys. Lett.* **107** 262103

[19] Ferry D K, Somphonsane R, Ramamoorthy H and Bird J P 2015 *J. Comp. Electron.* **15** 144

[20] Ferry D K, Goodnick S M and Bird J P 2009 *Transport in Nanostructures* 2nd edn (Cambridge: Cambridge University Press)

[21] Guilianai G F and Quinn J J 1982 *Phys. Rev.* B **26** 4421

[22] Ferry D K 2018 *An Introduction to Quantum Transport in Semiconductors* (Singapore: Jenny Stanford Publishing) section 8.3

[23] Chow E, Wei H P, Girvin S M and Shayegan M 1996 *Phys. Rev. Lett.* **77** 1143

[24] Prasad C, Ferry D K and Wieder H H 2004 *J. Vac. Sci. Technol.* B **22** 2059

[25] Hwang E H, Adam S and das Sarma S 2007 *Phys. Rev. Lett.* **98** 186806

[26] Wick F G 2008 *Phys. Rev. (Ser. I)* **27** 11

[27] Price P J 1979 *Semiconductors and Semimetals* vol 14, ed R K Willardson and A C Beer (New York: Academic) pp 249–308

[28] Mori H, Oppenheim I and Ross J 1962 in *Studies in Statistical Mechanics* ed J de Boer and G E Uhlenbeck (North Holland: Amsterdam)

[29] Chester G V 1963 *Rep. Prog. Phys.* **26** 411

[30] Barker J M, Ferry D K, Koleske D D and Shul R J 2005 *J. Appl. Phys.* **97** 063705

[31] Barker J M, Ferry D K and Goodnick S M 2004 *J. Vac. Sci. Technol.* B **22** 2045

[32] Gunn J B 1963 *Sol. State Commun* **1** 88

[33] Ridley B K and Watkins T B 1961 *Proc. Phys. Soc. London* **78** 293

[34] Ridley B K and Watkins T B 1961 *J. Phys, Chem. Sol.* **22** 155

[35] Ferry D K 2017 *Semicond. Sci. Technol.* **32** 085003

[36] He G *et al* 2017 *Sci. Reports* **7** 11256

[37] Ferry D K 2019 *Semicond. Sci. Technol.* **34** 044001

[38] Ridley B K 1963 *Proc. Phys. Soc. London* **82** 954

[39] Ferry D K, Heinrich H, Keeler W and Müller E A 1973 *Phys. Rev.* B **8** 1538

[40] Chynoweth A G and Pearson G L 1958 *J. Appl. Phys.* **29** 1103

[41] Senitsky B and Moll J L 1958 *Phys. Rev.* **110** 612

[42] Ancker-Johnson B, Cohen R W and Glicksman M 1961 *Phys. Rev.* **124** 1745

[43] Ferry D K and Dougal A A 1966 *IEEE Trans. Electron. Dev.* **13** 627

[44] Böer K W and Ovshinsky S R 1969 *J. Appl. Phys.* **41** 2675

[45] Edwards A H, Barnaby H J, Campbell K A, Kozicki M N, Liu W and Marinella M J 2015 *Proc. IEEE* **103** 1004

[46] Zwanzig R W 1961 *Lectures in Theoretical Physics* vol 3 ed W E Brittin, B Downs and J Downs (New York: Interscience)

[47] Hess K and Vogl P 1979 *Sol. State Commun.* **30** 807

[48] Ferry D K 2016 *Semicond. Sci. Technol.* **31** 11LT02

Lightning Source UK Ltd.
Milton Keynes UK
UKHW051856270122
397741UK00002B/117